Ecological Genetics

Ecological Genetics: Design, Analysis, and Application

Andrew Lowe, Stephen Harris, and
Paul Ashton

Blackwell
Publishing

BLACKWELL PUBLISHING
350 Main Street, Malden, MA 02148-5020, USA
9600 Garsington Road, Oxford OX4 2DQ, UK
550 Swanston Street, Carlton, Victoria 3053, Australia

The right of Andrew Lowe, Stephen Harris, and Paul Ashton to be identified as the Authors of this Work
has been asserted in accordance with the UK Copyright, Designs, and Patents Act 1988.

First published 2004 by Blackwell Publishing Ltd

5   2007

Library of Congress Cataloging-in-Publication Data

Lowe, Andrew, Dr.
    Ecological genetics : design, analysis, and application / Andrew
Lowe, Stephen Harris, and Paul Ashton.
        p. ; cm.
    Includes bibliographical references and index.
    1. Ecological genetics.   I. Harris, Stephen, 1966–
II. Ashton, Paul, Dr.   III. Title.
        [DNLM: 1. Ecology. 2. Genetics.   QH 456 L913e 2004]
    QH456.L69 2004
        576.5'8–dc21

                                              2003011404

ISBN-13: 978-1-4051-0033-5 (pbk. : alk. paper)

A catalogue record for this title is available from the British Library.

Set in 9/11pt Janson
by Graphicraft Ltd, Hong Kong

The publisher's policy is to use permanent paper from mills that operate a sustainable forestry policy, and
which has been manufactured from pulp processed using acid-free and elementary chlorine-free practices.
Furthermore, the publisher ensures that the text paper and cover board used have met acceptable
environmental accreditation standards.

For further information on
Blackwell Publishing, visit our website:
www.blackwellpublishing.com

Contents

Preface

Ecological genetics provides the framework by which the patterns of genetic variation can be investigated and mechanisms of speciation can be understood. The application of molecular markers in genetics and ecology over the last 20 years has revolutionized our understanding of the subject. This has led to an increasing utilization of ecological genetics approaches to a wide variety of biological problems. This text provides an introduction to the area through a clear exposition of the various concepts that may influence genetic variation in an organism and its constituent populations, along with a detailed consideration of the appropriate form of analysis for each investigation.

The concept for this book was formed during the annual Ecological Genetics Group (EGG) Conference at Edge Hill, UK, in Easter 2000, following a workshop conducted by the three authors. Our experiences at the workshop and as active researchers and teachers within the field of ecological and conservation genetics, have made it obvious to us that there is currently no single, readily available volume that addresses the experimental design, data analysis, and biological interpretation of molecular marker studies, despite the increasing significance of such studies at undergraduate and postgraduate levels. This book has therefore been written to address these important issues. However, we have not attempted any detailed description of molecular genetic methodologies since there are many texts already produced on this subject. The book is rather aimed to bridge the knowledge gap between the selection of genetic markers with which to address a problem and the biological interpretation of resultant data. The principal readership is upper undergraduate and postgraduate level and as an introduction for researchers in other areas.

Interest in genetic variation and its relationship to the environment is not new and seminal studies have been conducted on plants and animals by researchers for more than a century. In the fourth edition of his book *Ecological Genetics*, Ford (1975) stated that "if ecological genetics is to be successfully developed, we need new thinking, new methods, new material". Over the last 25 years, new thinking, new methods, and new material have indeed become available and are being successfully used to develop the field of ecological genetics, much of which has now been encompassed by the phrase "molecular ecology".

The contemporary approach of ecological genetics emphasizes the importance of population size and structure, the interaction among populations through migration and dispersal (i.e. gene flow), historical and contemporary contributions to population genetic structure, the interaction between local selection and genetic drift, and an expansion of the phenotype to include quantitative and qualitative features. Within this contemporary vision of ecological genetics, we have interpreted it as the investigation of the origin and maintenance of genetic variation within and between populations, which ultimately leads to adaptation and speciation.

Laboratory advancements in the utilization of molecular markers in the last 30 years have meant that vastly greater numbers of increasingly more variable loci can be identified and used; this shows no sign of slowing. In addition, there have been methodological advances in the statistical analysis of quantitative and phylogenetic characters. Furthermore, low-cost computing power and the concomitant development of analysis software, much of which is web-accessible, have opened up the possibilities of the complex analysis of vast amounts of empirical data. These developments mean that detailed genetic analyses are no longer restricted to an elite group of "model" organisms (e.g. mice, *Drosophila*, maize); but should, at least in theory, be open to any organism. Indeed genomic investigations, where the expression patterns of genetic variants in different environments are investigated, are being undertaken for an increasing range of species. Genetics is here to stay as an essential component of ecology, and ecology will become increasingly important for the interpretation of data emerging from genomics research programs.

Ecological genetics is of both pure and applied scientific interest, with a wide range of applications. The picture of modern ecological genetics that emerges is one of relevance to the interaction of large-scale geographic patterns of demography with genetic dynamics among small, partially isolated, and potentially locally adapted populations, or metapopulations. Ecological genetics offers contributions to a range of contemporary debates and issues, including:
- the impact of introduced alien species;
- release of genetically modified (GM) organisms;
- potential land-use change and threats to genetic resources;
- conservation genetic considerations for isolated and threatened species;
- assessing variation across the range of species, their evolutionary history and local adaptation considerations;
- assessing past range changes induced by climate change and how these may be used to predict future responses to global warming; and
- describing contemporary and ancient speciation events.

Special features

In this brave new world the new ecological genetic traveler needs a handbook to help highlight pitfalls and provide guidelines; this book equips the reader with the necessary tools and background information for such a journey of discovery:
- A general introduction to the area of ecological genetics (from genetic diversity, gene flow, and phylogeography to speciation) gives guidelines on the methods of analysis to be applied and an overview of the statistical procedures involved, so that students can be sure that the correct approach is being taken for their project.
- Guidelines for making decisions on materials sampling and emphasis on the biological knowledge of the study organism as a primary requisite, upon which reasonable project planning decisions can be made.
- An important aspect of the book is the emphasis on having an explicit hypothesis to test and to understand both the markers that may be applied and the analyses available.
- Summary sections on the most commonly used molecular techniques, their strengths and weaknesses, and their applicability to the analytical issues covered in each chapter.
- Illustration of the biological application of theoretical ecological genetic principles through boxes containing relevant examples, together with informative tables and helpful figures.
- A final chapter presents four case studies from butterflies, ragworts, bears, and oaks that integrate diverse aspects of ecological genetics from genetic diversity and differentiation, gene flow, and mating system, to phylogeography, hybridization, and speciation.
- The book aims to provide a reference handbook for use with the many population genetic software packages now available on the web, few of which informatively explain which statistics are most appropriate to apply to a particular biological situation. To help with this aspect, the book is accompanied by a Blackwell's hosted website detailing the software packages available and a decision tool for students to choose the most appropriate analysis methods for their particular case.

Acknowledgments

This book is dedicated to and is produced with very special thanks to Jude and Jacob, Carolyn and Filipe, and Janis, Layla, Sarah, Mark, and Robert. We are indebted to the thorough reviewing efforts of Cecile Bacles, Elizabeth Barrett, Mike Bruford, Sophie Gerber, Felix Gugerli, Geoff Oxford, and particularly to Pete Hollingsworth, whose input almost deserves recognition as a coauthor. We are also very grateful for additional comments from and discussion with Stephen Cavers, Chloe Galley, Ruth Eastwood, and Richard Ennos. We would also like to thank Victoria Annis for technical assistance, organizational help, and general encouragement in the latter stages of the work. Jan Jackson also helped throughout the project. The coherence of our view on ecological genetics almost certainly arises from the fact that we have all survived the rigorous training of the Richard Abbott school of ecological genetics at the University of St Andrews.

Abbreviations

A	adenine
AA	amino acid
AFLP	amplified fragment length polymorphism
AMOVA	analysis of molecular variance
ANOVA	analysis of variance
bp	base pair
C	cystosine
CAP	cleaved amplified polymorphism
CITES	Convention on International Trade in Endangered Species
cpDNA	chloroplast DNA
ddATP	dideoxyadenosine triphosphate
ddCTP	dideoxycytosine triphosphate
ddGTP	dideoxyguanosine triphosphate
$dd^{T}TP$	dideoxythymine triphosphate
DNA	deoxyribonucleic acid
F1	first filial generation
F2	second filial generation
G	guanine
g	gram
GM	genetically modified
IAM	infinite allele model
ISSR	inter SSR
Kb	kilobase
mtDNA	mitochondrial DNA
mya	million years ago
nDNA	nuclear DNA
ng	nanogram
PCR	polymerase chain reaction
RAMPO	random amplified microsatellite polymorphism
RAPD	randomly amplified polymorphic DNA
RE	restriction enzyme

RFLP	restriction fragment length polymorphism
RNA	ribonucleic acid
rRNA	ribosomal RNA

SCAR	sequence-characterized amplified region
SMM	stepwise mutation model
SNP	single nucleotide polymorphism
SSR	simple sequence repeat

| T | thymine |
| tRNA | transfer RNA |

| UNCED | United Nations Conference on Environment and Development |
| UPGMA | unweighted pair group method with arithmetic mean |

1

Ecological genetics

". . . the methods of genetics, diligently applied, obviously give one the power to replace loose speculation and guesswork by irrefutable inductions, and so to lay down a foundation upon which the evolutionist and taxonomist can build with safety."

Trow (1912)

Summary

1 Ecological genetic investigations have traditionally emphasized the integration of field ecology and laboratory genetics. Recent conceptual (e.g., phylogenetic analysis) and methodological (e.g., DNA markers) developments have revitalized interest in ecological genetics.

2 A contemporary vision of ecological genetics is presented as the investigation of the origin and maintenance of genetic variation within and between populations (in the broadest sense), which ultimately leads to speciation.

3 The outcomes of ecological genetic investigations have both pure and applied applications, for example, in the evolutionary consequences of habitat fragmentation, the construction of conservation policies for species reintroduction and for risk assessment of the release of genetically modified organisms.

4 New developments that allow high-throughput analysis of specific genomic loci, combined with the development of environmental genomic approaches to assessing adaptive variation, improved bioinformatics tools for data storage and access, and advanced statistical theory for enhanced data interpretation will provide exciting approaches to ecological genetics in the future.

1.1 What is ecological genetics?

The term ecological genetics was first used in print by Ford (1964) in his landmark book *Ecological Genetics*, although it had been used some years earlier when the Ecological Genetics Group held their first meeting in Aberystwyth in 1956. Ford considered that ecological genetics dealt with the "adjustments and adaptations of wild populations to their natural environment," whilst it was the genius of Darwin and Wallace that brought together ecological ("struggle for existence") and genetic (variation) concepts.

However, interest in genetic variation and its relationship to the environment was not new in the 1960s (Cain and Provine 1992), and had previously been a central element of Turesson's (1922, 1925, 1930) seminal work. Turesson grew populations of plants from different Swedish habitats in a common garden, and was one of the first to grasp the concept that habitat-correlated genetic variation was widespread among plant species. Extensions of this work by Clausen, Keck, and Hiesey (1940, 1948), investigated plant populations from an altitudinal gradient in central California. By growing plants in common gardens at different altitudes, they were able to analyze variation (i.e., local adaptation) that would otherwise have been masked by the use of a single common garden. This was the theme later taken up by gene ecologists (Briggs and Walters

1997). In animals, the link between ecological and genetic variation was thoroughly invest-igated in the banded snail, *Cepaea nemoralis* (Cain and Sheppard 1950, 1954). Each of the above case studies emphasize the two essential and inseparable aspects of Ford's view of ecological genetics, the combination of ecological fieldwork and laboratory genetics. It was through these types of studies that major syntheses of ecological genetics and evolutionary biology have been made (e.g., Dobzhansky 1970, Grant 1963, 1981, Stebbins 1950).

In the fourth edition of *Ecological Genetics*, Ford (1975, p. 11) stated that "if ecological genetics is to be successfully developed, we need new thinking, new methods, new material." Over the last 25 years, new thinking, new methods and new material have indeed become available and are being successfully used to develop ecological genetics (Berry, Crawford, and Hewitt 1992, Real 1994), much of which has now been encompassed by the phrase "molecular ecology". The renewed interest in the field can be measured in part by the increased incidence of meetings tackling ecological genetic issues and the success of a journal dedicated to the area, only initiated a decade or so ago.

The conceptual developments that have had substantial impacts on the development of ecological genetics have been in areas as diverse as the development of rigorous methods for phy-logenetic analysis (Avise 2000), the development of procedures for the detailed analysis of gene flow (Bossart and Prowell 1998) and the development of metapopulation theory (Hanski 1999). The greatest impact of new methods has been the development of protein and DNA markers for the analysis of genetic variation (Baker 2000). Such laboratory advances have meant that vastly greater numbers of increasingly more variable loci can be identified and used; this shows no sign of slowing. In addition, there have been methodological advances in the statistical ana-lysis of quantitative characters (Lynch and Walsh 1998). Furthermore, low-cost computing power and the concomitant development of analysis software, much of which is web-accessible, have opened up the possibilities of the complex analysis of vast amounts of empirical data. These developments mean that detailed genetic analyses are no longer restricted to an elite group of organisms (e.g., mice, *Drosophila*, maize); now ecological genetic procedures and analysis should, at least in theory, be open to any organism.

Classical ecological genetics was principally concerned with documenting and measuring the magnitude of selection in natural systems. Such investigations tended to focus on single popula-tions, in which population size was not an essential feature, and on aspects of the phenotype under simple gene control. In contrast, the contemporary approach emphasizes the importance of population size and structure, the interaction among populations through migration and dis-persal (i.e., gene flow), the interaction between local selection and genetic drift, and an expansion of the phenotype to include quantitative and qualitative features (Real 1994).

However, into this brave new world, Ford's (1975) argument, that the choice of material for ecological genetic investigations is of the utmost importance, is still valid, albeit with modifications in the light of nearly 30 years of additional research and development. Thus, it is necessary to have a clear, testable hypothesis. In addition, the ideal study taxon should: (i) have a known ecology; (ii) be easy to collect; (iii) have a short life-span so that numerous generations can be investigated; (iv) have genetic variation that is easily recognized and interpreted and is capable of statistical analysis; (v) be at a high enough density that it can easily be found; and (vi) be easily maintained in the laboratory or greenhouse. However, it is rarely possible to fulfill all of these criteria. Ford (1975, p. 9) indicated that the choice of material seems "rather an art than a science", and stated that "the more rewarding decisions are generally reached by a thorough . . . grasp of the essentials of the problem to be solved, estimated in the light of much knowledge and experience".

Lifetimes of research are often required for robust investigations of ecological genetics, and require a combination of detailed fieldwork and the application of genetic (now usually molecular) techniques. However, importantly ecological genetics requires experimentation, as emphasized by Ford (1964). Documentation of variation needs to be combined with the genera-tion of hypotheses that can be empirically tested. We have interpreted contemporary ecological genetics as the investigation of the origin and maintenance of genetic variation within and between populations, which ultimately leads to adaptation and speciation. Ecological genetics provides the means by which mechanisms of speciation can be understood and the patterns of genetic variation in nature investigated.

1.2 Why study ecological genetics?

The picture of modern ecological genetics that emerges is one of relevance to the interaction of large-scale geographic patterns of demography with genetic dynamics among small, partially isolated, and potentially locally adapted populations. Framed in such a context, ecological genetics is of both great pure and applied scientific interest. Adaptive radiation, the evolution of ecological diversity within a rapidly multiplying phylogenetic lineage, is one major research area where the combination of ecological and genetic analysis has been particular valuable (e.g., Schluter 2000).

The debate over the introduction of alien species and release of genetically modified (GM) organisms and their potential impacts on the environment and closely related species is an area where a combination of fieldwork and experimental work, within an ecological genetic framework, is likely to produce exciting pure and applied science. One of the major environmental concerns over GM crops relates to transgene movement into wild relatives, where the pattern of hybridization will affect the scale and rapidity of any ecological change, plus the feasibility of any necessary containment. In the UK, one of the species of most concern is the widely cultivated oilseed rape (*Brassica napus*), since it has numerous wild relatives (e.g., *Brassica rapa*) that are either native or naturalized. Using remote sensing, Wilkinson et al. (2000) identified possible sites of sympatry between *B. napus* and *B. rapa* across 15,000 km^2 of south-east England. In 1998, two sympatric populations were found over the entire survey area, and in 1999, every newly recruited plant in these populations was tested for hybridity using a combination of flow cytometry and nuclear DNA markers. Despite the size of this investigation, only one hybrid was observed among 505 plants screened in the *B. rapa* populations. Thus, it is possible to investigate the interaction between populations of domesticated and wild organisms.

As human impacts on habitat landscape increase, and habitats are fragmented, the interactions between wild and domesticated populations become more complex. Similarly, wild populations that were once large and connected by gene flow may become fragmented, and the population size diminished (Frankham, Ballou, and Briscoe 2002). Under such conditions, migration becomes critical for the reestablishment of local populations, so that gene flow may counter some of the detrimental effects of inbreeding depression. Similarly, changes in habitat management may have significant influences on the genetics of rare species. For example, in the UK, the Red Squirrel (*Scirius vulgaris*) is a nationally rare species, although it is locally common in the fragmented coniferous woodlands of northern England and the Scottish Borders. Using a combination of detailed analysis of habitat fragmentation and analysis of microsatellite variation in museum specimens of *S. vulgaris*, Hale et al. (2001) were able to show that the creation of a large coniferous plantation (since the 1960s) was enough to produce miscegenation of *S. vulgaris* populations up to 100 km apart. Thus, the genetic consequences of habitat change may occur hundreds of kilometers away from the site of change.

Any ecological genetics program that focuses on demographic and genetic variation across large-scale geographic variation has obvious implications for conservation. Populations that are small and locally adapted may also suffer from increased gene flow through the disruption of locally adapted genotypes. However, any migrants into these populations may not have equal effects on local population growth and fitness. Thus, knowledge of the genetic diversity that exists among migrants is essential when constructing conservation policies for species reintroduction or augmenting declining populations (Frankham, Ballou, and Briscoe 2002). For example, conservation organizations are increasingly interested in ensuring that only local seed is planted within specific areas. This, of course, raises the issue of what is meant by local: the same population, the same county, the same country or even the same continent.

Conservation biology and GM risk assessment are not the only areas where the results from ecological genetic investigations are a valuable applied tool. The long-term sustainability of ecological systems depends on the ability of organisms to adapt and respond to changes in their biotic and abiotic environment. The methods for studying adaptive evolution in contemporary populations provided by ecological genetics are equally important in assessing the ability of organisms to respond to both local and global environmental perturbations. Furthermore, it is now possible to investigate the genetic and ecological outcomes of major environmental changes, for example, the effects of the last glaciation on the recolonization of Europe by plant and animal species (e.g., Hewitt 2000, Taberlet et al. 1998). The accurate reconstruction of historical species' range shifts and correlation with past climate change offer great potential for improving predictions of floral and faunal adaptive response and gene flow under future rapid climate change scenarios.

New developments in DNA extraction technology have been incorporated in ecological genetic approaches. It is no longer necessary to collect whole organisms; all that is required is a small amount of suitable material. Furthermore, the use of preserved materials means that it is possible to add a temporal dimension (tens, even thousands of years) to investigate genetic diversity and structure. Also, hundreds of samples may be readily analyzed for large genomic segments. Thousands of markers are available, which are variable at many different levels. However, the next major development will be through environmental genomics initiatives, where the expression patterns of genetic variants in different environments can be investigated. Genetics is here to stay as an essential component of ecology, while ecology will become increasingly important for the interpretation of data emerging from genomics research programs.

REFERENCES

Avise, J.C. 2000. Phylogeography. *The History and Formation of Species*. Cambridge, MA: Harvard University Press.

Baker, A.J. 2000. *Molecular Methods in Ecology*. Oxford: Blackwell Science.

Berry, R.J., Crawford, T.J., and Hewitt, G.M. 1992. *Genes in Ecology*. Oxford: Blackwell Science.

Bossart, J.L. and Prowell, D.P. 1998. Genetic estimates of population structure and gene flow: limitations, lessons and new directions. *Trends in Ecology and Evolution*, 13: 202–6.

Briggs, D. and Walters, S.M. 1997. *Plant Variation and Evolution*. Cambridge: Cambridge University Press.

Cain, A.J. and Sheppard, P.M. 1950. Selection in the polymorphic land snail *Cepaea nemoralis* (L.). *Heredity*, 4: 275–94.

Cain, A.J. and Sheppard, P.M. 1954. Natural selection in *Cepaea*. *Genetics*, 39: 89–116.

Cain, A.J. and Provine, W.B. 1992. Genes and ecology in history. In R.J. Berry, T.J. Crawford, and G.M. Hewitt, eds. *Genes in Ecology*. Oxford: Blackwell Science, pp. 3–28.

Clausen, J., Keck, D.D., and Hiesey, W.M. 1940. Experimental studies in the nature of species, I. The effect of varied environments on western North American plants. *Publications of the Carnegie Institution*, 520.

Clausen, J., Keck, D.D., and Hiesey, W.M. 1948. Experimental studies in the nature of species, III. Environmental responses of climatic races of *Achillea*. *Publications of the Carnegie Institution*, 581.

Dobzhansky, T.G. 1970. *Genetics of the Evolutionary Process*. New York: Columbia University Press.

Ford, E.B. 1964. *Ecological Genetics*. London: Methuen.

Ford, E.B. 1975. *Ecological Genetics*. London: Chapman and Hall.

Frankham, R., Ballou, J.D., and Briscoe, D.A. 2002. *Introduction to Conservation Genetics*. Cambridge: Cambridge University Press.

Grant, V. 1963. *The Origin of Adaptations*. New York: Columbia University Press.

Grant, V. 1981. *Plant Speciation*. New York: Columbia University Press.

Hale, M.L., Lurz, P.W.W., Shirley, M.D.F., Rushton, S., Fuller, R.M., and Wolff, K. 2001. Impact of landscape management on the genetic structure of red squirrel populations. *Science*, 293: 2246–8.

Hanski, I. 1999. *Metapopulation Ecology*. Oxford: Oxford University Press.

Hewitt, G. 2000. The genetic legacy of the Quaternary ice ages. *Nature (London)*, 405: 907–13.

Lynch, M. and Walsh, B. 1998. *Genetics and Analysis of Quantitative Traits*. Sunderland, MA: Sinauer.

Real, L.A. 1994. *Ecological Genetics*. Princeton, NJ: Princeton University Press.

Schluter, D. 2000. *The Ecology of Adaptive Radiation*. Oxford: Oxford University Press.

Stebbins, G.L. 1950. *Variation and Evolution in Plants*. New York: Columbia University Press.

Taberlet, P., Fumagalli, L., Wust-Saucy, A.-C., and Cosson, J.-F. 1998. Comparative phylogeography and postglacial colonization routes in Europe. *Molecular Ecology*, **7**: 453–64.

Turesson, G. 1922. The genotypical reponse of the plant species to the habitat. *Hereditas*, **3**: 211–350.

Turesson, G. 1925. The plant species in relation to habitat and climate. *Hereditas*, **6**: 147–236.

Turesson, G. 1930. The selective effect of climate upon plant species. *Hereditas*, **14**: 99–152.

Wilkinson, M.J., Davenport, I.J., Charters, Y.M., Jones, A.E., Allainguillaume, J., Butler, H.T., Mason, D.C., and Raybould, A.F. 2000. A direct regional scale estimate of transgene movement from genetically modified oilseed rape to its wild progenitors. *Molecular Ecology*, **9**: 983–91.

2

Markers and sampling in ecological genetics

". . . we have concluded that all collecting trips to fairly unknown regions should be made twice; once to make mistakes and once to correct them."
John Steinbeck (1958) *Log from the Sea of Cortez*

Summary

1 Sampling methodology is important in ecological genetic studies and the exact strategy depends on the nature of the investigation. For an effective strategy, sampling within and among populations and sampling within and among genomes must be investigated.
2 Ecological genetic markers are derived from the nuclear, mitochondrial and, in the case of plants, chloroplast genomes. The characteristics of these three genomes vary and determine the questions that they may be used to answer.
3 Tissues for analysis may be derived from a wide range of different sources, including fresh and preserved materials. When analyses are conducted, the importance of voucher material for the verification of identification must not be overlooked.

2.1 Introduction

Ecological genetics is one of many fields of biological research that has benefited from the scientific revolutions in molecular biology and information technology. Surveying patterns of variation and testing evolutionary hypotheses about process have been two important approaches in empirical ecological genetics research. However, limited resources are generally available for such studies, thus efficient experimental design and sampling become crucial. In order to study natural populations one has to consider suitable strategies for sampling within and among both populations and genomes (i.e., intra- and inter-population and genomic variability). In addition, choices must be made over the markers to be studied. Where studies were once limited to a relatively small number of morphological or biochemical markers, tens and often hundreds of DNA markers can now be used; the limitation may now be access to experimental material. Ideally, project design should incorporate a pilot phase during which checks can be made on the adequacy of the sampling protocols and markers to answer the chosen problem.

Early in a study it is necessary to answer the questions: (i) is the proposed sampling of the organism or genome sufficient to test the proposed hypothesis? (ii) are the resources available for the strategy proposed? (iii) are other strategies available? and (iv) should the project be abandoned? The latter question is valid, since there is little point in undertaking an extensive piece of work only to discover later that the strategy was unable to test the original hypothesis because of inadequate sampling of either the organism or its genome. However, since sampling is both expensive and time consuming, there is also little point in sampling more than is necessary. Issues associated with sampling are important when planning an ecological genetics research project and form the basis of this chapter. First, genomes and molecular markers will be considered, followed by a brief overview of sampling strategies. The bulk of the chapter will be concerned with the determination of the sizes of within and among population samples necessary to address ecological genetics questions.

2.2 Methods of data generation

2.2.1 Genome types

Genetic markers occur in the nuclear and organelle (chloroplast and mitochondria) genomes; in animals, only the mitochondrial genome is available for analysis, whilst, in plants, both mitochondrial and chloroplast genomes are available. These three genomes differ in their evolutionary characteristics, for example, inheritance and sequence and structural mutation rates, which determine the types of genetic issues that they may be used to study. Furthermore, within each genome, the coding and non-coding regions evolve at different rates. Hence, choice of region for analysis within a genome will depend on the level of variation necessary for a particular investigation.

There has been considerable interest in the structure and evolution of the nuclear, mitochondrial, and chloroplast genomes. However, for the purposes of this section, only those features of the genomes that are of most relevance to their use, as genetic markers, will be introduced (Table 2.1). The inheritance of organelle genes differs from that of nuclear genes in that they show (Birky 2001): (i) vegetative segregation, where organelle segregation is determined by stochastic processes; (ii) uniparental inheritance, where organelles are transmitted from generation to generation through only one parent (most often the maternal parent but paternal transmission of organelles is known to occur); (iii) intracellular selection, where heteroplasmic cells become homoplasmic through selection of particular organelle types; and (iv) reduced recombination, where genomes from different lineages apparently never recombine (Fig. 2.1). However, to demonstrate strict uniparental inheritance of an organelle it is necessary to have large sample sizes (Milligan 1992).

The nuclear genome (nDNA) is the largest of the genomes available for genetic analysis, and varies in size from approximately 5×10^4 kb to 7×10^8 kb. However, this difference in genome size is not associated with either genetic or morphological complexity, thus, *Amoeba dubia* nDNA (ca. 6.7×10^8 kb) is approximately 200 times bigger than that of the human nDNA (3.3×10^6 kb). Nuclear genomes are packaged into varying numbers of chromosomes within a cell, for example, the angiosperm *Haplopappus gracilis* has only four chromosomes per mitotic cell and the fern *Ophioglossum reticulatum* has 1260 chromosomes. The difference between morphological complexity and genome size is known as the C-value paradox and is due to the presence of repetitive DNA. The nDNA is a mixture of coding and non-coding regions, which are divided into highly repetitive sequences, moderately repetitive sequences, and single copy sequences based on their relative abundances. Highly repetitive sequences are usually non-coding, whilst moderately repetitive sequences encode polypeptide-encoding sequences, such as histones, of which there may be several hundred copies, and transfer RNA (tRNA) and ribosomal RNA (rRNA). There are two classes of rRNA: 5S rRNA, encoded by 5S rDNA, and 18S-26S rRNA, encoded by 18S-26S rDNA (Jorgensen and Cluster 1988). Both of these sequences are clustered in the genome and are tandemly repeated, the coding region within each repeat unit separated by a non-coding region. Divergence across the region is associated with functionality of the region; non-coding regions diverge more than coding regions (Fig. 2.2; Jorgensen and Cluster 1988). There are two distinct types of repetitive DNAs: interspersed repeats, which consist of individual units of specified sequence scattered across the genome; and tandem repeats, which are typically short sequences repeated hundreds or thousands of times. Minisatellite DNA is one type of repetitive DNA that has been widely used as a source of DNA markers, particularly in studies of animals (Carter 2000).

Table 2.1

Characteristics of the nuclear (nDNA) and mitochondrial (mtDNA) genomes of animals and plants and the plant chloroplast (cpDNA) genome.

Feature	nDNA Animal	Plant	cpDNA	mtDNA Animal	Plant
Inheritance	Biparental	Biparental	Most angiosperms maternal; conifers paternal	Mostly maternal	Mostly maternal
Structure	Linear	Linear	Circular	Circular	Circular, linear and more complex shapes
Size (kb)	$4.90 \times 10^4 - 7.00 \times 10^8$	$5.00 \times 10^4 - 3.07 \times 10^8$	71–214	15–20	200–2400
Substitution rate (substitutions per synonymous site per year)*	3.5×10^{-9}	$4.1 - 5.7 \times 10^{-9}$	$0.86 - 1.20 \times 10^{-9}$	56×10^{-9}	$0.36 - 0.50 \times 10^{-9}$
Substitution rate compared to plant mtDNA	8.1	11.4	2.4	130.2	1.0
Foreign sequences	Common	Common	Rare	Rare	Common
Structural mutations	Common	Common	Rare	Rare	Common
Recombination	Yes	Yes	Intramolecular	No	Inter- and intramolecular

* These are mean values for the genome. Substitution rates will vary from region to region within a genome.
Values from Brown et al. (1982), Li (1997), Wolfe, Li, and Sharp (1987), and Wolfe, Sharp, and Li (1989).

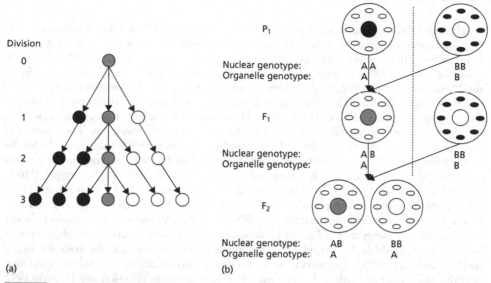

Fig. 2.1

Organelle genetics. (a) Vegetative segregation of an organelle mutation from a heteroplasmic cell (gray) at division 0 into homoplastic cells (black and white) (for the sake of simplicity only the organelle complement of cell is shown). (b) Uniparental maternal inheritance of organelles (ellipses) and biparental inheritance of nuclear DNA (circles).

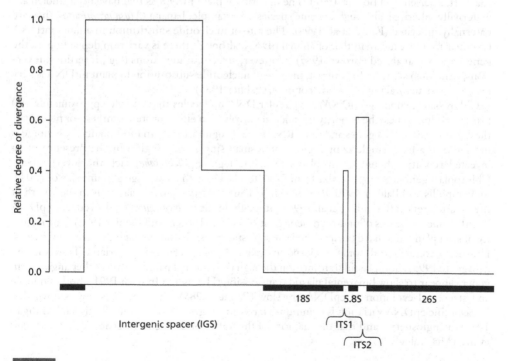

Fig. 2.2

Relative divergence of coding (thick line) and non-coding (thin line) regions across the nuclear 18S-26S ribosomal DNA repeat. ITS, internal transcribed spacer. Data from Appels and Dvorak (1982) and Jorgensen and Cluster (1988).

Single-copy DNA sequences, which include most polypeptides and their 3' and 5' untranslated regions and introns, occur once per haploid genome and represent only a fraction of the total genome.

In animals, the mitochondrial genome (mtDNA) is generally a small (15–20 kb) circular molecule (Boore 1999), although three of the four cnidarian Classes are known to have linear mtDNAs (Bridge et al. 1992). Eighty-seven complete metazoan mtDNA sequences are available and they almost all contain 37 genes, typically encoding 13 proteins for oxidative phosphorylation, two rRNAs and 22 tRNAs (Boore 1999). There is also a single large non-coding region that is known to contain controlling sequences for transcription and replication. The majority of animal mtDNAs are inherited through the maternal parent (Boore 1999), although in the bivalve genus *Mytilus*, mtDNAs are transmitted through both males and females, consequently within any one individual, tissues are heteroplasmic for two different mtDNA haplotypes (Hoeh, Blakley, and Brown 1991, Stewart et al. 1995). In higher plants, mtDNAs are much larger (200–2400 kb) and more complex than those of either animals or fungi (15–176 kb), and have much greater coding capacity. In addition to tRNAs and rRNAs, the genome encodes polypeptides involved in energy metabolism and polypeptides of unknown function: complete mtDNA sequences are available for two plants. The difference in size between the compact animal mtDNA and plant mtDNA appears to be mainly due to introns, intergenic sequences and duplications plus integrated chloroplast and nuclear DNA sequences (Hanson and Folkerts 1992, Palmer 1992, Wolstenholme and Fauron 1995). However, despite detailed investigations, plant mtDNA structure and organization remain poorly understood. The master circle hypothesis of mtDNA structure, which proposes that small circular mtDNAs are the product of recombination within large circular mtDNAs, may be too simplistic, since many plant mtDNAs appear to exist as dynamic equilibria between circular, linear, and more complex molecular arrangements (Backert, Nielsen, and Börner 1997). The majority of plant mtDNAs that have been studied are maternally inherited, although in some species, for example, banana (*Musa acuminata*), they are paternally inherited (Fauré et al. 1994). The rate of nucleotide substitution in animal mtDNA is similar to or greater than that of animal nDNA, although there is variation depending on the gene sequence analyzed (Brown 1985). However, structural mutations (e.g., insertions, inversions, deletions) are rare. In contrast, the rate of nucleotide substitution in plant mtDNA is very low, whilst structural mutations are common (Palmer 1985).

Chloroplast genomes (cpDNA) are circular DNA molecules that encode approximately 100 protein-coding genes (the majority of which are involved in either protein synthesis or photosynthesis), about 30 tRNA genes and four rRNA genes (Sugiura 1989). In most plants the genome is divided into a large single-copy region and a small single-copy region by the presence of an inverted repeat; although some plants (e.g., the legume *Vicia faba*) lack the invert repeat. Chloroplast genomes range in size from 71 kb to 214 kb, with an average genome size of 150 kb (dePamphilis and Palmer 1990, Palmer 1987). The majority of size variation is in the length of the inverted repeat (Goulding et al. 1996), although, in the case of *Epifagus*, the reduced cpDNA length is due to the loss of photosynthetic genes (Wolfe, Morden, and Palmer 1992). The majority of seed plants that have been studied show plastid transmission through the maternal parent. However, conifers and some angiosperms (e.g., *Actinidia chinensis*; Cipriani, Testolin, and Morgante 1995) show plastid transmission through the paternal parent, whilst other angiosperm genera appear to show biparental plastid transmission (Harris and Ingram 1991). Both sequence and structural evolution of cpDNA are slow (Palmer 1985), although it is now known that intraspecific cpDNA variation is common (Harris and Ingram 1992, Soltis, Soltis, and Milligan 1992) in angiosperms and that the majority of this variation is due to insertion/deletion (indel) events (McCauley 1995).

2.2.2 Techniques for data generation

Since the earliest ecological genetics studies were undertaken at the start of the twentieth century, genetic markers have been utilized to assign unambiguous genotypes to individual organisms

(Briggs and Walters 1997). Genetic markers are observable traits (the expression of which indicates the presence or absence of certain genes) that are classified into five broad groups: morphological, cytological, biochemical, protein, and DNA. Morphological traits, for example, flower color in peas and bristle arrangement in *Drosophila*, were the first markers to be used, although they may only be available in limited numbers. Cytological characteristics, for example, chromosome number and structure or chromosome pairing behavior, provided a means of detecting variation that was not visible at the morphological level (e.g., Stebbins 1971). Chemical markers, that is, secondary products (e.g., flavonoids and terpenes), offered particular promise as ecological genetic markers as they showed considerable intraspecific variation (e.g., Crawford 1990, Giannasi and Crawford 1986), although the analysis of chemical marker data has been criticized (Birks and Kanowski 1988, 1993). Other biochemical markers include immunological markers (Maxson and Maxson 1990), although these have largely been abandoned in favor of more reliable and easily utilized markers. However, protein and, more recently, DNA markers have revolutionized the availability of markers in ecological genetics studies. Markers were once limited to a small number of well-characterized genes that were available in only a small number of model organisms, for example, *Drosophila*, *Zea*, and *Arabidopsis*, where crosses could easily be made. Relatively little was known about the ecological genetics of more diverse organisms in natural environments. Protein and DNA markers changed this and allowed new evolutionary mechanisms to be discovered that would otherwise have gone unnoticed.

An ideal genetic marker for ecological genetic studies has six important characteristics (Weising et al. 1995):

• *Detect qualitative or quantitative variation.* The marker should be either present or absent, or the level of its expression should show discrete variation, that is, high versus low.

• *Show no environmental or developmental influences.* If an individual is translocated into three separate environments then it should display the same genotype irrespective of environment and if a marker is found in the juvenile it should also be present in the adult.

• *Show simple codominant inheritance.* In a diploid, both alleles at a locus should be visible in the heterozygote condition. In the dominant situation, one allele is present and it is impossible to distinguish between the dominant homozygote and the heterozygote condition.

• *Detect silent nucleotide changes.* The marker should be capable of detecting changes in the coding region of a genome that results in synonymous amino acid substitution, that is, mutations in codons that result in the incorporation of identical amino acids into a protein sequence. For example, the codons GTT, GTC, GTA, and GTG all code for the amino acid valine.

• *Detect changes in coding and non-coding portions of the genome.* The markers should be randomly distributed across the genome, and not restricted to just one class of DNA.

• *Detect evolutionary homologous changes.* The markers used for genetic analysis should be homologous, that is, similar due to descent from a common ancestor. However, loci and alleles may be defined in genetic studies in manners other than by descent, for example, origin or state (see Gillespie 1998).

None of the marker systems currently used in ecological genetics studies have all of these ideal characters. There are marker systems that are preferred for certain problems, for example, microsatellites will be the preferred markers for detailed analysis of gene flow within populations, whilst other problems may be studied equally effectively using different marker systems, for example, PCR-RFLPs and allozyme analysis would be equally useful for estimating genetic diversity within a population. However, the choice of a marker system is a compromise between the properties of the marker system and its availability. Thus, marker choice must be based on the hypothesis that is being tested, the properties of the marker system, and the resources that are available for the research program.

Protein and DNA markers are assumed to be selectively neutral, although, from an ecological genetics viewpoint, an equally important group of markers are adaptive. In many studies, adaptive markers have often been ignored, in favor of neutral markers, despite their evolutionary importance (Ennos, Worrell, and Malcolm 1998). The importance of both field and laboratory

Table 2.2

Summary of the main characteristics of major sources of molecular markers used in ecological genetics research.

	Allozymes	RFLP	RAPD	AFLP	PCR-RFLP	SSR	Sequencing
Basis	Detection of charged AA distribution differences	Detection of relative RE site positions	Distribution of random primers through genome	PCR of RE fragment subset using modified primers	RE digest of PCR products	PCR of simple sequence repeat regions	Direct sequencing of PCR products
Polymorphism	Charged AA substitutions	Nucleotide substitutions; indels; inversions	Nucleotide substitutions; indels; inversions	Nucleotide substitutions; indels; inversions	Nucleotide substitutions; indels; inversions	Repeat number changes	Nucleotide substitutions; indels
Abundance in genome	Low	High	Very high	High	High	Medium	High
Level of polymorphism	Low	Medium	Medium	Medium	Medium	High	Medium
Dominance	Usually codominant	Codominant	Dominant	Codominant/dominant	Codominant	Codominant	Codominant
Amount of material	Very little	2–10 µg DNA	10–25 ng DNA	1–2 µg DNA	50–100 ng DNA	50–100 ng DNA	10–25 ng DNA
Multiplex ratio	?	1–2	5–10	30–100	Low	1	?
Sequence information needed	No	No	No	No	Yes	Yes	Yes
Radioactive detection needed	No	Yes/No	No	Yes/No	No	No/Yes	No/Yes
Development costs	Low	Medium	Low	Medium	Medium/high	High	Medium
Start-up costs	Low	Medium/high	Low	Medium	High	High	High
Automation	No	Limited	Yes	Yes	Limited	Yes	Yes
Reproducibility	Medium/high	High	Low	Medium	High	High	High

AA, amino acid; AFLP, amplified fragment length polymorphism; PCR, polymerase chain reaction; PCR-RFLP, polymerase chain reaction-restriction fragment length polymorphism; RAPD, randomly amplified polymorphic DNA; RE, restriction endonuclease; RFLP, restriction fragment length polymorphism; SSR, simple sequence repeats or microsatellites.

work in ecological genetics studies is vividly demonstrated in the analysis of phosphoglucose iso-merase (PGI) variation in the butterfly species, *Colias eurytheme* and *Colias philodice eriphyle* (Watt 1977, 1991). PGI polymorphism influences viability, flight time, mating success, and fecundity, and different PGI genotypes are selected for or against depending on the ambient temperature, for example, homozygous individuals of genotype *Pgi-22* are favored in normal conditions but selected against in hot conditions (Watt 1983).

For variation to be of the greatest value in ecological genetics studies, it should be character-ized genetically. That is, the genetics of each of the putative loci investigated should be deter-mined through crossing experiments. However, for the majority of organisms it is impractical to attempt such programs, because of fecundity, physical crossing difficulties, or reproductive age. For example, to produce a segregating F_2 generation for a tropical forest tree may take 100 years since the plant may not reach reproductive age for 50 years. Such difficulties led Gillet and Gregorius (2000) to propose a method for testing marker inheritance from the genetic analysis of single tree progeny arrays. However, for the majority of studies, detailed genetic analyses are not undertaken, rather loci and alleles are interpreted based on expected patterns of segregation.

Six of the most commonly used types of protein and DNA markers (allozymes, restriction fragment length polymorphisms (RFLP), randomly amplified polymorphic DNA, amplified fragment length polymorphism (AFLP), microsatellites, and sequence analysis; Table 2.2) are described in 'Essential methods information,' pp. 36–44, together with their advantages, dis-advantages, and some of their applications (Table 2.3). However, numerous other types of marker systems have been proposed, for example, single nucleotide polymorphisms (SNPs; Cutler et al. 2001, Gibson 2002). Marker systems may be classified according to their modes of inheritance, that is, dominant (e.g., AFLPs) versus codominant (e.g., RFLPs), the numbers of putative loci that they detect, that is, few loci (e.g., allozymes) versus many loci (e.g., RAPDs), the numbers of alleles that they detect at a locus, that is, diallelic (e.g., RAPDs) versus multiallelic (e.g., SSRs), or their ease of use, that is, simple (e.g., RAPDs) versus complex (e.g., AFLPs).

Allozymes are the most widely used and understood of the marker systems currently applied in ecological genetic studies (Butlin and Tregenza 1998). The codominant expression, cost effectiveness, and simplicity of allozyme detection have ensured that there has been continued interest in these markers, although their use is declining because of the low number of alleles detected per locus, the absence of phylogenetic information, and the need to have access to suitable fresh material (Newbury and Ford-Lloyd 1993, Schaal et al. 1998, Wendel and Weeden 1990). In studies of animals, the use of allozyme markers in ecological studies has almost entirely given way to DNA markers, since allozyme extraction is usually destructive. DNA can be extracted from diverse dead materials, for example, dried leaves (e.g., Chase and Hills 1991), museum specimens (e.g., Iduica, Whitten, and Williams 2001) or feces (e.g., Kohn et al. 1999), whilst functional proteins must be extracted from either fresh or rapidly frozen material (Wendel and Weeden 1990). Combined with the ability to identify DNA markers in small amounts of DNA using the polymerase chain reaction (PCR), there has been a revolution in genetic studies. It is now possible to design PCR primers to almost any DNA region and for these regions to be consistently amplified across almost any organism, whilst high throughput sequence analysis means that variation can now be easily detected at the level of individual nucleotides rather than through secondary methods, for example, RFLPs.

Attempts have been made to compare the different techniques using measures of genetic diversity and two criteria have been proposed for technique comparison: information content and multiplex ratio (Rafalski et al. 1997). The greater a marker's information content the easier it is to detect polymorphism, whilst the multiplex ratio indicates the number of loci scored in each analysis. These criteria have been summarized into a single parameter, the marker index, which is highly correlated in AFLPs, RFLPs, and SSRs (Powell et al. 1996; Table 2.2). However, the assessment of information content, as defined by Rafalski et al. (1997), for anonymous markers (e.g., RAPDs, AFLPs) is problematic, since locus, and hence allele, identities are generally unknown. Furthermore, marker comparisons may not be legitimate since each marker system

Table 2.3

Utility of molecular markers in different types of ecological genetics studies of animals and plants. This table should be used in conjunction with Table 2.2 and Boxes 2.1 to 2.6, since each approach has its own advantages and disadvantages which must be taken into account when making a decision over which markers to use in an investigation.

	Allozymes	RFLP and PCR-RFLP nDNA	cpDNA	mtDNA (plant)	mtDNA (animal)	RAPD	AFLP	SSR	Sequencing nDNA	cpDNA	mtDNA (plant)	mtDNA (animal)
Genetic diversity	++	+++	+	+	+	++	++	++	+++	++	+	++
Population differentiation	+++	++	++	++	++	++	++	++	+++	++	++	+++
Gene flow	++	++	+	(+)	-	(+)	(+)	+++	+++	++	(+)	++
Polyploidy	+++	+++	++	-	-	-	-	+	++	++	-	-
Hybridization	++	+++	++	+	+	++	++	+	++	++	+	+
Phylogeny	(+)	+	++	(+)	++	-	-	(+)	+++	+++	(+)	+++
Individual genotyping	(+)	++	-	-	-	+++	+++	+++	+++	-	-	-
Phylogeography	-	?	++	(+)	++	-	-	-	?	+++	(+)	+++

RFLP, restriction fragment length polymorphism; PCR, polymerase chain reaction; RAPD, randomly amplified polymorphic DNA; AFLP, amplified fragment length polymorphism; SSR, simple sequence repeats. Method is equally useful in animals or plants if not indicated otherwise; +++, excellent; ++, good; +, OK; (+), has been used; −, unlikely to be useful or useless; ?, uncertain or not used.

detects a different type of genetic variation. For example, RFLPs may detect restriction site variation or indels depending on how they are scored, whilst variation at SSR loci is usually detected as indels.

2.3 Principles of sampling

2.3.1 Theory

In natural ecosystems, organisms show spatial and temporal structuring of variation, which can have profound influences on their genetics. Organisms may be distributed in two dimensions, for example, the distribution of the tree *Combretum molle* through tropical and southern Africa, or in three dimensions, for example, the distribution of barnacle larvae in the North Sea, or in four dimensions, for example, the distribution of migrating birds between Europe and Africa. Similarly, within a species, genetic variation is hierarchically structured, that is variation can occur within and among populations and regions. Thus ecological and genetic processes, such as patterns of gene flow and localized extinction, influence patterns of genetic variation, for example, outcrossing species tend to have the majority of their genetic variation distributed within populations, whilst inbreeding species tend to have the majority of their genetic variation distributed among populations (Loveless and Hamrick 1984). Furthermore, patterns of variation also exist within a genome with respect to the genes that are sampled (e.g., Hartl, Willing, and Nadlinger 1994). The most effective sampling strategy for the problem of interest would be designed based on knowledge of the structure of the genetic variation. However, in the majority of studies, this is not available *a priori* and thus other means of making sampling decisions must be considered, for example, morphological variation, preliminary genetic analyses, or knowledge of breeding system and gene flow. Clearly, in all but very restricted studies or where the organism is very rare, it is impossible to study every individual and population: in all organisms, it is currently impossible to study every gene.

It is necessary to have some means of sampling material and some objective criteria for doing so; the limits of which will be imposed by the study. Sampling strategies are driven by two factors: the objectives of the study and the resources available. The aim of a sampling strategy is to maximize efficiency, that is, to provide the best statistical estimates of parameters, with the narrowest confidence intervals, at the lowest possible cost.

At the start of any sampling strategy, the population must be defined. At least three different population definitions (statistical, ecological, and genetical) can be recognized in ecological genetics research, in addition to the population of genes contained within each individual. A statistical population is the universe of items that are under study, in contrast to the ecological population, which is the group of organisms of one taxon that occur in a particular area at a particular time (sometimes called the biological population or provenance; Turnbull and Griffin 1986). A genetical population is all of the individuals that are connected by gene flow (sometimes called the gene pool). These three population definitions may coincide (Fig. 2.3a), although more frequently they will not (Fig. 2.3b, c). In ecological genetics research, populations are usually ecologically defined and general conclusions drawn from these about the species or problem under investigation. However, such definitions may not comprise the statistical population, since only a limited part of a taxon's range may have been sampled or observations may only have been made at one point in time. Furthermore, since a sample of the genetical population usually provides the data, the problem of sampling more then one genetical population in an ecological population cannot be overlooked. Additional bias may also occur depending on the material sampled, for example, if tree seeds are sampled, not all of the individuals in a geographical area may produce seeds in one year and hence part of the gene pool may be ignored. Thus, consideration of the relationships between statistical, ecological, and genetical populations is an important part of designing an adequate sampling strategy.

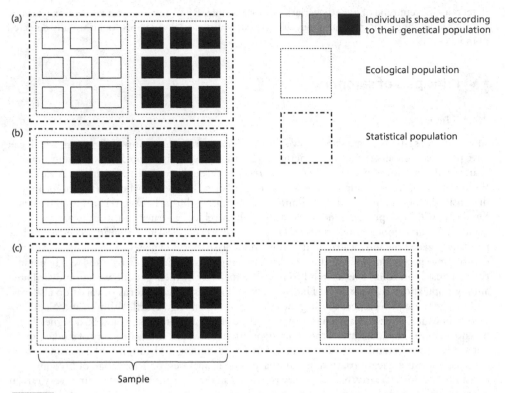

Fig. 2.3

Relationships between statistical, ecological, and genetical populations. (a) The sampled populations comprise the statistical population and the ecological and genetical populations coincide. (b) The sampled populations comprise the statistical population but the ecological and genetical populations do not coincide, as a result of either spatial or temporal factors. (c) The sampled populations do not comprise the statistical population but the ecological and genetical populations coincide.

Allied to population definition is the issue of population size. The number of individuals that occur in an area is the census population size. However, the definition of population size can be standardized in terms of its effective population size (N_e). The effective size of a population is the number of breeding adults in an idealized population of constant size, which would undergo the same rate of genetic drift as the actual population when gametes are sampled at random. The effective size of a population is usually smaller than the census sizes, and is affected by factors including wide fluctuations in population size, relatedness between individuals, inbreeding, unequal sex ratios, and wide variance in reproductive success (Frankham, Ballou, and Briscoe 2002).

The sampling unit in ecological genetics research may include individual populations, individuals within populations, or individual gametes. The definition of the sampling unit will vary with the investigation and must be clearly defined at the start of an investigation.

When statisticians consider sampling, it is assumed that samples are made according to the principles of probability sampling. That is, a set of distinct samples $S_1, S_2, S_3, \ldots S_n$ is defined to which sampling units are assigned. Each sample has a certain probability of selection and one of the S_i samples is selected using an appropriate random method. In practice, sampling for ecological genetics studies rarely meets these criteria of statistically rigorous sampling. For example, in studies of genetic diversity it is rare for the complete ecogeographical distribution of a taxon to be known, thus populations are usually opportunistically sampled based on an ecological definition. Similarly, sampling within populations usually takes place in a haphazard manner, based on either animal capture or a minimum distance of separation between plants. Furthermore, for example, in trees, sampling may be restricted to the most easily accessible branches.

Three basic sampling strategies may be of use in ecological genetics studies: simple random sampling, stratified random sampling, and systematic sampling. Each of these sampling strategies will be described in general terms, followed by a discussion of within-population and among-population sampling, although in specific situations other methods of sampling may be appropriate.

In random, stratified, and systematic sampling approaches, sample selection is made before the survey is conducted; none of the sampling decisions depend on the data that are gathered. However, sampling methods have been developed that make use of the data as they are collected, and these are known as adaptive sampling; this is ideally suited to populations that are highly clumped (Thompson 1992). For example, after analysis of genotype data, an allele at very low frequency might be observed in a population. There may be interest in knowing the distribution of the allele within the population and, hence, rather than sampling the population randomly, it may be more efficient to sample more intensively close to where the allele was observed.

Some methods of collecting samples mean that hierarchical samples are made. For example, a researcher interested in genetic diversity of ostracod species may use a plankton net to sample a very large number of ostracod individuals from a particular lake, and rather than analyze the whole sample a subsample is taken. This is a two-stage sampling process because there are two steps: a sample of units is selected and then elements within each unit are selected. In the example above, the unit is the sample from the lake and the elements are the smaller subsamples. This is analogous to the situation that occurs when populations from across the range of a species are sampled for genetic diversity research: the populations are the units and the individuals within each population are the elements. Clearly, subsampling may be undertaken at many different levels and hence this general approach to sampling is called multistage sampling (Bart, Fligner, and Notz 1998, Cochran 1977). Multistage sampling would be relatively simple if all sampling units contained the same number of elements. However, in the majority of situations, sample units are of unequal size and thus sampling is more complex (Bart, Fligner, and Notz 1998), for example, populations may contain different numbers of individuals.

In the majority of ecological genetics studies, the resources that are available to conduct the research determine sample size. However, there may be situations where alternative sampling methods are used, such that sample size is not fixed in advance but observations are made one at a time, and after each observation, the accumulated data are tested to determine whether a conclusion can be reached. In such sequential sampling procedures, the focus is on decision making, and has the great advantage that it minimizes sample size and hence reduces costs. These procedures have been discussed at length in Dixon and Massey (1983), Krebs (1999), Mace (1964), and Wetherill and Glazebrook (1983).

2.3.2 Simple random sampling

Theoretically, simple random sampling is one of the most straightforward probability sampling procedures and is defined as a statistical population consisting of N sampling units, from which n units are selected in such a way that every unit has an equal probability of being chosen (Fig. 2.4). Each possible sample unit is numbered from 1 to N and then n random numbers, between 1 and N, are drawn with or without replacement. Statistical details of this approach can be found in Bart, Fligner, and Notz (1998), Cochran (1977), and Krebs (1999). However, in practice these criteria are often difficult to follow, especially if the statistical population from which conclusions are to be drawn is greater than the study area.

In order to apply simple random sampling procedures in practice, it is necessary to know the distribution of all of the individuals within a population, for example, all trees in a population

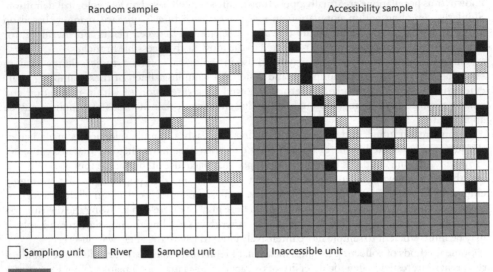

Random sample   Accessibility sample

☐ Sampling unit   ▦ River   ■ Sampled unit   ▨ Inaccessible unit

Fig. 2.4

Random versus accessibility sampling. Each population has been divided into 400 sampling units, of which 36 units were occupied by a river, and 36 units sampled.

would need to be labeled, and to know the ecogeographical distribution of all the populations of a species. However, for the majority of taxa, distribution patterns are poorly known, and may reflect collector bias rather than be true representations of distributions. For example, detailed distributions of even the most common organisms may be unknown in tropical regions, whilst the disjunct UK distribution of *Rosa caesia* ssp. *caesia* probably represents collecting activity, rather than a truly fragmented distribution (Graham and Primavesi 1993). Thus, for practical reasons, few ecological genetics investigations use simple random sampling, rather they use a series of methods that may be described as pseudorandom sampling and which Krebs (1999) has characterized as: (i) accessibility sampling; (ii) haphazard sampling; and (iii) judgmental sampling.

In accessibility sampling (Fig. 2.4), the sample is restricted to those units that are readily accessible (e.g., sampling in densely forested areas may only be practical where tracks or rivers occur; collecting permits may only be available for parts of a species' range). Haphazard sampling is similar, in that the sample is selected in a haphazard or opportunistic manner (e.g., samples are collected whenever the researcher is ready). Judgmental sampling involves the experience of the collector as to what constitutes a series of typical sampling units (e.g., sampling seed from individual trees that are >50 m apart in order to reduce the probability of sampling seed from closely related individuals). Whilst these methods may provide a good sample under appropriate conditions, they are rejected as statistically random samples since they cannot be evaluated by probability theory (Cochran 1977).

2.3.3 Stratified random sampling

Stratification of a population is a very powerful means of sampling, and is constantly done implicitly by researchers. In a stratified sample, a statistical population of N units is divided into L non-overlapping strata, which together comprise the whole population (i.e., $N = N_1 + N_2 + N_3 + \ldots + N_L$; Fig. 2.5). Stratification is usually based on geographical area and hence the area of each stratum is easily found. Once the strata have been determined, each stratum is sampled

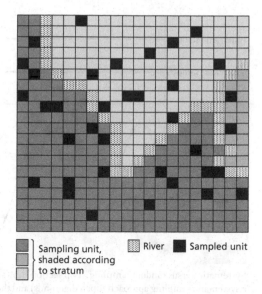

Fig. 2.5

Stratified random sampling. The population has been divided into 400 sampling units and three strata, of which 36 units were occupied by a river, and 36 units sampled.

Sampling unit, shaded according to stratum   River   Sampled unit

separately, and if strata are sampled using simple random sampling, the whole process is called stratified random sampling. Statistical details of this approach can be found in Bart, Fligner, and Notz (1998), Cochran (1977), and Krebs (1999).

One of the most important reasons for stratification of a sampling strategy is that if the strata are well defined there may be a gain in precision for estimates of the mean and confidence intervals of the whole population. However, stratification may be necessary if estimates of means and confidence intervals are needed for each stratum, there are more sampling problems in some areas than others (e.g., organisms may be more difficult to capture in some habitats than others), different groups of researchers sample different parts of the distribution, or populations extend over transnational boundaries, where different restrictions apply to sampling for biological research.

When planning a stratified sampling strategy it is necessary to determine the number of strata that are needed and how many sampling units should be measured in each stratum. The issue of the number of strata needed is not straightforward and is determined by the problem that is being investigated and the precision that is required. Cochran (1977) has argued that the number of strata should not exceed six, although fewer strata may be desirable. Stratification is usually based on ecological variables defined by the researcher, although other statistical methods of stratification are available (Iachan 1985).

Proportional allocation, constant allocation, and optimal allocation of resources are the main strategies for the allocation of sampling units among strata, although at least two individuals should be sampled per population so that variances can be estimated. The simplest method of stratified sampling is to allocate samples based on either a constant sampling fraction of each stratum (proportional allocation), for example, 5% of the sample units per stratum, or equal numbers of samples from each stratum (constant allocation), for example, 20 sample units per stratum. Proportional allocation weights each stratum according to its size relative to the total size and therefore tends to favor large strata, whilst constant allocation tends to favor small strata. The number of samples needed is determined by the acceptable confidence intervals for a particular study: the higher the level of precision the larger the sample needed.

When prior information is available, as, for example, after a pilot study, more powerful optimal allocation methods may be used. In particular, the cost of sampling may be minimized (e.g., Cochran 1977). In general, using cost functions, larger samples within a stratum are justified when the stratum is large and heterogeneous, and sampling is cheap (Krebs 1999). In

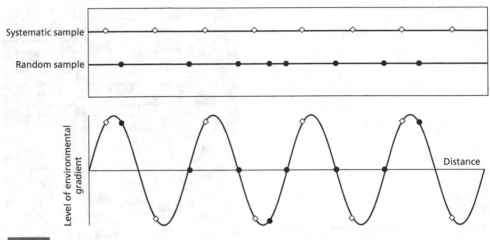

Fig. 2.6

Systematic versus random sampling. Top section shows two transects across a population, one taken using a systematic sampling approach (open diamonds) and the other taken using a random sampling approach (filled circles). The bottom section shows the distribution of these sample points across a population with a periodic environmental gradient.

addition to minimizing the total cost for a specified standard error, optimal allocation methods may be used to minimize the standard error of the stratified mean for a fixed total cost. Optimal allocation approaches to stratified sampling have been widely used by researchers interested in the maximization of genetic diversity within germplasm collections and the minimization of their associated costs (e.g., Brown 1989).

2.3.4 Systematic sampling

Systematic sampling is one of the most common and convenient methods of sampling in ecological genetics. That is, samples are made at fixed points on a line, grid, or physical feature (e.g., road or river). The main reasons for the use of this approach are the simplicity of its application in the field, the desire to sample evenly across a region, and to minimize sampling closely related individuals. Thus, this approach is particularly useful if a gradient or cline is suspected in a study, for example, a hybrid zone. One of the most common types of sampling procedure is to divide the region of interest into equal parts and sample from each part. However, periodic variation in the system being sampled is an important problem with systematic sampling, and will lead to bias in the estimation of the mean and variance of a population (Fig. 2.6; Krebs 1999). In practice, rather than periodic variation occurring, patterns tend to be highly clumped and irregular. Although statisticians condemn systematic sampling (e.g., Cochran 1977), Krebs (1999) concludes that systematic sampling can often be applied, and the resulting data treated as random sampling data, without bias. However, since there are concerns over data periodicity that may influence parameter estimates, it is necessary to be aware of these and, if there is a choice, always use random sampling.

2.4 Practice

The organism being investigated determines practical sampling, for example, approaches to sampling deep-sea crustaceans are unlikely to be of much use for sampling rainforest trees.

However, there are principles common to all methods of practical sampling, irrespective of the organism and problem being studied. For example, in addition to sampling design, the investigator must consider how an organism will be collected, what material will be sampled, how it will be transported between the field site and the laboratory, and ensure that sampling occurs in an appropriate legal and ethical manner. The design of effective sampling strategies is aided if the biology, for example, distribution, ecology, and reproductive biology, of the organism of interest is understood; it may even be appropriate to undertake an ecogeographical survey (e.g., Maxted and Guarino 1997). Furthermore, the sampling strategies used by researchers interested in similar problems will also be useful. Data on geographical distribution is particularly valuable for planning fieldwork, and can be found in libraries, especially taxonomic monographs, Floras, and Faunas, or summarized from biological collections. In addition to sampling in the field, alternative sources of material for ecological genetics studies include *ex situ* conservation collections (e.g., seed banks, zoological gardens, botanical gardens, sperm and egg banks) and museum collections (e.g., herbaria, skeleton collections, study skin collections, spirit collections). Specimens in such collections are, however, only as good as the information associated with them, and should be used with caution, for example, seed collected from plants in botanical gardens may be of hybrid origin. In addition, the methods used to sample specimens in the collection should be known, for example, seed banks may contain samples that are collected from individual plants in a population (family collections) or the whole population (bulk collections).

Numerous guidelines are available for practical sampling of different types of organisms and such works should be consulted at the planning stage of an investigation. These guidelines provide valuable information on minimum levels of documentation, safety in the field, and, for animals, appropriate humane methods of collection and, if necessary, killing. In addition, researchers must be aware of the appropriate methods for the preparation of high-quality voucher specimens. Voucher specimens should be collected and deposited in suitable, accessible biological collections, since a specimen is the only physical proof that a particular species was used in a study, and provides a link between the organism in the field and the data. Journal editors are increasingly demanding that ecological genetics data are supported by voucher material, especially in botanical and zoological journals. Ultimately, data are only as reliable as the accuracy and reliability of the observations; the linkage of data to a physical object and the documentation of such collections are of paramount importance (Goldblatt, Hoch, and McCook 1992). Errors continually creep into the literature, based on inadequate documentation and the absence of collections to support data. It is ironic that seed collections and germplasm conservationists have recognized this fact, yet many published studies that claim conservation value for their studies do not support their observations with the physical evidence of a specimen. Similarly, DNA sequence data that are deposited in sequence databases should be supported by reference back to a voucher specimen so that identifications can be checked in case of queries.

The type of material collected during sampling will depend on the problem being investigated and the marker chosen, for example, freshly collected material is generally necessary for allozyme studies and investigations that include the establishment of genetic crosses, whilst PCR-based sequence analysis may be used with prehistoric bone fragments (Scholz et al. 2000). However, an investigator has two main types of material available: living (or freshly collected) and preserved. Living materials would include whole organisms collected in the field (and transported to the laboratory), seeds and fruits and cuttings, whilst preserved material would include dried specimens and fluid-preserved samples. The collection of living materials or frozen material can lead to problems with transporting material from the field to the laboratory, particularly if field sites are remote. However, seeds are an ideal means of collecting living plant materials, although about 30% of seeds are recalcitrant and cannot be readily stored, subsequent germination of seeds may slow down the generation of data, and sampling occurs at the level of the progeny rather than of the mother. In contrast, with the development of PCR technology, small amounts of preserved materials can be used for DNA isolation, for example, DNA isolation from herbarium specimens, animal hairs, and entomological collections. Numerous methods for non-invasive sampling of genetic material have been developed (e.g., Taberlet and Waits 1998), whilst drying leaves in silica gel has become a favored method for collecting plant material (Chase and Hills 1991). A number of methods for the collection of genetic material and DNA isolation from animals and plants have also been developed (e.g., Doyle and Doyle 1987, Laulier et al. 1995, Lookerman and Jansen 1995).

As with all scientific research, ecological genetics programs must be conducted within appropriate legal and ethical frameworks. Among the many principles that were recognized by the United Nations Conference on the Environment and Development (UNCED), in Rio de Janeiro in 1992, was that individual countries have sovereign rights over the biological resources that occur in their territories (ten Kate and Laird 1999). Furthermore, individual countries have become aware of the diverse values of the biological resources of which they are custodians (Kunin and Lawton 1996). Thus, international treaties (e.g., Convention on International Trade in Endangered Species of wild fauna and flora (CITES)) and national legislation (e.g., UK Wildlife and Countryside Act 1981) now govern the utilization and study of biological resources. It is important that ecological genetics research is undertaken within these frameworks, that the necessary permits are obtained at the beginning of a research program, and that researchers comply with any restrictions imposed by the regulating authorities.

Restrictions on access to collection sites may mean that modifications must be made to sampling strategies or alternative sources of biological materials considered. In addition to over-coming problems of collecting restrictions, biological collections are a resource for ecological genetics research since temporal variation in population genetic structure, because of extinction or population genetic processes, may be studied and the costs of sampling materials can be reduced. However, the destructive sampling of specimens in collections should only be undertaken as a last resort. Unfortunately, specimens from biological collections are of varying quantity and quality. Collectors are biased in their choices of collection localities and tend to be attracted to areas where they know a particularly interesting species occurs, hence, there may be multiple collections from the same locality over many years. Such collections may be of great interest to understanding temporal variation, although there are rarely large numbers of collections from a single site collected at a single time. Collections will also represent areas that are relatively easily accessible, rather than be a true representation of the geographic distribution of a species (e.g., Stern and Eriksson 1996). In addition to spatially erratic collecting, there may also be temporally erratic collecting, where large amounts of collecting may have been undertaken during some periods, whilst during other periods there has been less collecting. Furthermore, attitudes to col-lecting have changed, for example, large numbers of British plant specimens were collected in the early years of the twentieth century but collecting has now been reduced to a fraction of this level. Specimens may be of varying quality, either as a result of the amount of label data that they carry or because of their treatment in the field or in the collection, for example, specimens may have been heat-treated, microwave-treated, or poisoned with heavy metals (e.g., Lookerman and Jansen 1995, Stern and Eriksson 1996). Since suitable biological collections may be scattered around the world, with different collection curators having different policies about destructive sampling in their collections, researchers who use biological collections as a source of material in ecological genetics studies should be aware of their responsibilities to both collections and future users.

2.5 Within-population sampling

2.5.1 Sampling to estimate allele frequency

Allele frequency is one of the most common values calculated in ecological genetics studies, when populations are compared. For a randomly sampled population, the sample size necessary to estimate allele frequency at a codominant locus with a given degree of confidence may

Box 2.1

Determination of sample size needed to estimate allele frequency at a codominant locus

In a randomly mating population of infinite size, it is desired to determine the size of sample necessary to estimate the frequency of an allele A at a diallelic locus within a specified margin of error. Let p be the frequency of allele A and q ($= 1 - p$) the frequency of allele B. Let ε be the acceptable margin of error to the estimate of p and α the probability of not achieving this margin of error. If the sample size is large ($n > 20$) and p is normally distributed then the estimate of the confidence interval is $\hat{p} \pm t_\alpha s_{\hat{p}}$, where $\hat{p}$ is the observed frequency, t_α is the normal deviate corresponding to a desired confidence probability and $s_{\hat{p}}$ is the standard error of $\hat{p}$ ($= \sqrt{\hat{p}(1 - \hat{p})/n}$). For 95% confidence intervals, the normal deviate is taken to be 2.00 (rather than 1.96, in order to render the estimate more conservative). Therefore, the desired margin of error, with a 5% chance of not achieving this error

level, is: $\varepsilon = 2 s_{\hat{p}} = 2\sqrt{\dfrac{\hat{p}(1 - \hat{p})}{n}}$, and thus

solving for n, the required sample size is:

$n_{codom} = \dfrac{4\hat{p}(1 - \hat{p})}{\varepsilon^2}$. Moderate departures

from normality do not affect estimation of confidence intervals. However, for small samples with skewed distributions

additional methods will be necessary (see Cochran 1977).

If a finite population is being sampled, then the finite population correction can be applied to the estimate of sample size needed. That is, $n^* \cong \dfrac{n}{1 + \dfrac{n}{N}}$, where n^* is

the estimated sample size required from a finite population of size N. However, if the sampling fraction does not exceed 5–10% then the finite population correction can be ignored.

For example, consider a study in which a researcher wishes to estimate the frequency of a particular codominant allele in a population within 5% of the true mean. Resources are limited and it is necessary to identify an optimal sample size, that is, $0.05 = 2 S_{\hat{p}}$. The worst-case scenario from this study would be when the allele frequency was 0.5, where a maximum sample size of 400 would mean that on 95% of occasions the true value of the frequency of p would be between 0.45 and 0.55. However, following a pilot study, additional information may be available that would enable the researcher to suggest that the frequency of the allele of interest was approximately 0.2, thus the necessary sample size would be 256. If the researcher can only accept an error level of ±0.01, then the sample size rises to 10,000 for the worst-case scenario and to 6400 for an estimated allele frequency of 0.2.

be determined using standard statistical methods (Box 2.1; Cochran 1977). In order to estimate sample size it is necessary to know the allele frequency. However, this information is rarely available at the start of an investigation, unless a pilot study has been undertaken. Therefore, sampling estimates may be based on the worst-case scenario, that is, the allele frequency where there is maximum variance. For a codominant, diallelic locus, this is where $p = 0.5$ (Fig. 2.7). Thus, with a 95% confidence interval the required sample size is: $\dfrac{1}{\varepsilon^2}$, which with a ±5% acceptable margin of error the maximum sample size becomes 400 independent gametes. As either the acceptable margin of error increases or the allele frequency changes from $p = 0.5$, the maximum necessary sample size decreases (Fig. 2.8a). Conversely, if one has a sample of 50 gametes (25 diploid individuals) then the margin of error, with a 5% chance of not achieving this error level, at an allele frequency of $p = 0.5$, will be ±14%.

However, a widely utilized group of ecological genetic DNA markers (e.g., RAPDs and AFLPs) show dominant inheritance. For these markers it is assumed that each amplification

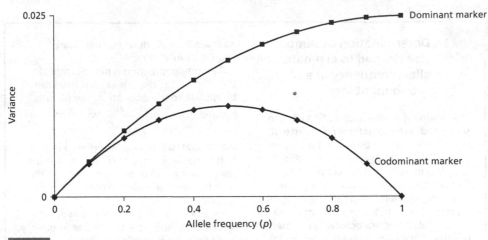

Fig. 2.7

Change in variance with allele frequency at a diallelic locus, in a population of 10 individuals. Individuals are fully outcrossed and only one progeny is sampled per individual mother.

product position on a gel represents a different locus and that there are only two alleles ("product present" and "product absent") possible at any locus. Thus, the only genotype that can be reliably identified is absent–absent: the present–present and present–absent genotypes will show identical product-present phenotypes. Let $\hat{x}$ be the estimated fraction of individuals in the population without the marker, that is, $\hat{x} = \hat{q}^2$, thus $\sqrt{\hat{x}}$ is the estimated frequency of product absence ($\hat{q}$). Using the notation and approach in Box 2.1, the number of individuals that must be sampled for a dominant marker (n_{dom}), with a 95% confidence interval, is given by:

$$n_{dom} = \frac{2(1 - \hat{q}^2)}{\varepsilon^2}$$

For dominant loci, variance increases as $\hat{q}$ is reduced (Fig. 2.7), and as the margin of error increases, the necessary sample size decreases (Fig. 2.8b). The ratio of n_{dom} to n_{codom} $\left(\dfrac{(1 + \hat{q})}{2\hat{q}}\right)$ is the proportional increase in the sampling that must be undertaken to achieve the same accuracy for dominant markers as for codominant markers (Fig. 2.9; Lynch and Milligan 1994).

So far, allele frequencies have been considered for a population that is randomly outcrossing, that is, the probability of selfing (s) is zero and only one progeny (k) is sampled per individual. However, studies of plants often use progeny arrays (seeds) to estimate population allele frequencies. However, when more than one progeny is analyzed from a single individual ($k > 1$) or when mating is non-random ($s > 0$), then the sample of gametes is no longer statistically independent, and, as a consequence, variance will increase (Fig. 2.10; Brown and Weir 1983). Increasing the size of a progeny array in an inbreeding species leads to little improvement in the accuracy of the estimate of allele frequency (Fig. 2.11), whilst for outbreeding species, there is little gain in accuracy beyond a sample size to progeny ratio of 0.2.

2.5.2 Sampling to estimate allele number

The mean number of alleles in a population sample is an important criterion that has been widely used in genetic conservation, particularly in the management of germplasm collections and

(a)

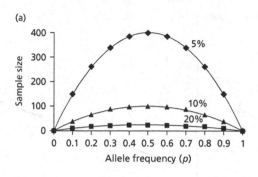

(b)

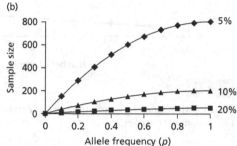

Fig. 2.8

Change in sample size with allele frequency at a diallelic locus (95% confidence interval) in an infinite population at three margins of error (5%, 10%, and 20%) for (a) codominant marker and (b) dominant marker.

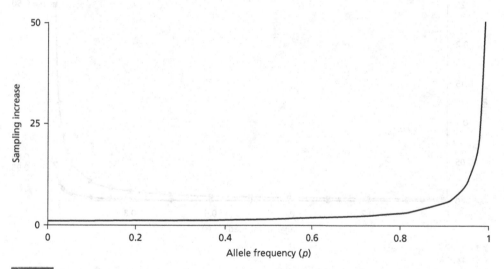

Fig. 2.9

Proportional increase in sampling needed to achieve the same accuracy for dominant markers and codominant markers in relation to allele frequency (p; $q = 1 - p$).

setting priorities for population conservation (Brown and Briggs 1991). Thus, there has been extensive discussion of the theory for the optimal sampling of allele numbers in samples for genetic conservation (e.g., Brown and Hardner 2000, Lawrence and Marshall 1997).

The number of alleles observed at a locus is a steadily increasing function that does not approach an asymptote, that is, for very large sample sizes it is possible to detect very rare alleles. The calculation of the theoretical allele profile for a population with two alleles at a locus is relatively simple (Marshall and Brown 1975). Consider a population with two alleles (A_1, A_2) at

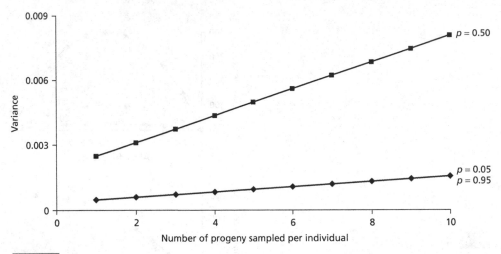

Variance when sampling is done from progeny arrays ($k = 1$ to 10). $N = 100$, $s = 0$, and $p = 0.05$, 0.5, and 0.95 for a codominant marker.

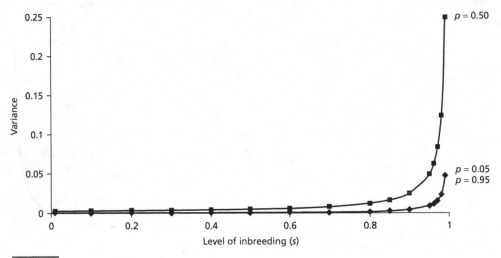

Variance when sampling is done at different levels of inbreeding; $s = 0$ is fully outcrossed and $s = 1$ is fully inbred. $N = 100$, $k = 1$, and $p = 0.05$, 0.5, and 0.95 for a codominant marker.

frequencies p_1 and p_2, respectively. The probability that a random sample of n gametes contains at least one copy of each allele ($P[A_1^+, A_2^+]$) is given by:

$$P[A_1^+, A_2^+] = 1 - (1 - p_1)^n - (1 - p_2)^n + (1 - p_1 - p_2)^n$$

If $p_1 = 0.95$ and $p_2 = 0.05$, then 59 gametes are required to obtain at least one copy of each allele with 95% certainty but if the frequency of the most common allele increases to 0.99 then a sample of 300 gametes is needed to ensure that at least one copy of each allele is obtained (Fig. 2.12).

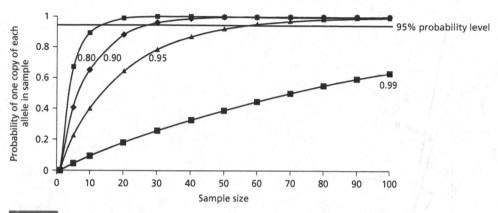

Fig. 2.12

The probability of sampling both alleles at a locus in samples of different size and with the frequency of the most common allele varying from 0.80 to 0.99.

That is, the sample size needed to have a 95% chance of sampling at least one copy of each allele is heavily dependent on the frequency of the rarest allele in the population. However, as the number of alleles per locus increases, the exact probability expression becomes more complex (Marshall and Brown 1975) but for the case of 20 alleles, each with a frequency of 0.05, a random sample of ca. 120 gametes will provide 95% certainty that one copy of each allele will be sampled, thus under most circumstances random samples of 50–100 gametes will be more than adequate. In practice, individual gametes are rarely sampled, rather individuals are sampled. Furthermore, in the case of plants it is usual to sample seeds from individual fruits.

Whilst the number of alleles that might be expected in a sample of given size is a function of allele frequency, if the alleles are neutral and follow Kimura and Crow's (1964) infinite neutral allele model (IAM), then Ewens (1972) has provided expressions for the numbers of alleles that can be expected in a sample of n gametes. In the IAM it is assumed that there is an equilibrium between the generation of alleles through mutation and their loss through genetic drift; both of which are stochastic processes. Furthermore, the IAM assumes each mutation produces a novel allele, not previously present in the population. The expected number of alleles (n_a) with a frequency between p and q ($0 < p < q < 1$) in an equilibrium population of effective size (N_e) is:

$$n_a = \theta \int_p^q \frac{(1-x)^{\theta-1}}{x} dx$$

The distribution of allele number in the population depends on a single parameter θ ($= 4N_e\mu$), where μ is the mutation rate. When θ is small (i.e., populations are small or mutation rate is low) most alleles are either very common or very rare. However, as the population size or mutation rate increases, a greater proportion of the alleles will occur at intermediate frequencies (Fig. 2.13). Thus, in very large populations it is expected that a virtually infinite number of alleles will occur, each at very low frequencies. Ewens (1972) showed that in a random sample of n gametes one would expect to capture n_a alleles ($E(n_a)$).

$$E(n_a) \approx \theta \, log_e[(n + \theta)/\theta] + 0.6$$

If θ is very small, $E(n_a) = 1$, whereas if θ is large, $E(n_a)$ approaches n, implying that if a sample is large enough and the mutation rate at the locus is high enough, then all of the sampled alleles will

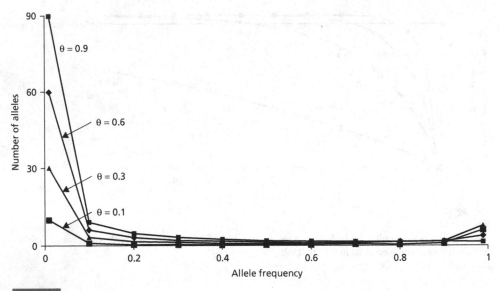

Fig. 2.13

Expected number of alleles under the infinite allele model at different allele frequencies for four values of θ (0.1, 0.3, 0.6, 0.9).

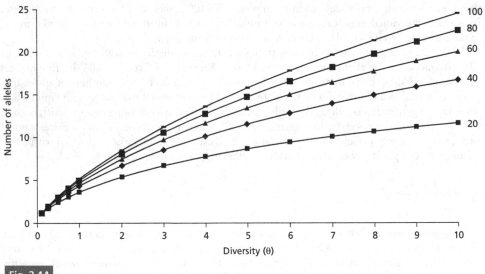

Fig. 2.14

The number of alleles (n_a) expected in samples of 20 to 100 gametes with varying values of θ ($= 4N_d\mu$).

be different (Fig. 2.14). Furthermore, there is a logarithmic increase in the proportion of alleles related to sample size but the effort associated with collecting samples from a population can be considered linear. Hence, a point is reached at which it is no longer economic to increase sample size in order to collect additional alleles.

2.5.3 Sampling to estimate gene diversity

Gene diversity, as formulated by Nei (1978), is one of the most widely used measures of genetic variation, since it does not depend on definitions of polymorphism, as the percentage of polymorphic loci does, and it can be explicitly formulated in terms of allele frequencies. Whilst the measure is dependent on sample size, the effect is very small compared to the mean number of alleles per locus, since Nei's measure of gene diversity is relatively unaffected by low-frequency alleles.

If the researcher is interested in accurate estimates of single locus gene diversities, Brown and Weir (1983) argue that it is necessary to sample approximately the same number of individuals as for accurate estimates of allele frequencies. However, Nei and Roychoudhury (1974) showed that when mean gene diversity is measured it is necessary to consider the sampling of both loci and alleles. That is, the variance of gene diversity ($V(\hat{h})$) is composed of a variance associated with sampling loci (interlocus variance; $V_l(h)$) and a variance associated with sampling alleles (intralocus variance; $V_a(h)$):

$$V(\hat{h}) = V_l(h) + V_a(h)$$

Increasing the number of individuals sampled will reduce intralocus variance; only by increasing the number of loci sampled will interlocus variance be reduced. Thus, in order to estimate mean gene diversity it is necessary to have optimal sampling of both individuals and loci. In the majority of investigations, $V_a(h)$ is much smaller than $V_l(h)$, unless the number of individuals sampled is very small, which led Nei and Roychoudhury (1974) and Nei (1978) to recommend that large numbers (>50) of codominant loci must be analyzed if reliable estimates of average gene diversity are to be obtained. However, for practical purposes, such large numbers of loci would need to be traded-off against the number of individuals sampled per population. Nei (1987) recommends that if about 25 loci are examined, then 20 to 30 individuals should be sampled per locus.

In practice, either the marker system or the study organism imposes technical limitations on the number of loci that can be studied. For allozymes, the number of markers may be limited by plant secondary chemistry that affects the ability to extract functional enzymes, whilst, in animals, the limitations may be imposed by access to suitable material for extraction. DNA markers allow large numbers of individuals to be screened, although marker systems may vary in the numbers of loci that they detect. Some marker systems identify few multiallelic loci (e.g., SSRs), whilst others identify many diallelic loci (e.g., AFLPs, RAPDs). However, Mariette et al. (2002), using computer simulations, showed that unless populations were in drift-migration equilibrium, there were substantial discrepancies between the diversities assessed using microsatellite or AFLP markers across the genome. Thus, intragenomic sampling must be considered if accurate estimates of gene diversity are to be obtained.

Natural populations are often subdivided into subpopulations, and thus it may be useful to study gene diversities within and between these subpopulations: a very common way of doing this is to use Nei's (1973) method. That is, gene diversity in the total population (H_T) can be decomposed into the gene diversities within (H_S) and between (D_{ST}) subpopulations:

$$H_T = H_S + D_{ST}$$

The relative magnitude of gene differentiation among subpopulations (G_{ST}) is measured as:

$$G_{ST} = \frac{D_{ST}}{H_T}$$

Pons and Petit (1995) investigated the optimal sampling strategy for G_{ST} at a haploid locus and showed that in order to minimize the variance of G_{ST} an optimal sample size of 2.5 individuals per population is necessary. Furthermore, in order to obtain reasonably precise estimates of G_{ST} it is important to have large numbers of populations sampled and that if the number of populations sampled is small then there is little point in sampling more than five individuals per population. Pons and Chaouche (1995) provide formulae for the estimation of optimal sample sizes for the analysis of population differentiation at diploid loci.

2.5.4 Sampling to estimate genetic distance

Molecular marker data are often used to infer similarities between species or populations using genetic distance, that is, the extent of genomic differences between two populations. Many genetic distance measures, for example, Rogers' genetic distance and Nei's genetic distance, are calculated based on allele frequencies (Nei 1987, Nei and Kumar 2000), and displayed as dendrograms (Swofford et al. 1996). Archie, Simon, and Martin (1989), working at the inter-specific level, conclude that four general properties of frequency data are likely to have sub-stantial effects on dendrogram stability: (i) overall patterns of genetic differentiation; (ii) levels of heterozygosity within species; (iii) patterns of gene identity among taxa and groups of taxa; and (iv) mean allele frequency and the degree of variability of frequencies for shared alleles. However, there is no *a priori* reason to think that these concerns are only the remit of interspecific invest-igations; they are also concerns of intraspecific investigations.

Nei (1978) showed that the standard error of genetic distance is high when gene diversity is high. Thus, when gene diversities of greater than 0.1 are expected it will be necessary to have relatively large numbers of individuals (>50) to construct robust dendrograms. As with meas-ures of gene diversity, two sampling processes are involved in the estimation of genetic distance: sampling of loci from the genome and sampling of individuals from the populations (Nei 1978, Nei and Roychoudhury 1974). However, despite recommendations that it is necessary to use >50 loci to obtain accurate genetic distance estimates, most allozyme studies, for example, use between 15 and 30 codominant marker loci, which is largely due to the difficulty of identifying suitable loci. The numbers of individuals sampled per locus usually vary between 20 and 30, although as few as one and as many as 1000 have been used in some studies (Archie, Simon, and Martin 1989). Furthermore, equal numbers of individuals are rarely sampled per locus.

One of the most influential papers in this area has been that of Gorman and Renzi (1979) who examined the effects of small sample size on dendrogram accuracy from two separate *Anolis* lizard data sets. These workers were specifically interested in the use of allozyme data for phylogeny reconstruction and came to the conclusion that only one or two individuals needed to be sampled in order to make a reasonable estimate of the similarities between taxa, a conclusion that was supported by the theoretical work of Nei (1978). However, Archie, Simon, and Martin (1989) reanalyzed Gorman and Renzi's data and concluded that their general conclusion was not sup-ported by their own data and that there can be a severe lack of stability in the case of small sample sizes. Archie, Simon, and Martin (1989) concluded that samples of at least 20 individuals are needed, from as many taxa as possible, although this does not guarantee accurate understanding of the differentiation found in poorly sampled taxa.

It is generally thought that loci with more alleles produce better estimates of genetic distance than loci with few alleles. However, loci with many alleles may be more difficult to score com-pared to those with fewer alleles, for example, due to alleles overlapping or only separating over a very short distance on a gel. Provided the divergence between populations is limited, Kalinowski (2002) showed, using simulation studies, that to estimate genetic distance it is unnecessary to have either highly polymorphic loci or many loci, rather sufficient numbers of alleles must be examined.

Sampling of the terminal nodes affects the structure of dendrograms generated from genetic distance measures to assess similarities between populations or to understand phylogenetic relationships (e.g., Baverstock and Moritz 1996, Lecointre et al. 1993, Swofford et al. 1996). However, the effects of terminal node sampling are incompletely understood and have only now started to be investigated in detail (Poe 1998). Poe (1998) has shown that for phylogenies with small numbers (ca. 20) of terminal taxa the effects appear to be small, although the effects on larger phylogenies are unknown. Investigations by Backlejau et al. (1996) of UPGMA trees have shown that ties within genetic distance data sets have major effects on dendrogram structures, although methods are available to test the statistical support for particular branching patterns (Felsenstein 1988 and Ritland 1989).

2.5.5 Sampling to estimate gene flow

The movement of genes among populations (gene flow; Chapter 4) is an important issue in ecological genetics. The majority of theoretical models are based on either continuous populations, using an isolation-by-distance approach, or island populations that become differentiated through mutation and genetic drift. The ease with which genetic structure statistics can be measured, and the widespread availability of biochemical and DNA markers has led to many studies that report estimates of gene flow (Neigel 1997). However, both the temporal and spatial scales of these studies violate the assumptions of gene flow models based on genetic structure (Chapter 4): such estimates reflect evolutionary history not ongoing processes. Thus, at the ecological level, the use of genetic structure to estimate gene flow is inadvisable (Bossart and Prowell 1998). An alternative approach to gene flow estimation is parentage analysis, in which parents (usually fathers) are identified and the pattern of gene movement quantified (Devlin and Ellstrand 1990, Smouse and Meagher 1994). Parental analysis provides a direct estimate of gene movement but it is usually based on only one or two reproductive episodes, rather than gene flow over a whole generation. Furthermore, parentage analysis-based gene flow estimates measure immigration into a circumscribed area that may or may not be a population.

To estimate the sample size needed to detect immigration of male gametes, it is necessary to know the exclusion probability of the marker (Box 2.2), the level of gene movement one wishes to

---

Box 2.2 Exclusion probabilities

An exclusion probability is the ability a marker to exclude a given relationship and is determined by the genotypes of the supposed relatives, allele frequencies at the loci, and the number of independent loci tested (Jamieson and Taylor 1997).

The exclusion probability can be increased in two ways: (i) using loci with many alleles per locus (Fig. 2a(i)); and (ii) increasing the number of loci investigated (Fig. 2a(ii)). Thus, for a codominant locus with n different alleles (e.g., SSR loci), the ith allele with population frequency p_i, the generalized paternity exclusion probability is:

$$P = 1 - 2\sum_{i=1}^{n} p_i^2 + 2\sum_{i=1}^{n} p_i^3 + 2\sum_{i=1}^{n} p_i^4 - 3\sum_{i=1}^{n} p_i^5$$
$$- 2\left[\sum_{i=1}^{n} p_i^2\right]^2 + 3\sum_{i=1}^{n} p_i^2 \sum_{i=1}^{n} p_i^3$$

However, P is maximized when all n alleles at the locus have a frequency of $1/n$. If K loci are investigated, each with an exclusion probability of P_i then the total paternity exclusion probability is:

$$P = 1 - \prod_{i=1}^{K}(1 - P_i)$$

Three types of exclusion probabilities can be calculated: (i) given two parents and one progeny, exclude a parent (i.e., paternity); (ii) given one parent and one progeny, exclude their relationship; and (iii) given two parents and one progeny, exclude both parents. For a codominant marker, the generalized paternity exclusion probability is given above. However, for one parent and one progeny, to exclude their relationship, the generalised exclusion probability is:

$$P = 1 - 4\sum_{i=1}^{n} p_i^2 + 2\left[\sum_{i=1}^{n} p_i^2\right]^2 + 4\sum_{i=1}^{n} p_i^3 - 3\sum_{i=1}^{n} p_i^4$$

whilst for two parents and one progeny, to exclude both parents the generalized exclusion probability is (Jamieson and Taylor 1997):

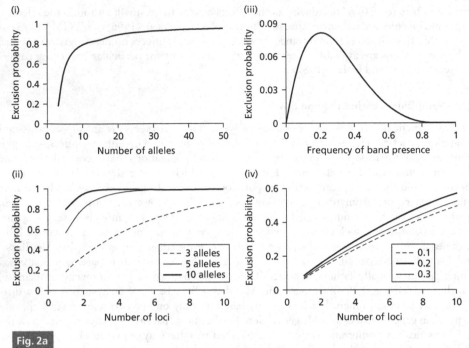

Paternity exclusion. (i) Effect of the number of alleles at a codominant locus on exclusion probability. (ii) Effect of the number of codominant loci with three, five, or 10 alleles per locus on exclusion probability. In each case, it is assumed that allele frequency is the reciprocal of the number of alleles per locus. (iii) Effect of the frequency of band presence on exclusion probability at a dominant locus. (iv) Effect of the number of loci on exclusion probability using a dominant marker with band presence frequencies of 0.1, 0.2, or 0.3.

$$P = 1 + 4\sum_{i=1}^{n} p_i^4 - 4\sum_{i=1}^{n} p_i^5 - 3\sum_{i=1}^{n} p_i^6 - 8\left[\sum_{i=1}^{n} p_i^2\right]^2$$

$$+ 8\sum_{i=1}^{n} p_i^2 \sum_{i=1}^{n} p_i^3 - 2\left[\sum_{i=1}^{n} p_i^3\right]^2$$

Exclusion probabilities can be extended to dominant marker systems (e.g., RAPDs and AFLPs), although the equations are very different to those generated using codominant markers (Gerber et al. 2000, Weir 1996). Consider a dominant marker with two phenotypes, band presence (+) and band absence (–). If p is the frequency of band presence, then in a panmictic population, the frequency of the – phenotype is $(1 - p)^2$ and that of the + phenotype is $p(2 - p)$. In the case of paternity, when the mother is known, the only combination of genotypes that can exclude a father is when the mother's phenotype is – (frequency is $(1 - p)^2$), that

of the progeny + (frequency is p) and the father – (frequency is $(1 - p)^2$). Therefore, the paternity exclusion probability is: $P = (1 - p)^2(1 - p)^2 p = p(1 - p)^4$. In the case of a single parent–progeny comparison, the exclusion probability is zero, that is, dominant markers cannot be used to address this type of question. If a pair of parents is compared to a progeny, the only case that can be excluded is when both parents have the – phenotype (frequency is $(1 - p)^2$) and the progeny have a + genotype (frequency is $p(2 - p)$), thus the exclusion probability is: $P = p(2 - p)(1 - p)^4$. Gerber et al. (2000) showed that both AFLP and SSR markers can produce high exclusion probabilities, although for dominant markers those with allele frequencies in the range of 0.1–0.4 were the most informative. However, as expected, codominant markers were more efficient at reconstructing parentage.

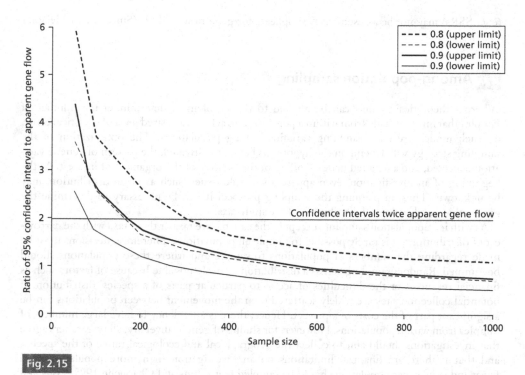

Effect of seed sample number on the relative size of a 95% confidence interval about an estimated rate of apparent gene flow to the value of the estimate itself. Assuming that total rate of gene flow equals 5%, that there are only five possible parents, and the probability of paternity exclusion is 0.80 or 0.90. Data from Sork et al. (1998).

detect, and the number of potential fathers within the study population. If the maternal genotype is known, then Sork et al. (1998; Fig. 2.15) have estimated that for a marker with an exclusion probability of 0.8, where 5% of gene movement is to be estimated and there are only five potential fathers in the population, then approximately 300 progeny must be sampled to ensure that the 95% confidence intervals are less than twice that of the apparent gene flow. If the exclusion probability of the markers increases to 0.9 then the sample size is approximately 200. It is clear that unless the populations are very small, then high exclusion probabilities are necessary to ensure that progeny sample sizes are empirically realistic.

For any mother, the number of progeny sampled should exceed the total number of potential fathers. In plants, sampling mothers close together will reduce the number of potential fathers that need to be sampled, a strategy that is useful for multi-site investigations. However, analysis of neighboring mothers may not be adequate for studies of population dynamics if there is high variation in female gene flow patterns within a population. For estimates of individual male fertility, many progeny per mother should be collected. Most models of parentage analysis assume that progeny are sampled at random from the available progeny pool. Therefore, to avoid non-random sampling in plants, one seed per fruit should be sampled. For species with singly sired, multi-seeded fruits, information on correlated matings may be desirable and therefore multiple seeds (full-sib progeny array) from the same fruit should be collected.

The best genetic markers for gene flow analysis are those loci with alleles in equal proportions since they give the highest exclusion probabilities, thus very diverse loci with many rare alleles (e.g., SSRs) may not be as useful as first appears for gene flow analysis (Smouse and Meagher 1994).

2.6 Among-population sampling

Whereas theoretical models can be applied to the problem of determining the number of samples that must be made from within a population to achieve a desired level of statistical error, no such models exist for sampling variation among populations. The optimization of the sampling strategy will depend on the hypothesis being investigated, the pattern of genetic variation expected, and a detailed understanding of the biology of the organism of interest. At the beginning of an investigation, even apparently simple issues, such as taxon distribution, may be unknown. Thus, in planning the sampling protocol it will be necessary to determine the geographical and ecological distribution of the study taxa.

As with intrapopulation sampling, except in the case of rare taxa or those taxa with the narrowest of distributions, it is rarely possible to sample all populations. Therefore, decisions must be made regarding the numbers of populations to sample and where these populations should be sampled. Random sampling across a distribution is rarely possible because of factors such as financial resources or the difficulties of access to particular parts of a species' distribution. If potential collection sites are widely scattered then the movement between populations can be a significant part of the costs of a project. Hence, there is a tendency to have large numbers of samples from within populations. However, for studies of genetic diversity, all researchers agree that investigations should aim to collect the geographical and ecological range of the species, and that if there are financial limitations on an investigation then more populations and fewer individuals per population should be sampled (e.g., Pons and Chaouche 1995, Pons and Petit 1995). In order to sample genetic diversity, Willan, Hughes, and Lauridsen (1990) have suggested that a minimum of five populations should be sampled, one from the center and four from the periphery of the distribution. However, the optimal allocation of limited resources between within- and among-population sampling must be determined by the biology of the organism being studied and the problem being investigated.

2.7 Power analysis

As ecological genetic research programs, particularly those that use molecular markers, move from phenomenological research, for example, the description of patterns of diversity and the recording of variation, investigations must incorporate the concept of statistical power into experimental design (Krebs 1999, Shrader-Frechette and McCoy 1992). Most simply, statistical power is the probability of getting a statistically significant result given that there is a biologically real effect in the population being studied. If a hypothesis is tested, and the result is found to be non-significant, the question remains as to whether this is because there is no effect or because the experimental design makes it unlikely that any biologically real effects will be detected.

There are four variables that will affect the outcome of a statistical analysis: (i) sample size; (ii) the probability of a Type I error (α; false positive); (iii) the probability of a Type II error (β; false negative); and (iv) the magnitude of the effect (Fig. 2.16). If three of these are fixed then the fourth can be calculated. Traditionally, experimental biologists have been concerned about Type I errors, although there is growing evidence that they should also concern themselves with Type II errors (Thomas and Juanes 1996). There are many computer programs now available to conduct power analysis, which have been reviewed by Thomas and Krebs (1997).

	Decision	
	Do not reject Ho	**Reject Ho**
Ho is true	P (correct negative) = $(1 - \alpha)$	False positive P (Type I error) = α
Ho is false	False negative P (Type II error) = β	P (correct positive) = $(1 - \beta)$

'Truth'

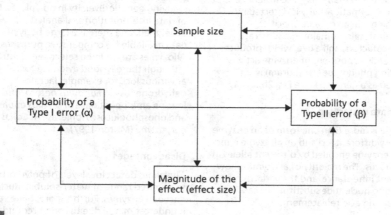

Fig. 2.16

Relationship between sample size, the magnitude of the expected effect and Type I and Type II errors in power analysis.

2.8 Further reading

Introductions to genome structure are provided by Alberts et al. (1994), Kubis, Thomas, and Heslop-Harrison (1998), Lewin (2000), and Lodish et al. (1995), whilst the structure of animal and plant organelle genomes is reviewed by Backert, Nielsen, and Börner (1997), Boore (1999), Hanson and Folkerts (1992), Palmer (1992), and Wolstenholme (1992). Protein and DNA marker technologies have been described in detail by Baker (2000), Hillis, Moritz, and Mable (1996), Hoelzel (1998) and Karp et al. (1998), whilst Karp et al. (1998), Rafalski et al. (1987), and Wolfe and Liston (1998) make recommendations about the applications of particular marker systems to particular types of problem. Sampling design has been the subject of extensive statistical research and is discussed in detail by Cochran (1977), Cox (1958), and Winer, Brown, and Michels (1991), whilst Krebs (1999) summarizes the application of experimental design in ecology. Recommended practical sampling guidelines for the different types of organisms include FAO (1995), McGavin (1997), and Southwood and Henderson (2000). Detailed treatments of power analysis can be found in Cohen (1988), Kraemer and Thiermann (1987), and Lipsey (1990), whilst excellent introductions are provided by Cohen (1992) and Muller and Benignus (1992), who consider how to determine the necessary power for a particular study.

Allozymes

Basis of technique

Non-denatured proteins, with different net charges, differentially migrate through a gel when an electrical current is applied. Protein charge characteristics result primarily from five amino acids (aspartic acid, arginine, glutamic acid, histidine, lysine). The net charge of a protein determines a protein's relative movement, although size and shape may also influence migration. Enzymatic activity is detected by soaking the gel in an enzyme-specific histochemical stain, containing a substrate, necessary cofactors, and a dye, which produces discrete bands at positions of enzyme activity. The banding phenotype (zymogram) is interpreted as a genotype (see Fig. 2b).

Type of data generated

An isozyme is more than one form of an enzyme encoded by different loci and an allozyme is more than one enzyme encoded by different alleles at the same locus. The majority of allozymes show codominant inheritance, and the variants are attributed to nucleotide substitutions causing charged amino acid replacement.

Data scoring

Data are usually interpreted as loci and alleles based on knowledge of enzyme structure and ploidy. Complications arise because of polyploidy, gene duplication, interlocus interactions, and null alleles (Weeden and Wendel 1990). Banding patterns may be complex and only capable of interpretation following crossing experiments. In such cases, either band positions or enzyme phenotypes may be compared, although information quality will be reduced compared to genotypic interpretations (see Brochmann, Soltis, and Soltis 1992).

Advantages

Allozymes are easily, safely, and cheaply detected. About 100 enzyme detection systems have been developed, mostly for animals (Murphy et al. 1996). In any study, it is unlikely that all available enzymes will be resolved: most studies use 10 to 30 enzymes, encoded by about 15 to 50 loci. Large numbers of individuals (ca. 40–50) can be analyzed at one time for many different enzymes (5–8). Allozyme analysis is currently the only easily used marker system for studying genetic diversity in polyploids. Methods of data interpretation and analysis are well developed, and there is a large body of existing data available for comparative purposes. Allozymes are assumed selectively neutral, although there is good evidence for selection at various loci, for example, lactate dehydrogenase in *Fundulus heteroclitus*, leucine aminopeptidase in *Mytilus edulis*, and phosphoglucose isomerase in *Colias eurytheme* (Mitton 1997).

Disadvantages

Allozymes detect low levels of polymorphism in a limited range of water-soluble, nuclear-encoded enzymes. Furthermore, gene variation is underestimated due to codon redundancy and synonymous nucleotide substitutions (Nei 1987), although isoelectric focusing may identify additional polymorphisms (Sharp, Desai, and Gale 1988). Problems are associated with environmental and ontogenetic expression with some enzyme systems, for example, alcohol dehydrogenase is induced by anoxia (Wendel and Weeden 1990). Furthermore, enzymes can be grouped according to the metabolic pathways they catalyze; the different classes having different variation patterns and hence different studies may not be comparable (Hartl, Willing, and Nadlinger 1994).

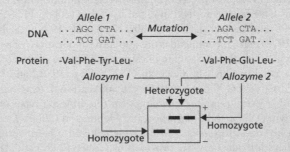

Fig. 2b

Allozyme variation. Relationship between allozyme variation at one locus, revealed on a starch gel (phenotype), and variation in the monomeric protein sequence and DNA sequence. Mutation of C in Allele 1 to A in Allele 2 results in the incorporation of glutamic acid rather than tyrosine into the protein and hence the allozymes migrate different distances in an electrical field.

Practical protocols and data analysis background

Large numbers of extraction buffers and electrophoretic conditions have been recommended for the detection of allozyme variation, and these have been summarized by Harris and Hopkinson (1976), Richardson et al. (1986), Herbert and Beaton (1989), Wendel and Weeden (1990), Kephart (1990), and Murphy et al. (1996). Discussions of the analysis of allozyme data can be found in Weeden and Wendel (1990), Murphy et al. (1996), and Richardson, Baverstock, and Adams (1986).

Applications of allozyme markers

The applications of allozyme markers include the estimation of gene diversity and population structure, and investigations of hybridization, introgression, gene flow, and polyploidy. However, enzyme markers have limited phylogenetic value (e.g., Buth 1984, Murphy et al. 1996; cf. Soltis, Soltis, and Gottlieb 1987).

Restriction fragment length polymorphisms (RFLPs)

Basis of technique

Restriction enzymes are used to detect variation in primary DNA structure. The number of bases in the restriction site (four base-cutting enzymes produce more fragments than six base-cutting enzymes) and the genome base composition determine the number of restriction sites identified in a genome. DNA variation is visualized, following Southern blotting of digested DNA, using DNA fragments (probes) that have been radioactively or chromatically labeled (see Fig. 2c). If the genome of interest is small and easily purified (e.g., animal mtDNA), variation can be detected directly (e.g., Dowling et al. 1996).

Type of data generated

RFLP analysis measures DNA variation that affects the relative positions of restriction sites. Thus, banding patterns are either compared directly or interpreted in terms of restriction site or DNA length mutations (indels), more rarely as inversions.

Data scoring

RFLP probes are usually considered to represent loci and alleles are defined by a specific probe–enzyme combination. RFLP data have also been scored in terms of the presence or absence of restriction fragments, as fragment direct analysis and fragment occurrence analysis, and the analysis of the position of restriction sites, as site occurrence analysis or site mutation analysis (Bremer 1991). The use of site data means that RFLP data can be ordered and hence capable of phylogenetic interpretation.

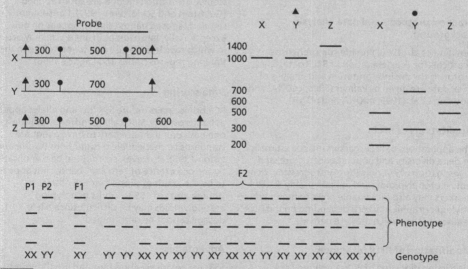

Fig. 2c

RFLP variation. Restriction sites from two restriction enzymes (indicated by triangle and circle) are mapped in three individuals (X, Y, Z). The relative positions of the restriction sites are shown in the left-hand diagram and the resulting pattern of restriction fragments on an agarose gel is shown in the right-hand diagram. The bottom of the diagram shows segregation of RFLP markers (identified using the circle enzyme) in the F2 generation when two homozygous parents are crossed.

Advantages

RFLP markers are codominant and it is possible to detect nDNA and organelle DNA polymorphisms in total DNA extracts. Furthermore, depending on the means of RFLP scoring, the data are ordered and can be interpreted phylogenetically. Results are highly repeatable and large amounts of variation can be detected depending on the probe and restriction enzyme used in the analysis. Furthermore, variation can be detected that occurs outside the region of interest, for example, the occurrence of site mutations outside the probe region (Jansen, Wee, and Millie 1998).

Disadvantages

RFLP analyses require large amounts of DNA, there is usually a need for access to radioisotopes and limited numbers of suitable nDNA markers are available (Dowling et al. 1996). Furthermore, RFLPs are expensive, time consuming to detect, and data from different laboratories are difficult to combine. Since DNA fragments migrate logarithmically on a gel, small changes in large fragments are more difficult to detect than similar size changes in small fragments. Although it is possible to use large numbers of enzymes to detect variation, this is often limited due to cost. Further-more, some enzymes have variable cutting depend-ing on DNA methylation, hence variation between RFLP patterns may be due to difference in gene activation (Gardiner-Garden, Sved, and Frommer 1992, Hepburn, Belanger, and Matthius 1987).

Practical protocols and data analysis background

Dowling et al. (1996) provide comprehensive protocols for the generation of RFLP data. Protocols for the interpretation and analysis of RFLP data are given by Palmer (1986; cpDNA) and Dowling et al. (1996; mtDNA and nDNA).

Applications of RFLPs

The applications of RFLP markers include estimation of gene diversity and population structure, and investigation of hybridization, introgression, gene flow, autopolyploidy, and allopolyploidy. RFLP markers may also be valuable phylogenetic and phylogeographic markers, depending on the DNA sequence from which they are derived.

Modification of RFLP technique

Restriction digestion of PCR-amplified DNA fragments (cleaved amplified polymorphic sequences; CAPS) means that populations can be quickly and cheaply screened for polymorphisms identified during either shotgun digestion or sequence analysis (Rafalski et al. 1997, Wolfe and Liston 1998). This approach increases the number of sequences that can be studied using RFLPs but

reduces the information that may be detected compared to standard RFLP approaches, since mutations outside the relatively small, amplified region will not be found. The approach eliminates methylation problems associated with some restriction enzymes, since PCR products are unmethylated. Suitable primers may be designed from sequence databases, analysis of low copy number random clones, and universal cpDNA, mtDNA, and nDNA sequences.

Microsatellites (SSRs)

Basis of technique

Microsatellites are short (10–50 copies) tandem repeats of mono- to tetra-nucleotide repeats, for example, $(AT)_n$ and $(CAG)_n$, which are assumed to be randomly distributed throughout the nDNA, cpDNA, and mtDNA (Goldstein and Schlötterer 1999, Jarne and Lagode 1996, Powell et al. 1996, Provan, Powell, and Hollingsworth 2001). Primers are designed to conserved regions flanking the variable SSR. SSR polymorphism is detected using PCR and separation of products on agarose, polyacrylamide, or automated DNA sequencing gels (see Fig. 2d).

Type of data generated

SSRs detect length variation that results from changes in the number of repeat units, to which stepwise mutation models are often applied (Goldstein and Schlötterer 1999). Consequently, regularly spaced bands (alleles) appear on gels. Exceptionally, large jumps in allele size may occur, in which case this may be due to mutations in the flanking regions (Jarne and Lagode 1996).

Data scoring

PCR primer pairs define SSR loci and alleles appear as bands on gels. Since alleles differ by as little as one base pair it is necessary to have adequate standards to check allele designations (Wolfe and Liston 1998). However, scoring can be complicated by the occurrence of "stutter" bands that appear to be the result of slipped-strand misrepair (Hauge and Litt 1993), whilst genotyping autopolyploids may be difficult since allele dosage cannot be determined.

Advantages

SSRs are relatively abundant and are thought to have a uniform coverage across the genome, although SSRs appear to be more common in animals than in plants (Wang et al. 1994). SSR markers are codominant and it is possible to detect both nDNA and organelle DNA polymorphisms in total DNA extracts. Mutation rates are high in SSRs compared to other DNA markers, making

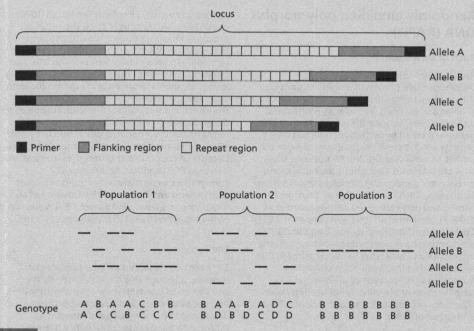

Fig. 2d

Microsatellite variation. The structures of four alleles at a typical microsatellite locus are shown at the top of the diagram and in the bottom of the diagram segregation of alleles is shown in three populations of seven individuals.

them useful markers for intrapopulation studies. Furthermore, locus definition based on defined primer pairs facilitates information exchange between laboratories.

Disadvantages

The initial identification of SSRs is expensive and requires cloning and sequencing, whilst SSR primer pairs tend to be species-specific. Cross-species amplification of SSRs has been shown, although SSR variability appears to be reduced (e.g., White and Powell 1997). The occurrence of "stutter" bands may make gels difficult to interpret, whilst the pattern of SSR mutation means that homoplasy between alleles may be high (Doyle et al. 1998, Goldstein and Pollock 1997). Relatively few loci are studied in SSR-based studies, which may be a disadvantage when investigating genetic distance, although Kalinowski (2002) has shown that rather than large numbers of loci what is important is to have large enough numbers of alleles.

Practical protocols and data analysis background

Karp, Isaac, and Ingram (1998) provide detailed protocols for the generation of SSR data, whilst Provan, Powell, and Hollingsworth (2001) and Zane

et al. (2002) provide details of methods for the isolation of SSRs from genomic DNA. Karp, Isaac, and Ingram (1998), Goldstein and Schlötterer (1999) and Scribner and Pearce (2000) provide introductions to methods of SSR data analysis.

Applications of SSR technique

The applications of SSR markers include estimation of gene diversity and population structure. Since SSR markers show a high number of alleles per locus they are ideally suited to the analysis of gene flow. However, SSRs may have limited phylogenetic value because of the problems of homoplasy and the recurrent generation of alleles of identical length (Doyle et al. 1998, Goldstein and Pollock 1997).

Modifications of SSRs

Inter-SSR (ISSR) is a modification of the SSR approach that makes use of the random distribution of SSRs across the genome, such that if two SSRs with a common motif are within a amplification distance sufficiently frequently then single primer amplification will yield a high degree of polymorphic bands (Tsumura, Ohba, and Strauss 1996). ISSR primers are designed to occur within the SSR itself and may or may not be anchored (Charters et al. 1996).

Randomly amplified polymorphic DNA (RAPD)

Basis of technique

RAPDs utilize single, arbitrary decamer oligonucleotide primers to amplify regions of the genome using the polymerase chain reaction (Williams et al. 1993). Venugopal, Mohapatra, and Salo (1993) suggest that there are sites in the genome flanked by perfect or imperfect invert repeats which permit multiple annealing of the primer to occur (see Fig. 2e). Primer annealing sites are scattered throughout the nuclear and cytoplasmic genomes, in all classes of DNA from single-copy DNA to multiple-copy DNA, and in coding and non-coding regions (Aagaard et al. 1995, Thormann et al. 1994, Williams et al. 1993). Complex, stochastic and dynamic equilibria mechanisms have been proposed to explain the interactions between the variously primed DNA molecules that involve selection of primer annealing sites followed by differential amplification of the initial amplification products (Caetano-Anollés 1993, Caetano-Anollés, Bassam, and Gresshoff 1991, 1992).

Type of data generated

Each band position on a gel is assumed to represent a diallelic locus (band present–absent). This locus definition means that RAPDs are dominant markers, that is, present–present homozygotes cannot be distinguished from present–absent heterozygotes at the phenotypic level. Bands are detected using either stains (e.g.,

ethidium bromide) or radioactive labels (Williams et al. 1993). RAPDs appear to be the result of processes including nucleotide substitutions that create or abolish primer sites, formation of secondary structures between priming sites and insertion, deletion or inversion of either priming sites or segments between priming sites (Bowditch et al. 1993, Williams et al. 1993). Furthermore, the effect of a nucleotide substitution appears to depend on its position in the primer recognition region (Caetano-Anollés, Bassam, and Gresshoff 1992). Polymorphism in product intensity occurs because of copy number differences, competition between PCR products, heterozygosity, comigration, or partial mismatching of primer sites (Adams and Demeke 1993, Dowling et al. 1996, Lorenz, Weihe, and Bšrner 1994, Venugopal, Mohapatra, and Salo 1993, Williams et al. 1993).

Data scoring

Data are usually scored as band presence or absence, although this is complicated by the occurrence of bands that vary in intensity within and between lanes. Some researchers have advocated the scoring of band intensity in addition to band presence–absence (Demeke, Adams, and Chibbar 1992), although this has not been widely accepted (Ritland and Ritland, 2000). Reproducibility of RAPD scoring is improved by scoring within a specific size range between monomorphic bands and independent replicate scoring of gels. Issues of product homology mean that RAPD markers should be used cautiously and that homology of RAPD products should be tested by hybridization of cloned products, RFLP analysis

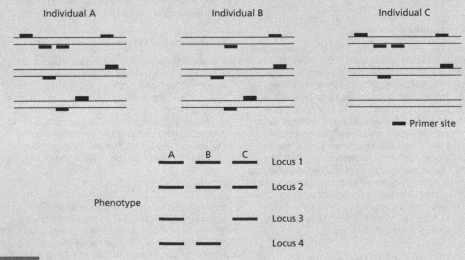

Fig. 2e

RAPD variation. Distribution of RAPD priming sites is shown across the DNA of three individuals at the top of the diagram. The phenotype resulting from the separation of PCR products on an agarose gel is shown at the bottom of the diagram, together with the usual designation of loci.

of gel-isolated products, or comparative genetic mapping (Hurme and Savolainen 1999, Rieseberg 1996, Thormann et al. 1994). Comparison with independent data sets may also prove useful (Adams and Rieseberg 1998).

Advantages

The technique is cheap, simple, requires no sequence information, is PCR-based, and a large number of putative loci may be screened (Weising et al. 1995). Furthermore, the technique is non-invasive and very small amounts of tissue are necessary, enabling it to be used easily with endangered species.

Disadvantages

The technique has been extensively criticized on technical (Jones et al. 1997) and theoretical (Harris 1999) grounds. These criticisms include issues associated with reproducibility, primer structure, marker dominance, product competition, product homology, allelic variation, genome sampling, and non-independence of loci.

Practical protocols and data analysis background

Weising et al. (1995) provide detailed protocols for RAPD analysis, along with other methods of DNA "fingerprinting". Other protocol compilations may be found in Williams et al. (1993) and Wolfe and Liston (1998). Guidelines for the analysis of RAPD data can be found in Weising et al. (1995) and Williams et al. (1993), whilst Lynch and Milligan (1994) discuss the use of RAPDs as dominant markers for population genetic analysis.

Applications of RAPDs

The application of RAPDs has been widespread, including estimation of gene diversity, population structure, and clonality, with more limited uses in investigations of hybridization, introgression, and gene flow. RAPDs are useful at the initial stages of an investigation but they have been superseded by other techniques (e.g., AFLPs) for more rigorous investigations. RAPD markers have also been used in phylogenetic and phylogeographic studies, although their use is controversial since RAPD markers cannot be ordered (Harris 1999).

Modifications of RAPDs

Criticisms of RAPDs associated with dominance, reproducibility, and product homology have led to modifications of the technique, including sequence characterized amplified regions (SCARs; Paran and Michelmore 1993) and randomly amplified microsatellite polymorphism (RAMPO; Ramser et al. 1997). RAPD products that are fixed between clones, populations, and species may be diagnostic and useful markers in studies of

hybridization and gene flow. Specific primers can be designed to such products by cloning and sequencing, so that SCARs can be developed. RAMPO markers behave like SSRs, except that the polymorphism detected may not be in an SSR region. All modifications of the technique compromise both the broad-range and simplicity of RAPDs.

Amplified fragment length polymorphism (AFLP)

Basis of technique

AFLP analysis involves the selective amplification of an arbitrary subset of restriction fragments generated by double digestion of DNA with a frequently cutting and a rarely cutting restriction enzyme (Vos et al. 1995). Fragment ends are modified by the addition of double-stranded adapters, which provide the primer sites for subsequent PCR amplification. Two phases of PCR amplification are involved. In the preselective amplification, primers are used which are complementary to the adaptors but have an additional base pair. Selective amplification uses the preselective PCR product as a template for amplification with selective primers that are identical to the preselective primers except for the addition of one to three additional selective bases and either a radioactive or fluorescent label. Fragments are separated on either polyacrylamide gels or an automated sequencer. The effect of adaptor ligation and the preselective and selective amplifications is to reduce the numbers of fragments detected (see Fig. 2f).

Type of data generated

The number of bands generated in an AFLP reaction is determined by the number of bases in the variable part of the selective primer (primers with one selective base produce more fragments than primers with either two or three selective bases) and genome complexity (Vos et al. 1995). Single nucleotide changes will be detected by AFLP analysis when the restriction sites are affected or nucleotides adjacent to the restriction sites are affected, causing the AFLP primers to mispair, thus preventing amplification. In addition, deletions, insertions, and rearrangements affecting the presence or size of restriction fragments will result in AFLPs.

Data scoring

Most AFLP markers are scored as diallelic markers, where alleles are detected as a band presence or absence, meaning that the markers are dominant. However, codominant AFLP markers may be detected because of small insertions or deletions in the restriction fragments (Vos et al. 1995).

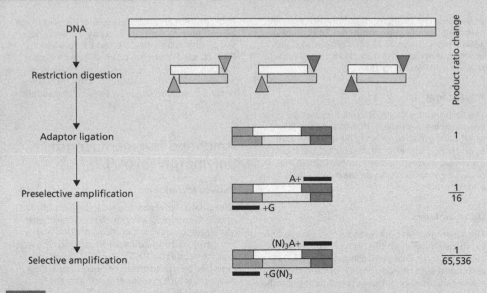

Fig. 2f

Steps in the detection of amplified fragment length polymorphisms. Native DNA is digested with two different restriction enzymes, shown by the differently shaded triangles, to produce a population of fragments that have similar or dissimilar ends. The population of fragments with the dissimilar ends is selected by ligation of appropriate adaptors. If the relative population of this class of fragments is defined as one then the preselective amplification will reduce this population to 1/16 of the original. The reduction in the case of selective amplification is determined by the number of selective bases in the primer: three bases will result in 1/65,536 of the original population being amplified, whilst for one and two base selective primers the changes are 1/256 and 1/4096, respectively.

Advantages

AFLPs are expected to be highly polymorphic, either dominant or codominant (although allelic relations may not be immediately obvious) and require no prior sequence knowledge (Rafalski et al. 1997). Since long amplification primers (ca. 17–21 nucleotides long) are used in AFLP analysis and these anneal perfectly to their target sequences, the AFLP technique is very reliable and robust compared to other techniques that use random amplification techniques (Vos et al. 1995), particularly when products are separated on an automated DNA sequencer. Furthermore, automated DNA sequencers mean that product profiles can be automatically scored. High marker densities can be obtained with a typical AFLP fingerprint containing between 50 and 100 amplified products, of which up to 80% may serve as markers.

Disadvantages

The technique requires a high degree of technical skill and relatively large amounts of high quality DNA. As with other random amplification techniques, similar problems of product homology determination exist (Robinson and Harris 2000), and without detailed genetic analysis, locus designation may be equivocal.

Practical protocols and data analysis background

Karp, Isaac, and Ingram (1998) and Vos et al. (1995) provide detailed protocols for the generation of AFLPs, whilst Karp, Isaac, and Ingram (1998) discuss the analysis of these data.

Applications of AFLPs

The majority of AFLP applications have been for genome mapping and breeding studies, although the application of AFLPs in ecological genetics is becoming widespread, especially for studies of gene diversity, population structure, and clonality. AFLPs have also been applied to gene flow analysis but since parentage assignment is limited to the exclusion of progeny markers within the potential parent population, large numbers of markers must be screened (Gerber et al. 2000). AFLP markers have also been used in phylogenetic and phylogeographic studies, although their use is controversial since AFLP markers cannot be ordered (Robinson and Harris 2000).

Modifications of AFLPs

Criticisms of AFLPs associated with dominance and product homology have led to modifications of the AFLP technique to produce sequence-characterized products (Meksem et al. 2001).

DNA sequence analysis

Basis of technique

Specific DNA regions are amplified by PCR and then subjected to cycle sequencing. During cycle sequencing, dideoxynucleotides, nucleotide analogs that prevent further extension of the DNA strand, are incorporated into the extending DNA strand and once incorporated, DNA synthesis is terminated. This creates a population of DNA sequences of different lengths, which can then be separated on polyacrylamide gels (Brown 1994). In manual sequencing, DNA strands are radioactively labeled, whilst in automated sequencing each of the dideoxynucleotides is labeled with a different colored fluorescent dye (Adams, Fields, and Venter 1994, Brown 1994). (See Fig. 2g.)

Type of data generated

In DNA sequence analysis, the order of nucleotides in a piece of DNA is determined. In addition to nucleotide substitutions, comparison between amplification products from different samples means that it is possible to detect deletions and duplications; the amount of variation detected and its type depend on the regions being sampled (Bishop and Rawlings 1997).

Data scoring

Data are scored directly as the separate nucleotide bases. These are either read from autoradiographs, in the case of manual sequencing, or as peaks on an automated DNA sequencer trace.

Advantages

Direct sequencing of DNA produces easily scored, high-quality information, whilst automated sequencing and high-powered computer facilities mean that large amounts of data can be generated. Since DNA sequence data are being generated, comparisons between taxa can be quickly and easily made, whilst universal sequence primers mean it is possible to sequence most taxa with no knowledge of their DNA sequence (e.g.,

Baldwin 1992, Demesure, Sodzi, and Petit 1995, Palumbi 1996).

Disadvantages

DNA sequence analysis is expensive for general diversity surveys since loci are screened one at a time. Some DNA samples may prove very difficult to sequence due to the occurrence of secondary structures, for example, the occurrence of adenine repeats, whilst the presence of too much variation may make alignment of sequences between samples difficult. Direct sequencing means that heterozygosity at individual nucleotide positions may be detected and it may be necessary to clone PCR products to eliminate this problem. Cloning is essential for the analysis of single or low copy number nuclear genes in order to distinguish between paralogy and orthology (Doyle and Gaut 2000).

Practical protocols and data analysis background

Numerous protocols have been published for the generation of DNA sequence data, for example, Adams, Fields, and Venter (1994), Ansorge, Vos, and Zimmermann (1997), Brown (1994), Hillis et al. (1996), Roe, Crabtree, and Khan (1996) and other potentially useful sequences are continually published in the literature or may be found by searching on-line databases. Methods for DNA sequence alignment can be found in Bishop and Rawlings (1997) and Miyamoto and Cracraft (1991), whilst methods for the analysis of sequence data have been summarized by Hall (2001), Nei and Kumar (2000), and Percus (2002).

Applications of technique

DNA sequence analysis can be used in applications that include estimation of gene diversity and population structure, and investigation of hybridization, introgression, and gene flow. However, the approach has found its greatest value for phylogenetic and phylogeographic analyses, where it is necessary to have ordered characters.

Modifications of DNA sequence analysis

Once polymorphism has been identified with DNA sequence analysis, it is possible to incorporate these data with other approaches to overcome problems of cost. DNA polymorphism may be located to restriction sites and hence restriction enzymes may be used to generate PCR-RFLPs (Wolfe and Liston 1998). Single nucleotide polymorphism (SNP; Gibson 2002) analysis potentially provides a means of screening large numbers of individuals for single nucleotide polymorphisms.

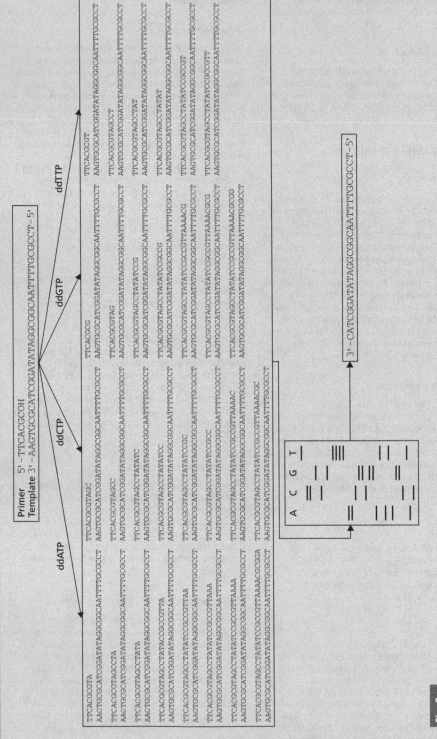

Fig. 2g

Dideoxynucleotide sequencing. The DNA strand for sequence analysis is synthesized using a specific primer and four dideoxynucleotides that prevent additional nucleotide incorporation when they are incorporated into the DNA strand. The result is a population of differently sized DNAs, which when separated on polyacrylamide gels, can be used to infer nucleotide order. Manual sequencing analysis involves the incorporation of radionucleotides into the synthesized DNA strand, whilst automated sequence analysis makes use of dideoxynucleotides that are labeled with different colored fluorescent dyes.

REFERENCES

Aagaard, J.E., Vollmer, S.S., Sorensen, F.C., and Strauss, S.H. 1995. Mitochondrial DNA products among RAPD profiles are frequent and strongly differentiated between races of Douglas-fir. *Molecular Ecology*, **4**: 441–7.

Adams, M.D., Fields, C., and Venter, J.C. 1994. *Automated DNA Sequencing and Analysis*. London: Academic Press.

Adams, R.P. and Rieseberg, L.H. 1998. The effects of non-homology in RAPD bands on similarity and multivariate statistical ordination in *Brassica* and *Helianthus*. *Theoretical and Applied Genetics*, **97**: 323–6.

Adams, R.P. and Demeke, T. 1993. Systematic relationships in *Juniperus* based on random amplified polymorphic DNAs (RAPDs). *Taxon*, **42**: 553–71.

Alberts, B., Bray, D., Lewis, J., Ruff, M., Roberts, K., and Watson, J.D. 1994. *Molecular Biology of the Cell*. New York: Garland.

Ansorge, W., Vos, W., and Zimmermann, J. 1997. DNA Sequencing Strategies. *Automated and Advanced Approaches*. Heidelberg: Spektrum Akademischer Verlag.

Appels, R. and Dvorak, J. 1982. Relative rates of divergence of spaces and gene sequences within the rDNA region of species in the Triticeae. Implications for the maintenance of homogeneity of a repeated gene family. *Theoretical and Applied Genetics*, **63**: 361–5.

Archie, J.W., Simon, C., and Martin, A. 1989. Small sample size does decrease the stability of dendrograms calculated from allozyme frequency data. *Evolution*, **43**: 678–83.

Backeljau, T., de Bruyn, L., de Wolf, H., Jordaens, K., van Dongen, S., and Winnepenninckx, B. 1996. Multiple UPGMA and neighbor-joining trees and the performance of computer packages. *Molecular Biology and Evolution*, **13**: 309–13.

Backert, S., Nielsen, B.L., and Börner, T. 1997. The mystery of the rings: structure and replication of mitochondrial genomes from higher plants. *Trends in Plant Science*, **2**: 477–83.

Baker, A.J. 2000. *Molecular Methods in Ecology*. Oxford: Blackwell Science.

Baldwin, B.G. 1992. Phylogenetic utility of the internal transcribed spacers of nuclear ribosomal DNA in plants: an example from the Compositae. *Molecular Phylogenetics and Evolution*, **1**: 3–16.

Bart, J., Fligner, M.A., and Notz, W.I. 1998. *Sampling and Statistical Methods for Behavioural Ecologists*. Cambridge: Cambridge University Press.

Baverstock, P.R. and Moritz, C. 1996. Project design. In D.M. Hillis, C. Moritz, and B.K. Mable, eds. *Molecular Systematics*. Sunderland, MA: Sinauer, pp. 17–27.

Birks, J.S. and Kanowski, P.J. 1988. Interpretation of the composition of coniferous resin. *Silvae Genetica*, **37**: 29–39.

Birks, J.S. and Kanowski, P.J. 1993. Analysis of resin compositional data. *Silvae Genetica*, **42**: 340–50.

Birky, C.W. 2001. The inheritance of genes in mitochondria and chloroplasts: laws, mechanisms, and models. *Annual Review of Genetics*, **35**: 125–48.

Bishop, M.J. and Rawlings, C.J. 1997. *DNA and Protein Sequence Analysis*. Oxford: IRL Press.

Boore, J.L. 1999. Animal mitochondrial genomes. *Nucleic Acids Research*, **27**: 1767–80.

Bossart, J.L. and Prowell, D.P. 1998. Genetic estimates of population structure and gene flow: limitations, lessons and new directions. *Trends in Ecology and Evolution*, **13**: 202–6.

Bowditch, B.M., Albright, D.G., Williams, J.G.K., and Braun, M.J. 1993. Use of amplified polymorphic DNA markers in comparative genome studies. *Methods in Enzymology*, **224**: 294–309.

Bremer, B. 1991. Restriction data from chloroplast DNA for phylogenetic reconstruction: is there only one accurate way of scoring? *Plant Systematics and Evolution*, **175**: 39–54.

Bridge, D., Cunningham, C.W., Schierwater, B., Desalle, R., and Buss, L.W. 1992. Class-level relationships in the phylum Cnidaria: evidence from mitochondrial genome structure. *Proceedings of the National Academy of Sciences USA*, **89**: 8750–3.

Briggs, D. and Walters, S.M. 1997. *Plant Variation and Evolution*. Cambridge: Cambridge University Press.

Brochmann, C., Soltis, D.E., and Soltis, P.S. 1992. Electrophoretic relationships and phylogeny of Nordic polyploids in *Draba* (Brassicaceae). *Plant Systematics and Evolution*, **182**: 35–70.

Brown, A.H.D. 1989. Core collections: a practical approach to genetic resources management. *Genome*, **31**: 818–24.

Brown, A.H.D. and Briggs, J.D. 1991. Sampling strategies for genetic conservation in *ex situ* collections of endangered plant species. In D.A. Falk and K.E. Holsinger, eds. *Genetics and Conservation of Rare Plants*. Oxford: Oxford University Press, pp. 99–119.

Brown, A.H.D. and Hardner, C.M. 2000. Sampling the gene pools of forest trees for *ex situ* conservation. In A. Young, D. Boshier, and T. Boyle, eds. *Forest Conservation Genetics. Principles and Practice*. Wallingford, UK: CABI Publishing, pp. 185–96.

Brown, A.H.D. and Weir, B.S. 1983. Measuring genetic variability in plant populations. In S.D. Tanksley and T.J. Orton, eds. *Isozymes, Plant Genetics and Breeding*. Amsterdam: Elsevier, pp. 219–39.

Brown, T.A. 1994. *DNA Sequencing. The Basics*. Oxford: IRL Press.

Brown, W.M. 1985. The mitochondrial genome of animals. In R.J. MacIntyre, ed. *Molecular Evolutionary Genetics*. London: Plenum Press, pp. 95–130.

Brown, W.M., Prager, E.M., Wang, A., and Wilson, A.C. 1982. Mitochondrial DNA sequences of primates: tempo and mode of evolution. *Journal of Molecular Evolution*, **18**: 225–39.

Buth, D.G. 1984. The applications of electrophoretic data in systematic studies. *Annual Review of Ecology and Systematics*, **15**: 501–22.

Butlin, R.K. and Tregenza, T. 1998. Levels of genetic polymorphism: marker loci versus quantitative traits. *Philosophical Transactions of the Royal Society, Series B*, **353**: 187–98.

Caetano-Anollés, G. 1993. Amplifying DNA with arbitrary oligonucleotide primers. *PCR Methods and Applications*, **3**: 85–94.

Caetano-Anollés, G., Bassam, B., and Gresshoff, P.M. 1991. DNA amplification fingerprinting using very short arbitrary primers. *Biotechnology*, **9**: 553–7.

Caetano-Anollés, G., Bassam, B., and Gresshoff, P.M. 1992. Primer–template interactions during DNA amplification fingerprinting with single arbitrary oligonucleotides. *Molecular and General Genetics*, **235**: 157–65.

Carter, R.E. 2000. DNA fingerprinting using minisatellite probes. In A.J. Baker, ed. *Molecular Methods in Ecology*. Oxford: Blackwell Science, pp. 113–35.

Charters, Y.M., Robertson, A., Wilkinson, M.J., and Ramsay, G. 1996. PCR analysis of oilseed rape cultivars (*Brassica napus* L. ssp. *oleifera*) using 5'-anchored simple sequence repeat (SSR) primers. *Theoretical and Applied Genetics*, **92**: 442–7.

Chase, M.W. and Hills, H.H. 1991. Silica gel: an ideal material for field preservation of leaf samples for DNA studies. *Taxon*, **40**: 215–20.

Cipriani, G., Testolin, R., and Morgante, M. 1995. Paternal inheritance of plastids in interspecific hybrids of the genus *Actinidia* revealed by PCR-amplification of chloroplast DNA fragments. *Molecular and General Genetics*, **247**: 693–7.

Cochran, W.G. 1977. *Sampling Techniques*. New York: Wiley.

Cohen, J. 1988. *Statistical Power Analysis for the Behavioral Sciences*. Hillsdale, NJ: Erlbaum.

Cohen, J. 1992. A power primer. *Psychological Bulletin*, **112**: 155–9.

Cox, D.R. 1958. *Planning of Experiments*. New York: Wiley.

Crawford, D.J. 1990. *Plant Molecular Systematics: Macromolecular Approaches*. London: Wiley.

Cutler, D.J., Zwixk, M.E., Carrasquillo, M.M., Yohn, C.T., Tobin, K.P., Kashuk, C., Mathews, D.J., Shah, N.A., Eichler, E.E., Warrington, J.A., and Chakravarti, A. 2001. High-throughput variation detection and genotyping using microarrays. *Genome Research*, **11**: 1913–25.

Demeke, T., Adams, R.P., and Chibbar, R. 1992. Potential taxonomic use of random amplified polymorphic DNA (RAPD): a case study in *Brassica. Theoretical and Applied Genetics*, **84**: 990–4.

Demesure, B., Sodzi, N., and Petit, R.J. 1995. A set of universal primers for amplification of polymorphic non-coding regions of mitochondrial and chloroplast DNA in plants. *Molecular Ecology*, **4**: 129–31.

dePamphilis, C.W. and Palmer, J.D. 1990. Loss of photosynthetic and chlororespiratory genes from the plastid genome of a parasitic flowering plant. *Nature*, **348**: 337–9.

Devlin, B. and Ellstrand, N.C. 1990. The development and application of a refined method for estimating gene flow from angiosperm paternity analysis. *Evolution*, **44**: 248–59.

Dixon, W.F. and Massey, F.J.J. 1983. *Introduction to Statistical Analysis*. New York: McGraw-Hill.

Dowling, T.E., Moritz, C., Palmer, J.D., and Rieseberg, L.H. 1996. Nucleic acids III: analysis of fragments and restriction sites. In D.M. Hillis, C. Moritz, and B.K. Mable, eds. *Molecular Systematics*. Sunderland, MA: Sinauer, pp. 249–320.

Doyle, J.J. and Doyle, J.L. 1987. A rapid DNA isolation procedure for small amounts of fresh leaf tissue. *Phytochemical Bulletin*, **19**: 11–15.

Doyle, J.J. and Gaut, B.S. 2000. Evolution of genes and taxa: a primer. *Plant Molecular Biology*, **42**: 1–23.

Doyle, J.J., Morgante, M., Tingey, S.V., and Powell, W. 1998. Size homology in chloroplast microsatellites of wild perennial relatives of soybean (*Glycine* subgenus *Glycine*). *Molecular Biology and Evolution*, **15**: 215–18.

Ennos, R.A., Worrell, R., and Malcolm, D.C. 1998. The genetic management of native species in Scotland. *Forestry*, **71**: 1–23.

Ewens, W.J. 1972. The sampling theory of selectively neutral alleles. *Theoretical Population Biology*, **3**: 87–112.

FAO. 1995. Collecting woody perennials. In L. Guarino, V. Ramanatha Rao, and R. Reid, eds. *Collecting Plant Genetic Diversity: Technical Guidelines*. Wallingford, UK: CAB International, pp. 485–509.

Fauré, S., Noyer, J.-L., Carreel, F., Horry, J.-P., Bakry, F., and Lanaud, C. 1994. Maternal inheritance of chloroplast genome and paternal inheritance of mitochondrial genome in bananas (*Musa acuminata*). *Current Genetics*, **25**: 265–9.

Felsenstein, J. 1988. Phylogenies from molecular sequences: inference and reliability. *Annual Review of Genetics*, **22**: 521–65.

Frankham, R., Ballou, J.D., and Briscoe, D.A. 2002. *Introduction to Conservation Genetics*. Cambridge: Cambridge University Press.

Gardiner-Garden, M., Sved, J.A., and Frommer, M. 1992. Methylation sites in angiosperm genomes. *Journal of Molecular Evolution*, **34**: 219–30.

Gerber, S., Mariette, S., Streiff, R., Bodénès, C., and Kremer, A. 2000. Comparison of microsatellites and amplified fragment length polymorphism markers for parentage analysis. *Molecular Ecology*, **9**: 1037–48.

Giannasi, D.E. and Crawford, D.J. 1986. Biochemical systematics II: a reprise. *Evolutionary Biology*, **20**: 25–248.

Gibson, G. 2002. Microarrays in ecology and evolution: a preview. *Molecular Ecology*, **11**: 17–24.

Gillespie, J.H. 1998. *Population Genetics. A Concise Guide*. Baltimore: The Johns Hopkins University Press.

Gillet, E. and Gregorius, H.R. 2000. Qualified testing of single-locus codominant inheritance using single tree progenies. *Biometrics*, **56**: 801–7.

Goldblatt, P., Hoch, P.C., and McCook, L.M. 1992. Documenting scientific data: the need for voucher specimens. *Annals of the Missouri Botanical Garden*, **79**: 969–70.

Goldstein, D.B. and Schlötterer, C. 1999. *Microsatellites. Evolution and Applications*. Oxford: Oxford University Press.

Goldstein, D.G. and Pollock, D.D. 1997. Launching microsatellites: a review of mutation processes and methods of phylogenetic inference. *Journal of Heredity*, **88**: 335–42.

Gorman, G.C. and Renzi, J. 1979. Genetic distance and heterozygosity estimates in electrophoretic studies: effects of sample size. *Copeia*, **2**: 242–9.

Goulding, S.E., Olmstead, R.G., Morden, C.W., and Wolfe, K.H. 1996. Ebb and flow of the chloroplast inverted repeat. *Molecular and General Genetics*, **252**: 195–206.

Graham, G.G. and Primavesi, A.L. 1993. *Roses of Great Britain and Ireland*. London: Botanical Society of the British Isles.

Hall, B.G. 2001. *Phylogenetic Trees Made Easy. A How-to Manual for Molecular Biologists*. Sunderland, MA: Sinauer.

Hanson, M.R. and Folkerts, O. 1992. Structure and function of the higher plant mitochondrial genome. *International Review of Cytology*, **141**: 129–72.

Harris, H. and Hopkinson, D.A. 1976. *Handbook of Enzyme Electrophoresis in Human Genetics*. Amsterdam: North Holland.

Harris, S.A. 1999. RAPDs in systematics – A useful methodology? In P.M. Hollingsworth, R.M. Bateman, and R.J. Gornall, eds. *Advances in Molecular Systematics*. London: Taylor & Francis, pp. 211–28.

Harris, S.A. and Ingram, R. 1991. Chloroplast DNA and biosystematics: the effects of intraspecific diversity and plastid transmission. *Taxon*, **40**: 393–412.

Hartl, G.B., Willing, R., and Nadlinger, K. 1994. Allozymes in mammalian population genetics and systematics: indicative functions of a marker system reconsidered. In B. Schierwater, B. Streit, G.P. Wagner and R. DeSalle, eds. *Molecular Ecology and Evolution: Approaches and Applications*. Basel: Birkhauser Verlag, pp. 299–310.

Hauge, X.Y. and Litt, M. 1993. A study of the origin of "shadow bands" seen when typing dinucleotide repeat polymorphisms by the PCR. *Human Molecular Genetics*, **2**: 411–15.

Hepburn, A.G., Belanger, F.C., and Matthius, J.R. 1987. DNA methylation in plants. *Developmental Genetics*, **8**: 475–93.

Herbert, P.D.N. and Beaton, M.J. 1989. *Methodologies for Allozyme Analysis using Cellulose Acetate Electrophoresis. A Practical Handbook*. Beaumont, TX: Helena Laboratories.

Hillis, D.M., Mable, B.K., Larson, A., Davis, S.K., and Zimmer, E.A. 1996. Nucleic acids IV: sequencing and cloning. In D.M. Hillis, C. Moritz, and B.K. Mable, eds. *Molecular Systematics*. Sunderland, MA: Sinauer, pp. 321–81.

Hillis, D.M., Moritz, C., and Mable, B.K. 1996. *Molecular Systematics*. Sunderland, MA: Sinauer.

Hoeh, W.R., Blakley, K.H., and Brown, W.M. 1991. Heteroplasmy suggests limited biparental inheritance of *Mytilus* mitochondrial DNA. *Science*, **251**: 1488–90.

Hoelzel, A.R. 1998. *Molecular Genetic Analysis of Populations: A Practical Approach*. Oxford: IRL Press.

Hurme, P. and Savolainen, O. 1999. Comparison of homology and linkage of random amplified polymorphic DNA (RAPD) markers between individual trees of Scots pine (*Pinus sylvestris* L.). *Molecular Ecology*, **8**: 15–22.

Iachan, R. 1985. Optimum stratum boundaries for shellfish surveys. *Biometrics*, **41**: 1053–62.

Iudica, C.A., Whitten, W.M., and Williams, N.H. 2001. Small bones from dried mammal museum specimens as a reliable source of DNA. *Biotechniques*, **30**: 732–6.

Jamieson, A. and Taylor, S.C.S. 1997. Comparisons of three probability formulae for parentage exclusion. *Animal Genetics*, **28**: 397–400.

Jansen, R.K., Wee, J.L., and Millie, D. 1998. Comparative utility of chloroplast DNA restriction site and DNA sequence data for phylogenetic studies in plants. In D.E. Soltis, P.S. Soltis, and J.J. Doyle, eds. *Molecular Systematics of Plants II. DNA Sequencing*. London: Kluwer Academic Press, pp. 87–100.

Jarne, P. and Lagode, P.J.L. 1996. Microsatellites, from molecules to populations and back. *Trends in Ecology and Evolution*, **11**: 424–9.

Jones, C.J., Edwards, K.J., Castaglione, S., Winfield, W.O., Sala, F., van de Wiel, C., Bredmeijer, G., Vosman, B., Matthes, M., Daly, A., Brettschneider, R., Bettini, P., Buiatti, M., Maestri, E., Malcevschii, A., Marmiroli, N.,

Aert, R., Volckaert, G., Rueda, J., Linacero, R., Vazquez, A., and Karp, A. 1997. Reproducibility testing of RAPD, AFLP and SSR markers in plants by a network of European laboratories. *Molecular Breeding*, **3**: 381–90.

Jorgensen, R.A. and Cluster, P.A. 1988. Modes and tempos in the evolution of nuclear ribosomal DNA: new characters for evolutionary studies and new markers for genetic and population studies. *Annals of the Missouri Botanical Garden*, **75**: 1238–47.

Kalinowski, S.T. 2002. How many alleles per locus should be used to estimate genetic distances? *Heredity*, **88**: 62–5.

Karp, A., Isaac, P.G., and Ingram, D.S. 1998. *Molecular Tools for Screening Biodiversity*. London: Chapman and Hall.

Kephart, S.R. 1990. Starch gel electrophoresis of plant isozymes: a comparative analysis of techniques. *American Journal of Botany*, **77**: 693–712.

Kimura, M. and Crow, J.F. 1964. The number of alleles that can be maintained in a finite population. *Genetics*, **49**: 725–38.

Kohn, M.H., York, E.C., Kamradt, D.A., Haught, G., Sauvajot, R.M., and Wayne, R.K. 1999. Estimating population size by genotyping faeces. *Proceedings of the Royal Society London, Series B*, **266**: 657–63.

Kraemer, H.C. and Thiermann, S. 1987. *How Many Subjects? Statistical Power Analysis in Research*. Newbury Park, CA: Sage.

Krebs, C.J. 1999. *Ecological Methodology*. Menlo Park, CA: Benjamin Cummings.

Kubis, S., Thomas, S., and Heslop-Harrison, J.S. 1998. Repetitive DNA elements as a major component of plant genomes. *Annals of Botany*, **82** (Suppl. A): 45–55.

Kunin, W.E. and Lawton, J.H. 1996. Does biodiversity matter? Evaluating the case for conserving species. In K.J. Gaston, ed. *Biodiversity. A Biology of Numbers and Difference*. London: Blackwell Science, pp. 283–308.

Laulier, M., Pradier, E., Bigot, Y., and Periquet, G. 1995. An easy method for preserving nucleic acids in field samples for later molecular and genetic studies without refrigerating. *Journal of Evolutionary Biology*, **8**: 657–63.

Lawrence, M.J. and Marshall, D.F. 1997. Plant population genetics. In N. Maxted, B.V. Ford-Lloyd, and J.G. Hawkes, eds. *Plant Genetic Conservation. The in situ Approach*. London: Chapman and Hall, pp. 99–113.

Lecointre, G., Philippe, H., Le, H.L.V., and Le, G.H. 1993. Species sampling has a major impact on phylogenetic inference. *Molecular Phylogenetics and Evolution*, **2**: 205–24.

Lewin, B. 2000. *Genes VII*. Oxford: Oxford University Press.

Li, W.-H. 1997. *Molecular Evolution*. Sunderland, MA: Sinauer.

Lipsey, M.W. 1990. *Design Sensitivity. Statistical Power for Experimental Research*. London: Sage.

Lodish, H., Baltimore, D., Berk, A., Zipurskey, S.L., Matsudaira, P., and Dornall, J. 1995. *Molecular Cell Biology*. New York: Scientific American Books.

Lookerman, D.J. and Jansen, R.K. 1995. The use of herbarium material for DNA studies. In T.F. Stuessy, ed. *Sampling the Green World*. New York: Columbia University Press, pp. 205–20.

Lorenz, M., Weihe, A., and Bõrner, T. 1994. DNA fragments of organellar origin in random amplified polymorphic DNA (RAPD) patterns of sugar beet (*Beta vulgaris* L.). *Theoretical and Applied Genetics*, **88**: 775–9.

Loveless, M.D. and Hamrick, J.L. 1984. Ecological determinants of genetic structure in plant populations. *Annual Review of Ecology and Systematics*, **15**: 65–95.

Lynch, M. and Milligan, B.G. 1994. Analysis of population genetic structure with RAPD markers. *Molecular Ecology*, **3**: 91–9.

Mace, A.E. 1964. *Sample-size Determination*. New York: Reinhold.

Mariette, S., le Corre, V., Austerlitz, F., and Kremer, A. 2002. Sampling within the genome for measuring within-population diversity: trade-offs between markers. *Molecular Ecology*, **11**: 1145–56.

Marshall, D.R. and Brown, A.H.D. 1975. Optimum sampling strategies in genetic conservation. In O.H. Frankel and J.G. Hawkes, eds. *Crop Genetic Resources for Today and Tomorrow*. Cambridge: Cambridge University Press, pp. 53–80.

Maxson, L.R. and Maxson, R.D. 1990. Proteins II: immunological techniques. In D.M. Hillis and C. Moritz, eds. *Molecular Systematics*. Sunderland, MA: Sinauer, pp. 127–55.

Maxted, N. and Guarino, L. 1997. Ecogeographic surveys. In N. Maxted, B.V. Ford-Lloyd, and J.G. Hawkes, eds. *Plant Genetic Conservation. The in situ Approach*. London: Chapman and Hall, pp. 69–87.

McCauley, D.E. 1995. The use of chloroplast DNA polymorphism in studies of gene flow in plants. *Trends in Ecology and Evolution*, **10**: 198–202.

McGavin, G. 1997. *Insects and Other Terrestrial Arthropods*. London: Royal Geographical Society.

Meksem, K., Ruben, E., Hyten, D., Triwitayakorn, K., and Lightfoot, D.A. 2001. Conversion of AFLP bands into high-throughput DNA markers. *Molecular Genetics and Genomics*, **265**: 207–14.

Milligan, B.G. 1992. Is organelle DNA strictly maternally inherited? Power analysis of a binomial distribution. *American Journal of Botany*, **79**: 1325–8.

Mitton, J.B. 1997. *Selection in Natural Populations*. Oxford: Oxford University Press.

Miyamoto, M.M. and Cracraft, J. 1991. *Phylogenetic Analysis of DNA Sequences*. Oxford: Oxford University Press.

Muller, K.E. and Benignus, V.A. 1992. Increasing scientific power with statistical power. *Neurotoxicology and Teratology*, 14: 211–19.

Murphy, R.W., Sites, J.W., Buth, D.G., and Haufler, C.H. 1996. Proteins: isozyme electrophoresis. In D.M. Hillis, C. Moritz, and B. Mable, eds. *Molecular Systematics*. Sunderland, MA: Sinauer, pp. 51–120.

Nei, M. 1973. Analysis of genetic diversity in subdivided populations. *Proceedings of the National Academy of Sciences USA*, 70: 3321–3.

Nei, M. 1978. Estimation of average heterozygosity and genetic distance from a small number of individuals. *Genetics*, 89: 583–90.

Nei, M. 1987. *Molecular Evolutionary Genetics*. New York: Columbia University Press.

Nei, M. and Kumar, S. 2000. *Molecular Evolution and Phylogenetics*. Oxford: Oxford University Press.

Nei, M. and Roychoudhury, A.K. 1974. Sampling variance of heterozygosity and genetic distance. *Genetics*, 76: 379–90.

Neigel, J.E. 1997. A comparison of alternative strategies for estimating gene flow from genetic markers. *Annual Review of Ecology and Systematics*, 28: 105–28.

Newbury, H.J. and Ford-Lloyd, B.V. 1993. The use of RAPD for assessing variation in plants. *Plant Growth Regulation*, 12: 43–51.

Palmer, J.D. 1985. Evolution of chloroplast and mitochondrial DNA in plants and algae. In R.J. MacIntyre, ed. *Molecular Evolutionary Genetics*. London: Plenum Press, pp. 131–240.

Palmer, J.D. 1986. Isolation and structural analysis of chloroplast DNA. *Methods in Enzymology*, 118: 167–86.

Palmer, J.D. 1987. Chloroplast DNA evolution and biosystematic uses of chloroplast DNA variation. *American Naturalist*, 130: 6–29.

Palmer, J.D. 1992. Chloroplast and mitochondrial genome evolution in land plants. In R.G. Herrmann, ed. *Cell Organelles*. Berlin: Springer-Verlag, pp. 99–133.

Palumbi, S.R. 1996. Nucleic acids II: the polymerase chain reaction. In D.M. Hillis, C. Moritz, and B.K. Mable, eds. *Molecular Systematics*. Sunderland, MA: Sinauer, pp. 205–47.

Paran, I. and Michelmore, R.W. 1993. Development of reliable PCR based markers linked to downy mildew resistance genes in lettuce. *Theoretical and Applied Genetics*, 85: 989–93.

Percus, J.K. 2002. *Mathematics of Genome Analysis*. Cambridge: Cambridge University Press.

Poe, S. 1998. Sensitivity of phylogeny to taxonomic sampling. *Systematic Biology*, 47: 18–31.

Pons, O. and Chaouche, K. 1995. Estimation, variance and optimal sampling of gene diversity. II. Diploid locus. *Theoretical and Applied Genetics*, 91: 122–30.

Pons, O. and Petit, R.J. 1995. Estimation, variance and optimal sampling of gene diversity. I. Haploid locus. *Theoretical and Applied Genetics*, 90: 462–70.

Powell, W., Morgante, M., Andre, C., Hanafey, M., Vogel, J., Tingey, S., and Rafalski, A. 1996. The comparison of RFLP, RAPD, AFLP and SSR (microsatellite) markers for germplasm analysis. *Molecular Breeding*, 2: 225–38.

Provan, J., Powell, W., and Hollingsworth, P.M. 2001. Chloroplast microsatellites: new tools for studies in plant ecology and evolution. *Trends in Ecology and Evolution*, 16: 142–7.

Rafalski, J.A., Vogel, J.M., Morgante, M., Powell, W., Andre, C., and Tingey, S.V. 1997. Generating and using DNA markers in plants. In B. Birren and E. Lai, eds. *Non-mammalian Genomic Analysis: A practical Guide*. New York: Academic Press, pp. 75–134.

Ramser, J., Weising, K., Lopez, P.C., Terhalle, W., Terauchi, R., and Kahl, G. 1997. Molecular marker based taxonomy and phylogeny of guinea yam (*Dioscorea rotundata* – *D. cayenensis*). *Genome*, 40: 903–15.

Richardson, B.J., Baverstock, P.R., and Adams, M. 1986. *Allozyme Electrophoresis: A Handbook for Animal Systematics and Population Structure*. Sydney: Academic Press.

Rieseberg, L.H. 1996. Homology among RAPD fragments in interspecific comparisons. *Molecular Ecology*, 5: 99–105.

Ritland, C. and Ritland, K. 2000. DNA-fragment markers in plants. In A.J. Baker, ed. *Molecular Markers in Ecology*. Oxford: Oxford University Press, pp. 208–34.

Ritland, K. 1989. Genetic differentiation, diversity, and inbreeding in the mountain monkeyflower (*Mimulus caespitosus*) of the Washington Cascades. *Canadian Journal of Botany*, 67: 2017–24.

Robinson, J.P. and Harris, S.A. 2000. Amplified fragment length polymorphism and microsatellites: a phylogenetic perspective. In E. Gillett, ed. *Which Marker for Which Purpose?* Hamburg: Institut für Forestgenetik und Forestpflanzenzüchtung, pp. 95–121.

Roe, B.A., Crabtree, J.S., and Khan, A.S. 1996. *DNA Isolation and Sequencing*. New York: Wiley.

Schaal, B.A., Hayworth, D.A., Olsen, K.M., Rauscher, J.T., and Smith, W.A. 1998. Phylogeographic studies in plants: problems and prospects. *Molecular Ecology*, 7: 465–74.

Scholz, M., Bachmann, L., Nicholson, G.J., Bachmann, J., Giddings, I., Rueschoff, T.B., Czarnetzki, A., and Pusch, C.M. 2000. Genomic differentiation of Neanderthals and anatomically modern man allows a fossil-DNA-based classification of morphologically indistinguishable hominid bones. *American Journal of Human Genetics*, 66: 1927–32.

Scribner, K.T. and Pearce, J.M. 2000. Microsatellites: evolutionary and methodological background and empirical applications at individual, population and phylogenetic levels. In A.J. Baker, ed. *Molecular Methods in Ecology.* Oxford: Oxford University Press, pp. 235–73.

Sharp, P.J., Desai, S., and Gale, M.D. 1988. Isozyme variation and RFLPs at the ß-amylase loci in wheat. *Theoretical and Applied Genetics,* **76**: 691–9.

Shrader-Frechette, K.S. and McCoy, E.D. 1992. Statistics, costs and rationality in ecological inference. *Trends in Ecology and Evolution,* **7**: 96–9.

Smouse, P.E. and Meagher, T.R. 1994. Genetic analysis of male reproductive contributions in *Chamaelirium luteum* (L.) Gray (Liliaceae). *Genetics,* **136**: 313–22.

Soltis, D.E., Soltis, P.S., and Milligan, B.G. 1992. Intraspecific chloroplast DNA variation: systematic and phylogenetic implications. In P.S. Soltis, D.E. Soltis, and J.J. Doyle, eds. *Molecular Systematics of Plants.* London: Chapman and Hall, pp. 117–50.

Soltis, P.S., Soltis, D.E., and Gottlieb, L.D. 1987. Phosphoglucomutase gene duplication in *Clarkia* (Onagraceae) and their phylogenetic implications. *Evolution,* **41**: 667–71.

Sork, V.L., Campbell, R., Dyer, J., Fernandez, J., Nason, J., Petit, R., Smouse, P., and Steinberg, E. 1998. Gene flow in fragmented, managed and continuous populations. Research Paper No. 3. National Center for Ecological Analysis and Synthesis, Santa Barbara, California.

Southwood, T.R.E. and Henderson, P.A. 2000. *Ecological Methods.* Oxford: Blackwell Science.

Stebbins, G.L. 1971. *Chromosomal Evolution in Higher Plants.* London: Edward Arnold.

Stern, M.J. and Eriksson, T. 1996. Symbioses in herbaria: recommendations for more positive interactions between plant systematists and ecologists. *Taxon,* **45**: 49–58.

Stewart, D.T., Saavedra, C., Stanwood, R.R., Ball, A.O., and Zouros, E. 1995. Male and female mitochondrial DNA lineages in the Blue Mussel (*Mytilus edulis*) species group. *Molecular Biology and Evolution,* **12**: 735–47.

Sugiura, M. 1989. The chloroplast chromosomes in land plants. *Annual Review of Cell Biology,* **5**: 51–70.

Swofford, D.L., Olsen, G.J., Waddell, P.J., and Hillis, D.M. 1996. Phylogenetic inference. In D.M. Hillis, C. Moritz, and B.K. Mable, eds. *Molecular Systematics.* Sunderland, MA: Sinauer, pp. 407–514.

Taberlet, P. and Waits, L.P. 1998. Non-invasive genetic sampling. *Trends in Ecology and Evolution,* **13**: 26–7.

ten Kate, K. and Laird, S.A. 1999. *The Commercial Use of Biodiversity. Access to Genetic Resources and Benefit-Sharing.* London: Earthscan.

Thomas, L. and Juanes, F. 1996. The importance of statistical power analysis: an example from *Animal Behaviour. Animal Behaviour,* **52**: 856–9.

Thomas, L. and Krebs, C.J. 1997. A review of statistical power analysis software. *Bulletin of the Ecological Society of America,* **78**: 126–39.

Thompson, S.K. 1992. *Sampling.* New York: Wiley.

Thormann, C.E., Ferreira, M.E., Camargo, L.E.A., Tivang, J.G., and Osborn, T.C. 1994. Comparison of RFLP and RAPD markers to estimating genetic relationships within and among cruciferous species. *Theoretical and Applied Genetics,* **88**: 973–80.

Tsumura, Y., Ohba, K., and Strauss, S.H. 1996. Diversity and inheritance of inter-simple sequence repeat polymorphisms in Douglas-fir (*Pseudotsuga menziesii*) and sugi (*Cryptomeria japonica*). *Theoretical and Applied Genetics,* **92**: 40–5.

Turnbull, J.W. and Griffin, A.R. 1986. The concept of provenance and its relationship to infraspecific classification. In B.T. Styles, ed. *Infraspecific Classification of Wild and Cultivated Plants.* Oxford: Clarendon Press, pp. 157–89.

Venugopal, G., Mohapatra, S., and Salo, D. 1993. Multiple mismatch annealing: basis for random amplified polymorphic DNA fingerprinting. *Biochemical and Biophysical Research Communications,* **197**: 1382–7.

Vos, P., Hogers, R., Bleeker, M., Reijans, M., van de Lee, T., Hornes, M., Frijters, A., Pot, J., Peleman, J., Kuiper, M., and Zabeau, M. 1995. AFLP: a new technique for DNA fingerprinting. *Nucleic Acids Research,* **23**: 4407–14.

Wang, Z., Weber, J.L., Zhong, G., and Tanksley, S.D. 1994. Survey of plant short tandem DNA repeats. *Theoretical and Applied Genetics,* **88**: 1–6.

Watt, W.B. 1977. Adaptation at specific loci. 1. Natural selection in phosphoglucose isomerase of *Colias* butterflies: biochemical and population aspects. *Genetics,* **87**: 177–94.

Watt, W.B. 1983. Adaptation at specific loci. 2. Demographic and biochemical elements in the maintenance of the *Colias* PGI polymorphism. *Genetics,* **103**: 691–724.

Watt, W.B. 1991. Biochemistry, physiological ecology, and population genetics – the mechanistic tools of evolutionary biology. *Functional Ecology,* **5**: 145–54.

Weeden, N.F. and Wendel, J.F. 1990. Genetics of plant isozymes. In D.E. Soltis and P.S. Soltis, eds. *Isozymes in Plant Biology.* London: Chapman and Hall, pp. 46–72.

Weir, B.S. 1996. *Genetic Data Analysis II.* Sunderland, MA: Sinauer.

Weising, K., Nybom, H., Wolff, K., and Meyer, W. 1995. *DNA Fingerprinting in Plants and Fungi.* London: CRC Press.

Wendel, J.F. and Weeden, N.F. 1990. Visualisation and interpretation of plant isozymes. In D.E. Soltis and P.S. Soltis, eds. *Isozymes in Plant Biology*. London: Chapman and Hall, pp. 5–45.

Wetherill, G.B. and Glazebrook, K.D. 1986. *Sequential Methods in Statistics*. London: Chapman and Hall.

White, G. and Powell, W. 1997. Cross-species amplification of SSR loci in the Meliaceae family. *Molecular Ecology*, 6: 1195–7.

Willan, R.L., Hughes, C.E., and Lauridsen, E.B. 1990. Seed collections for tree improvement. In N. Glover and N. Adams, eds. *Tree Improvement of Multipurpose Species*. Arlington, VA: Winrock International Institute for Agricultural Development, pp. 11–37.

Williams, J.G.K., Hanafey, M.K., Rafalski, J.A., and Tingey, S.V. 1993. Genetic analysis using random amplified polymorphic DNA markers. *Methods in Enzymology*, 218: 704–40.

Winer, B.J., Brown, D.R., and Michels, K.M. 1991. *Statistical Procedures in Experimental Design*. New York: McGraw-Hill.

Wolfe, A.D. and Liston, A. 1998. Contributions of PCR-based methods to plant systematics and evolutionary biology. In D.E. Soltis, P.S. Soltis, and J.J. Doyle, eds. *Molecular Systematics of Plants II. DNA Sequencing*. London: Kluwer Academic, pp. 43–86.

Wolfe, K.H., Morden, C.W., and Palmer, J.D. 1992. Small single-copy region of plastid DNA in the non-photosynthetic angiosperm *Epifagus virginiana* contains only two genes. Differences among dicots, monocots and bryophytes in gene organization at a non-bioenergetic locus. *Journal of Molecular Biology*, 223: 95–104.

Wolfe, K.H., Li, W.-H., and Sharp, P.M. 1987. Rates of nucleotide substitution vary greatly among plant mitochondrial, chloroplast, and nuclear DNAs. *Proceedings of the National Academy of Sciences USA*, 84: 9054–8.

Wolfe, K.H., Sharp, P.M., and Li, W.-H. 1989. Rates of synonymous substitution in plant nuclear genes. *Journal of Molecular Evolution*, 29: 208–11.

Wolstenholme, D.R. 1992. Animal mitochondrial DNA: structure and evolution. *International Review of Cytology*, 141: 172–216.

Wolstenholme, D.R. and Fauron, C.M.-R. 1995. Mitochondrial genome organisation. In C.S. Levings and I.K. Vasil, eds. *The Molecular Biology of Plant Mitochondria*. Dordrecht: Kluwer, pp. 1–59.

Zane, L., Bargelloni, L., and Patarnello, T. 2002. Strategies for microsatellite isolation: a review. *Molecular Ecology* 11: 1–16.

3

Genetic diversity and differentiation

"No-one supposes that all the individuals of the same species are cast in the very same mould. These individual differences are highly important for us, as they afford materials for natural selection to accumulate. . . ."

Charles Darwin (1859)

Summary

1 Genetic variation can be described as having three main components: genetic diversity (the amount of genetic variation); genetic differentiation (the distribution of genetic variation among populations); and genetic distance (the amount of genetic variation between pairs of populations).

2 Methods of measuring genetic diversity and its various components are available for codominant, dominant, and haploid markers. Similarly various statistics for measuring genetic differentiation between populations for a range of markers are available.

3 The Hardy–Weinberg equilibrium is fundamental to the interpretation of genetic variation data, and sophisticated methods are available to analyze deviations from this equilibrium.

4 The possible biological applications of the measurements of genetic variation include estimating gene flow, taxonomy, identifying genetic bottlenecks, and informing *ex situ* conservation sampling strategies.

3.1 Introduction

Variation is present, in some form, in natural populations of all organisms. This variation is particularly prominent when one looks at species that have been artificially selected by man are considered, for example, dogs or dahlias. The observed variation, the phenotype, may be reflected in genetic variation, the genotype. However, the genotype interacts with the environment to produce the phenotype. For instance, plants of identical genotype may produce different phen types when grown in different environments. Furthermore, variation may be associated with relatively simple characters (qualitative variation), for example, yellow and green peas in Mendel's classic experiment, or complex traits (quantitative variation), such as, the length of a mammalian femur.

Genetic variation, the raw material upon which natural selection acts, is continually being created by mutation and at the same time eroded by selection and drift. The ability of a species to respond to selection is dependent upon the presence of heritable variation. If genetic variation is present within a species, any alterations in selective pressures due to environmental changes will allow some individuals to survive and reproduce. For example, the celebrated case of the phenotypic changes observed in the peppered moth (*Biston betularia*) in Britain in the eighteenth century following the Industrial Revolution. The melanic variant, already present at low frequency in the pre-industrial population, was not conspicuous to bird predators against the background of soot-polluted tree trunks. This type became the predominant form over the pale and stippled variant, which had been common in pre-industrial times due to its camouflaged appearance against unpolluted trees, but was highly visible against polluted trees. This variation allowed the species to respond to the selective pressure imposed by increased soot in industrialized areas. Thus the absence of genetic variation may result in a species lacking the adaptive capacity to respond to environmental perturbations, which will ultimately lead to extinction.

The genetic variation within a species is therefore a fundamental concept for ecological geneticists, and has three components: (i) genetic diversity (the amount of genetic variation); (ii) genetic differentiation (the distribution of genetic variation among populations); and (iii) genetic distance (the amount of genetic variation between pairs of populations). Furthermore, genetic variation can be considered in terms of that which is neutral with respect to natural selection, that is, all alleles at a locus have equal effects on the individual carrying them, and that which is adaptive, that is, individuals that have particular alleles at a locus have greater fitness than other individuals possessing different alleles at the locus. From the point of view of ecological genetics, it is those differences between individuals that are due to genes, and not the environment, which are the most important, whether they be neutral or adaptive variation. Molecular markers are used to describe and estimate genetic variation. However, markers for neutral loci only reveal one aspect of a species' genetic variation (i.e., the neutral variation), the other important part of genetic diversity is adaptive variation, which is currently difficult to estimate using molecular markers (Ennos, Worrell, and Malcolm 1998), although many advances are being made in the fields of environmental and functional genomics and deal with polymorphisms at adaptive loci. This chapter will deal mainly with analyses of neutral genetic variation.

This chapter is concerned with describing genetic variation within and among populations. Following an introduction to the biological and genetic factors that influence genetic variation, the various approaches to the estimation of genetic diversity, differentiation, and diversity are considered. Finally, the uses of genetic variation measures are investigated in the context of ecological genetics.

3.2 Factors influencing diversity and differentiation

3.2.1 Characteristics of the organism and its environment

Population size

Random changes in allele frequency are related to population size: the smaller the population, the more likely chance events are to change allele frequencies. This random process of allele frequency change is called genetic drift, and is a result of random sampling of gametes. At its most extreme, drift can lead to the extinction of alleles and the loss of polymorphism such that a locus becomes fixed for a single allele. Thus, in order to eliminate the effects of drift populations must be large. Importantly, drift is independent of natural selection.

Reductions in population size can occur though events such as colonization of a new habitat by a small number of individuals (founder effects) or through processes such as habitat fragmentation where previously widespread populations are reduced in size (bottlenecks). However, population size and the effects of genetic drift are also phenomena associated with sampling of populations for either experimental or conservation purposes. Except in the rarest of species it is not possible to sample every individual in a population, hence stochastic variation due to drift may result. One of the expected outcomes of either founder effects or genetic bottlenecks is that the newly created population or the remaining population may have a different allele frequency to that of the source or original population (Box 3.1). The level of heterozygosity is reduced from one generation to the next at a rate proportional to the population size. This can be described by the equation $H_e = 1 - 1/(2N_e)^t$ where H_e is the level of heterozygosity, N_e is the effective population

Box
3.1

Chance and the Hardy–Weinberg equilibrium

Five factors lead to deviation from the Hardy–Weinberg equilibrium. Three of these, mutation, natural selection, and assortative mating, are expanded within this chapter, whilst a fourth, gene flow, is the subject of a subsequent chapter (chapter 4). An example of the remaining factor – chance events due to small population size – is considered here.

The effect of **chance events** was demonstrated by Byrne and Nichols (1999) in the mosquito *Culex pipiens* on the London Underground. *Culex pipiens* is found in the tunnels of the Underground, where it is an intermittent irritant for maintenance workers, and it was a more notable problem during the Second World War when it found a ready source of food in the humans that sought shelter during air raids. *Culex pipiens* is also found above

ground in London, where it feeds upon birds, and there are other ecological differences between the two groups of mosquitoes. From a survey of 20 allozyme loci of both above-ground and subterranean populations, the subterranean populations showed much lower levels of genetic diversity than those found on the surface (Table 3a) and there are very close similarities within the subterranean populations with virtually no differentiation between them (Table 3b). Further comparisons with Mediterranean and laboratory-bred populations indicated that the London Underground population had most likely been founded by a single colonization event from above-ground populations. This single chance event led to the variation present in the underground having a subset of the diversity found on the surface, and is an example of an extreme founder effect.

Table 3a

Genetic diversity measures for populations of *Culex pipiens* from underground, overground, and laboratory origins in London. Subterranean populations show consistently lower levels of genetic variation from those above ground. (Data is modified from that of Byrne and Nichols available at www.qmw.ac.uk/nugbt112.) Diversity indices are explained in section 3.4.

	Underground	Overground	Laboratory
No. of populations	8	11	1
Mean no. of alleles per locus (A)	2.45	3.5	1.15
% polymorphic locus (P)	0.6	0.9	0.15
Mean no. of alleles per polymorphic locus (AP)	3.6	3.8	2.3
Mean heterozygosity	0.07	0.27	0.05

Table 3b

Mean (and range) of Nei's (1972) D showing difference between mosquito populations of different origins. Calculation of D is described further in section 3.6.3 and Box 3.12. Overground population is on average twice as different from each other as underground population is from each other. Data from Byrne and Nichols (1999).

	Underground	Overground
Underground	0.025 (0.0001–0.004)	
Overground	0.20 (0.07–0.23)	0.05 (0.17–0.1)
Laboratory	0.23 (0.025–0.26)	0.20 (0.13–0.27)

size and t is the number of generations. So in one generation an effective population of 10 individuals loses $1 - 1/(2 \times 10)^1 = 0.05$ of the variation present in the initial population.

Population size, and the stochastic loss of alleles, is an important issue in conservation biology. This raises the question of how large a population needs to be before the loss of alleles through genetic drift becomes unimportant. Theoretically, drift will always occur unless the population is infinitely large. Since infinitely large populations do not exist, this has led to the development of the concept of minimum viable populations (MVP). The aim of this approach is to identify an effective population size (N_e), which is large enough to retain sufficient genetic variation to suffer the vicissitudes of demographic and environmental stochasticity without becoming extinct. The "50/500" rule states that 50 individuals are necessary to conserve genetic variation in the short term (i.e., a few generations) and 500 individuals are necessary to prevent long-term loss of genetic diversity (Franklin 1980, Soule 1980). However, these values have been widely criticized (Lande 1995) as they appear to underestimate the numbers of individuals needed to maintain long-term viability of a population. The effective population size may fluctuate from generation to generation and the numbers required to maintain levels of genetic variation are also highly dependent on the life history of the organism under study.

Gene flow

Gene flow is the proportion of newly immigrant genes moving into a population (chapter 4). Gene flow comprises both immigration of alleles into a population and migration of alleles out of the population. For stationary organisms, for example, most plants, gene flow comprises movement of gametes (e.g., pollen) and movement of zygotes (e.g., seeds). In contrast, in many social animals (e.g., lions, *Panthera leo*) immigration of gametes may occur through movement of males, whilst females remain in the group to which they were born. Such patterns of allele movement can have profound impacts on the structure of genetic diversity (chapter 4). The extent of gene flow is determined by the mobility of the species, dispersal ability of propagules/gametes, and the degree of isolation of populations, whether that is physical, ecological, or temporal.

Reproductive system

The simplest assumption about a reproductive system is that any single allele carried by a gamete is equally likely to fuse with any other allele, that is, alleles fuse at random (panmixia or non-assortative mating). However, this may not occur, in which case mating is described as assortative. Positive assortative mating (like breeding with like) is when mating between individuals of the same type occurs more often than expected at random, whilst negative assortative mating (like not breeding with like) is when mating between individuals of the same type occurs less often than expected. Assortative mating does not change allele frequencies between generations but it will influence the proportion found in a population of different genotypes. Positive assortative mating leads to a reduction in the expected proportion of heterozygous loci relative to panmixia, whilst negative assortative mating leads to more heterozygous loci than expected. At its most extreme, positive assortative mating occurs where an individual is completely self-fertilized.

Natural selection

The majority of molecular markers utilized in ecological genetics work are considered to be selectively neutral (chapter 2). However, some supposedly neutral loci have been shown to have adaptive significance or be closely linked to genes under selection. For example, adaptive characteristics have been demonstrated for the enzyme phosphoglucose isomerase in butterflies (e.g., *Colias*; Watt 1977, 1983) and leucine aminopeptidase in *Mytilus edulis* (Koehn and Hilbish 1987). More recently, comparisons of genetic variation derived from coding versus non-coding genomic regions of *Crassostrea virginica* (Karl and Avise 1992) and *Passerella iliaca* (Zink 1986,

1994) have revealed different levels of population differentiation. Significant population heterogeneity in DNA markers was not reflected in allozyme variation, implying that markers located within the coding regions are subject to selection. Furthermore, interaction of marker loci with other regions of the genome (pleitropy) that are subject to selection can also influence allele frequency (Charlesworth, Nordberg, and Charlesworth 1997).

3.2.2 Genome considerations

Mutation

Mutation occurs continually. However, the rate of mutation is likely to be extremely low, although this will depend on the genome and marker investigated (chapter 2). The rate of nuclear mutation per gene per cell generation is generally considered to be of the order of 1×10^{-5} to 1×10^{-6} mutations per 100,000 cells, although this is variable between loci (Skelton 1993). For instance, the yellow-body mutation in *Drosophila melanogaster* has a mutation rate of 1.2×10^{-5} mutations per 100,000 cells, whilst the sugary seed mutation in *Zea mays* occurs at a rate of 2.4×10^{-7} mutations per 100,000 cells (Skelton 1993). In situations where the mutation rate of the markers investigated is likely to be high (e.g., nuclear microsatellites) then appropriate statistical analyses are needed to incorporate this (Goldstein and Schlötterer 1999).

Polyploidy

Polyploidy, the occurrence of more than two copies of an entire nuclear genome within a cell, can have effects on the calculation and interpretation of genetic diversity statistics, depending on the nature of the polyploidy event (Grant 1981, Levin 2000). At its simplest, polyploidy arises through multiplication of genomes within a species (autopolyploidy) or by genome multiplication of interspecific crosses (allopolyploidy). However, it is important to distinguish the timescale over which these events occur, that is, whether the polyploid event is recent (neopolyploidy) or much older (ancient polyploidy; Levin 2000, Soltis and Soltis 1993, 1999; chapter 6).

Linkage

Linked genes are genes that are found on the same chromosome. When large numbers of loci are utilized it is inevitable that linked genes will be incorporated into the data set. However, linkage only becomes a consideration in the analysis of genetic data if the genes are very close together or recombination rates are very low. In such situations, linkage disequilibrium occurs, where an allele from one gene locus is found to be associated with an allele from another locus more frequently than expected under random association (Box 3.2).

3.2.3 Molecular marker suitability

The properties of different molecular marker types need to be considered carefully to minimize violating assumptions of the particular statistic to be used to describe genetic variation. The important characteristics of the different marker classes that can be considered when choosing which diversity, differentiation, or distance measure to apply are outlined below (see chapter 2 for a consideration of techniques).

Isozymes

Many diversity and differentiation statistics have been developed specifically with isozymes in mind (e.g., Kimura 1953, Slatkin 1985). Whilst isozymes do not show particularly high levels of variation, their typical codominance, known genomic origin (usually nuclear), and widespread application makes them very suitable for estimating diversity and differentiation. The drawbacks

Box 3.2

Understanding linkage disequilibrium

Linked genes are genes that are located on the same chromosome (Fig. 3a). If two alleles from different genes on the same chromosome tend to be associated in different individuals at a greater frequency than that expected due to random association, linkage disequilibrium is said to exist between those two genes.

Consider two loci (A and B) located closely together on the same chromosome, each with two alleles (*A* and *a* and *B* and *b*, respectively). Each allele is present in the population at a frequency of 0.5. The expected frequency of each combination of alleles on any particular chromosome is shown in Table 3c.

Figures in italics show observed frequencies. Any observed deviation from expectation in such a 2 × 2 contingency table can be tested statistically using a chi-squared test or Fisher's exact test. Even without statistical testing, it can be seen that the frequency of genotypes *AB* and *ab* are much higher than anticipated, with corresponding reductions in the expected frequencies of genotypes *Ab* and *aB*. This is due to linkage disequilibrium.

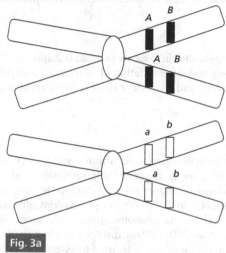

Fig. 3a

Homologous chromosome pairing in prophase I. Two genes are shown on each chromosome arm. As illustrated alleles *A* and *B* are linked, as are alleles *a* and *b*.

Table 3c

Expected frequency of pairwise combinations of linked alleles in the absence of linkage disequilibrium in bold. Figures in italics are observed frequencies and are very different from expected values, they indicate linkage disequilibrium.

Population allele frequency	*A*	*a*
	0.5	0.5
B	*AB*	*aB*
0.5	**0.25**	**0.25**
	0.42	*0.10*
b	*Ab*	*ab*
0.5	**0.25**	**0.25**
	0.08	*0.40*

---

of isozymes are that they are unordered and only low levels of polymorphism may be found, thus limiting their application to estimate genetic distance (although unordered methods relying purely on allele-sharing matches are available to estimate distance).

Microsatellites

Microsatellite markers are highly polymorphic, codominant loci and are abundant in most species' genomes. However, the high number of alleles per locus causes some bias in diversity and differentiation estimates due to increased heterozygosity levels. Other methods of analysis that use a step-wise mutation model, applied to repeat number variation to infer evolutionary relationships between alleles, can also be used effectively to calculate genetic distance and differentiation measures. The high mutation rate also means that microsatellite loci suffer from homoplasy problems (Schlötterer et al. 1998) and a large number of loci (more than 10) are recommended if such an evolutionary distance basis is used. A potentially high mutation rate and high allelic number may also violate some diversity models by increasing the within-population component of variation.

Dominant markers

This class of dominantly expressed markers (e.g., AFLPs, RAPDs, and ISSRs) are multilocus, and scattered throughout the entire genome (chapter 2). The application of dominant markers to diversity and differentiation estimates is, however, problematic due to the unknown proportion of heterozygotes in a population. Several methods can be used to circumvent this problem, either by using robust estimators or by working with classes of fragment where the error between observed and expected heterozygosity is reduced (e.g., Lynch and Milligan 1994). Due to the nature of the loci being scattered throughout the genome, such dominant markers are very useful for assessing whole genome genetic variation. Data are unordered, although the large potential number of loci that can be surveyed allow for powerful derivation of unordered estimates of distance. Inappropriate methods of analysis may significantly overestimate these parameters in natural populations.

Animal organellar DNA markers

Variation in the animal mitochondrial genome is high, and the data are ordered (Chapter 2). It is therefore an excellent marker for assessing diversity and differentiation. The main advantage of this genome results in the application of phylogenetic and coalescent models to differentiation and distance estimates.

Plant organellar DNA markers

Chloroplast genomes tend to harbor much lower levels of variation than animal mtDNA (Wolfe et al. 1987), although the mutations are ordered (chapter 2). Diversity, differentiation, and distance measures are therefore possible and the phylogenetic information content of the molecule can also be utilized to examine historical processes (chapter 5). Within populations, diversity tends to be low, although microsatellite loci in the chloroplast genome appear to be more variable than indel and sequence mutations (Provan et al. 2001). Plant mtDNA has low levels of unordered variation (Palmer 1992). When variation is found it can be used to assess population diversity and differentiation, although it is likely to be of limited use.

nDNA variation

Variation at neutral loci within the nuclear genome offers great potential for diversity and differentiation studies, particularly if data can be ordered (Zhang and Hewitt 2003). Data are codominant and thus a range of diversity and differentiation statistics can be used. Some problems exist with the application of uncharacterized nuclear loci, including parology (see chapter 2 for a full explanation), mutation rate, and selection, but with further development useful markers are becoming available, and should allow application of genetic distance estimates to such loci.

3.3 The Hardy–Weinberg equilibrium

The Hardy–Weinberg equilibrium, developed independently by British mathematician Godfrey H. Hardy and the German physician Wilhelm Weinberg in 1908, is a central concept to many genetic diversity and differentiation models. Given initial allele frequencies within a population, the model predicts the proportion of diallelic genotypes in the next generation following random mating. The model operates under conditions of no natural selection, no mutation, no gene flow into or out of an infinitely large population, and that mating is non-assortative. The influence of forms of non-assortative mating is explored in Boxes 3.3 and 3.4.

Box
3.3

The effect of positive assortative mating on genotype frequencies

Positive assortative mating is the phenomenon of an individual being more likely to mate with another individual of the same phenotype than would be expected due to chance. Self-fertilization, where an individual mates only with itself, is the most extreme form of positive assortative mating.

In a one-gene, two-allele system, the most variable scenario is when the two alleles are at the same frequency, that is,

p (frequency of A) = 0.5, q (frequency of a) = 0.5. With these allele frequencies and under Hardy–Weinberg equilibrium, the expected frequency of the three genotypes is as follows: AA: $0.5^2 = 0.25$; Aa: $2 \times 0.5 \times 0.5$ = 0.5; aa: $0.5^2 = 0.25$.

If we assume complete self-fertilization, the expected frequencies of the genotypes will change. In each generation self-fertlization of each homozygote will produce more homozygotes. However, self-fertilization of heterozygotes due to random fusion of gametes at fertilization will produce all three phenotypes, only half of which will be heterozygous.

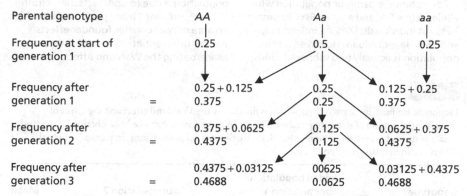

Parental genotype	AA	Aa	aa
Frequency at start of generation 1	0.25	0.5	0.25
Frequency after generation 1 =	0.25 + 0.125 0.375	0.25 0.25	0.125 + 0.25 0.375
Frequency after generation 2 =	0.375 + 0.0625 0.4375	0.125 0.125	0.0625 + 0.375 0.4375
Frequency after generation 3 =	0.4375 + 0.03125 0.4688	00625 0.0625	0.03125 + 0.4375 0.4688

Thus at each generation with complete selfing half of the heterozygotes will be lost (Fig. 3b).

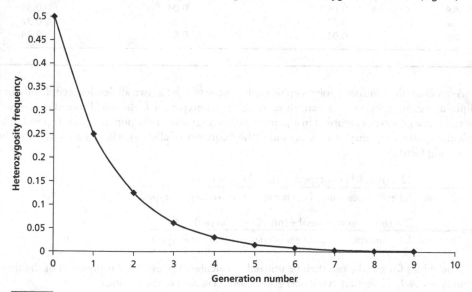

Declining frequency of heterozygotes in a two-allele system in a population where reproduction is completely by self-fertilization. Note the population starts with the maximum number of heterozygotes possible.

The Wahlund effect

Another possible cause of deviation from Hardy–Weinberg equilibrium, through a form of non-random mating, is the Wahlund effect. First described by Wahlund in 1928, this phenomenon occurs when an area is sampled as a single population but is actually composed of two or more distinct populations, which are differentiated and have little or no gene flow between them.

For instance, a panmitic population with alleles A (p = 0.5) and a (q = 0.5) will contain 25% AA individuals, 50% Aa individuals, and 25% aa individuals. However, if this population is actually two subpopulations of equal size, with frequencies of p = 0.8 and q = 0.2 in the first subpopulation and p = 0.2 and q = 0.8 in the second subpopulation, this gives expected genotype numbers in the two subpopulations as shown in Table 3d.

Note that the expected frequency of heterozygotes is higher when the sample is treated as a single population than two subpopulations (0.5 compared to 0.32).

Under such circumstances the deficiency of heterozygotes within the sampled population is due to undetected structuring. Similarly, smaller "true" populations within an area may also suffer founder effects further differentiating each of them and exacerbating the Wahlund effect.

Table 3d

Genotype frequencies in two subpopulations divided by the Wahlund effect and the effect of calculating the mean of these frequencies. The mean is the figure that would be obtained if the study area was treated as one population rather than two distinct subpopulations. In particular note the difference in homozygote frequencies.

Genotype	Divided population Subpopulation 1	Subpopulation 2	Mean
AA	0.64	0.04	0.34
Aa	0.32	0.32	0.32
aa	0.04	0.64	0.34

As an example, consider a polymorphic nuclear gene which has two alleles denoted A and a in a diploid organism, hence there are three possible genotypes (AA, Aa, aa). From the numbers of the three genotypes recorded in a population (or a sample of the population) it is possible to calculate p (the frequency of allele A) and q (the frequency of allele a), where $p + q = 1$, using the following formulae:

$$p = \frac{[2 \times (\text{no. } AA \text{ genotypes})] + (\text{no. } Aa \text{ genotypes})}{2 \times (\text{no. } AA \text{ genotypes} + \text{no. } Aa \text{ genotypes} + \text{no. of } aa \text{ genotypes})}$$

$$q = \frac{[2 \times (\text{no. } aa \text{ genotypes})] + (\text{no. } Aa \text{ genotypes})}{2 \times (\text{no. } AA \text{ genotypes} + \text{no. } Aa \text{ genotypes} + \text{no. } aa \text{ genotypes})}$$

The values for p and q can then be utilized to calculate the expected proportion of the three genotypes (AA, Aa, and aa) in the next generation according to the equation:

$$p^2 + 2pq + q^2 = 1$$

The number of each genotype is then given by: (i) p^2N = number of homozygous AA individuals in the population; (ii) $2pqN$ = number of heterozygous Aa individuals in the population; and

The Hardy–Weinberg equilibrium in practice

Senecio vulgaris (Common groundsel) is an annual plant of open and disturbed ground. Its capitulum (flower head) is polymorphic with some individuals having a rayed capitula and others having rayless capitula (chapter 7). There is also an intermediate form with short, stubby rays. This character is under the genetic control of a single gene with two alleles, ray (*R*) and rayless (*r*), that show incomplete dominance. The three genotypes give three different phenotypes: *RR* – rayed capitula (long rays 7–10 mm); *Rr* – short-rayed capitula; *rr* – rayless capitula.

A field survey in Ormskirk (Lancashire, UK) in 2002 revealed the following numbers of the three phenotypes: 152 *RR*; 12 *Rr*; and 354 *rr*. The frequency of the *R* allele (*p*) is calculated from $[(2 \times 152) + 12]/[2 \times (152 + 12 + 354)]$ as 0.31. Similarly, the frequency of the rayless allele is calculated as 0.69 (note that $p + q = 1$). Using these allele frequencies the expected numbers of the three phenotypes in the total population (*N*) according to Hardy–Weinberg principles and assuming random mating of genotypes are calculated as: expected number of *RR* ($p^2N = 50$); expected number of *Rr* ($2pqN = 222$); and the expected number of *rr* ($q^2N = 246$).

Comparison of the observed and expected numbers of genotypes in the population can be used to establish whether the population is in Hardy–Weinberg equilibrium (Table 3e).

Table 3e

Observed and expected genotype numbers in a population of *Senecio vulgaris*.

Genotype	Observed numbers	Expected numbers
RR	152	50
Rr	12	222
rr	354	246

Even without recourse to statistics it is clear that there is a great discrepancy between the observed and expected numbers of genotypes, with an abundance of homozygotes and a corresponding lack of heterozygotes in the wild. A chi-squared test reveals that the difference between observed and expected numbers is highly significant ($p < 0.0001$).

The cause of deviation from the Hardy–Weinberg expectations needs careful consideration of the conditions for the equilibrium along with the detailed knowledge of the species' biology. Given that the species is predominantly self-fertilizing, the most extreme form of positive assortative mating, non-random mating is the most likely reason for the results observed. However, while mutation is extremely unlikely to produce the observed result, changes through natural selection (homozygote advantage), gene flow (wind-borne seed), and drift (small population size) would need to be considered.

(iii) q^2N = number of homozygous *aa* individuals in the population, where *N* is the total population size.

The null hypothesis that the expected numbers of the three genotypes are equal to the observed numbers can be tested using a χ^2 (chi-squared) test or other statistical tests (e.g., Fisher's exact). A significant difference between observed and expected values (i.e., rejection of the null hypothesis) implies that one or more of the conditions of the Hardy–Weinberg equilibrium have not been met. A good understanding of both the conditions necessary for the equilibrium to be maintained and of the biology of the species being considered allows deviation from the equilibrium to be explained (Box 3.5).

In a three-allele system, the frequency of alleles A_1, A_2, and A_3 are denoted by p, q, and r, respectively. Hence the six possible genotype frequencies are calculated thus (where $p^2 + q^2 + r^2 + 2pq + 2pr + 2qr = 1$): $A_1A_1 = p^2$; $A_2A_2 = q^2$; $A_3A_3 = r^2$; $A_1A_2 = 2pq$; $A_1A_3 = 2pr$; and $A_2A_3 = 2qr$.

Conversely, if the genotype frequencies are known then the allele frequencies can be calculated using the same approach as that outlined for two alleles. Thus:

$$p = \text{frequency of allele } A_1 = \frac{[2 \times (\text{no. } A_1A_1 \text{ types})] + (\text{no. } A_1A_2 \text{ types}) + (\text{no. } A_1A_3 \text{ types})}{2 \times \text{total no. of individuals in the population}}$$

3.4 Genetic diversity

Genetic diversity is a commonly used expression to describe the heritable variation found within biological entities and can be measured at the individual, population, and species level. At any particular locus diversity may be present within an individual, for example an individual may be heterozygous. Diversity may also be present within a population, when the alleles present at a variable locus are found in different individuals. If there are more than two alleles and the organism is diploid, no individuals will exhibit the total variation present within the population. In this case the population diversity is described as being partitioned among individuals. In addition, diversity may be present within species and allelic variation may change between populations across the range of the species where some populations exhibit alleles that others lack and vice versa. In this case diversity is said to be partitioned among populations of a species. A species, population, or individual may be monomorphic at a particular locus, in which case all have the same allele. Alternatively, a species, population, or individual may be polymorphic at a locus, in which case different alleles are present. The distinction between the proportion of heterozygosity and the level of polymorphism, at the individual, population, and species levels, are two of the values that diversity statistics attempt to quantify.

Diversity in a broad sense has two components: richness and evenness. Richness is a measure of abundance, while evenness indicates how the variation is distributed. Richness can be related to the number of polymorphic loci present, or it can record the number of alleles at a locus. Evenness, on the other hand, can be represented by the mean number of alleles at a locus within a population or species, and requires an assessment of variance.

The approaches to quantifying genetic variation can be broadly divided into three categories, based on: (i) observations; (ii) assumptions; or (iii) similarities. Approaches based on observations are the simplest methods of assessing diversity, and within a single value attempt to record the breadth of diversity found within a group (e.g., the number of alleles) and are typically associated with measures of allelic richness. They make no assumptions about species biology. Several other measures are based on the assumptions of the Hardy–Weinberg equilibrium. Thus deviations from the Hardy–Weinberg equilibrium can be tested. The primary use of this approach is in determining the total amount of genetic diversity within a species and its degree of compartmentalization, within and between populations. Similarity approaches (e.g., Simpson's and Shannon's indices) are based upon the degree of allele sharing between populations and whilst they can be used with a range of molecular marker methods, they are usually associated with the analysis of dominant data. Similarity statistics measure the degree of difference between pairs of populations or species.

A further consideration in the choice of statistic is the number of copies of each gene within an individual (the ploidy level) and the possibility that molecular markers will highlight a number of forms or copies of each loci, some of which are not homologous (the problem of parology, chapter 2). A range of the most commonly applied diversity statistics is outlined below.

3.4.1 Measures of allelic richness

The most straightforward approach to measuring genetic diversity is to produce an estimate of allelic diversity, which involves measuring the number of alleles per locus or the number of polymorphic loci. An example of the use of allelic richness is given in Box 3.6.

Box
3.6

Allelic richness in elephant populations

Hunting of elephants (*Loxodonta africana africana*) by white settlers in southern Africa has caused both population fragmentation and severe reduction in numbers. Due to conservation efforts population numbers have recovered to some degree since 1900, when they were probably at their lowest.

Within South Africa the largest population (>8000 elephants) is currently in Kruger National Park. This has developed from only 10 elephants reported in the area in 1905, the increase occurring through successful breeding and immigration from Mozambique.

At the Addo National Park the elephant population numbers 325. This park was set up in 1931 to conserve the 11 elephants present in the area at that time. The reserve is isolated so no population increase by immigration has occurred and the current population stems from between seven and nine individuals that produced progeny following the establishment of the reserve. Hence this population has suffered a major bottleneck.

Whitehouse and Harley (2001) used 10 microsatellite loci to compare the genetic variation in Addo and Kruger National Parks (Fig. 3c). The DNA for Addo elephants came from biopsy darting and incorporated 105 individuals (approximately 40% of the current population). In addition samples from skins of two Addo elephants killed in the 1920s were assayed. The 108 Kruger elephant samples were randomly chosen from the dried hides of culled animals stored at the park.

Genetic diversity was compared using an allelic richness approach plus observed heterozygosity (H_O) and expected heterozygosity (H_T) (Table 3f).

The Addo population showed lower levels of genetic diversity than the Kruger population for all measures. Despite the very small sample size the level of variation in the Addo museum specimens is closer to the levels of variation found in the Kruger population.

This difference between the Addo museum specimens and the current population suggests that the bottleneck that occurred due to hunting has reduced genetic diversity in the population, a conclusion further endorsed by the

Fig. 3c

Location of South African national parks sampled for genetic variation in elephants (from Whiteh and Harley 2001).

Measures of allelic richness and gene diversity in two South African elephant populations.

Population	Mean sample size	P	AP	H_O	H_T
Addo	105	7	1.89	0.192	0.180
Kruger	108	9	3.89	0.422	0.444
Addo Museum	2	7	1.89	0.500	0.333

comparison of the Kruger and current Addo populations.

It is also likely that genetic diversity has been reduced due to population bottlenecks in the Kruger population. Consideration of identical loci between this study and that of Ugandan populations by Nyakaana and Arctander (1999) shows a lower number of alleles per polymorphic loci in the South African population (3.25 in Kruger National Park, 5.92 in Uganda). Loss of variation through bottlenecks is explained further in section 3.8.3.

Allelic diversity or allelic richness (A)

This is the mean number of alleles per locus, and includes monomorphic loci (although some loci only have a maximum of two alleles by definition, e.g., AFLPs). It may be calculated per population or per species. The statistic is sensitive to sample size (Nei 1987), although statistical standardization approaches are available (e.g., see section 3.7.1).

Percentage of polymorphic loci (P)

A locus is defined as polymorphic according to a predetermined criterion, for example, the frequency of the most common allele at the locus is <0.95. This measure is sensitive to the arbitrary definition of the criterion (Nei 1987). However, in contemporary literature loci are often treated as polymorphic if they exhibit any variation whatsoever.

Mean number of alleles per polymorphic locus (AP)

This is calculated as for *P* but includes only polymorphic loci.

Mean observed heterozygosity (H$_O$)

This value is the mean number of heterozygotes recorded at a particular locus expressed as a proportion of the total number of loci surveyed. This measure is widely applied to codominant data for diploid organisms. Obviously, haploid organisms cannot be heterozygous. It can be utilized with polyploids, albeit interpretation needs careful consideration (Nei 1987). However, this measure is inappropriate for dominant data, since the proportion of heterozygous individuals within a population cannot be easily identified.

3.4.2 Nei's gene diversity statistics

The term gene diversity was first used by Nei (1973), although Marshall and Allard (1970) used the same method to calculate their "polymorphic index". Whilst frequently referred to as expected heterozygosity, Weir (1990) considers gene diversity a more appropriate term to be

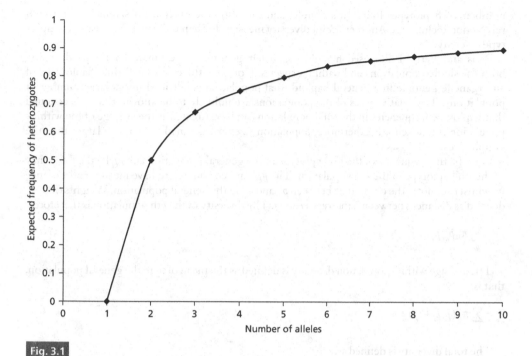

Fig. 3.1

The effect of increasing the number of alleles on Nei's gene diversity, H_T(expected diversity), in an infinitely large population when all alleles in a population are equally common.

used with the range of reproductive systems encountered (particularly those that self-fertilize) when describing diversity and its apportionment within and between populations.

Nei's (1973) measure of gene diversity uses the expected heterozygosity across the total species (H_T) defined as $1 - \sum_{i=1}^{i=K} \bar{p}_i^2$, where $\bar{p}$ is the mean frequency of the ith of K alleles across all populations surveyed. Nei's gene diversity is widely used but it does have its drawbacks (Nei 1987). For example, it varies between 0 and 1, and as the frequencies of alleles at a locus approach equality, the measure becomes less sensitive. In addition, the value is heavily dependent upon the frequency of the two most common alleles.

One effect of increasing the number of alleles at a locus is that the number of heterozygotes also increases. For instance, in a two-allele system (diallelic) with both alleles at equal frequency (0.5) the expected frequency of heterozygotes is 0.5. In a three-allele system with each allele at equal frequency, the expected frequency of heterozygotes is 0.6733; this trend continuing with increasing number of alleles (Fig. 3.1).

Thus, if multiallelic markers are used, then the population level heterozygosity increases purely as a function of the marker system used, not necessarily because of an actual increase in diversity. As long as markers with comparable levels of allelic richness are compared, then relative measures are permissible. However, where within-population heterozygosity increases due to high allele number, there will be a reduction in the amount of between-population diversity that can be partitioned, and this will bias measures of genetic differentiation.

3.4.3 Considerations for haploid genomes

Haploid genomes can be treated in much the same way as other genomes for the estimation of genetic diversity. For example, estimates of haplotype richness are based on counting the numbers of haplotypes found in a sample, and can also be corrected for sample size through rarefaction techniques. Another useful diversity measure for haploid data is a measure of haplotype diversity.

Pons and Petit (1996) used haploid organellar genomes to estimate haplotype diversity, both for single populations and within a species. Consider the case of I distinct haplotypes of an organelle genome in a general haploid total population subdivided into a large number of populations. The relative sizes of the populations are unlikely to be known, thus it is assumed that haplotype frequencies in the total population can be estimated as the average of the within-population frequencies. Furthermore, population sizes are assumed to be much larger than the sample sizes.

Let p_i be the frequency of the ith haplotype in the general population and p_{ki} be the frequency of the ith haplotype in the kth population. The p_{ki}'s are considered to have mean p_i and variance v_i, whilst c_{ij} denotes the covariance between p_{ki} and p_{kj} in the general population. Weights (π_{ij}) are defined as distances between haplotypes i and j. The diversity of the kth population is therefore:

$$v_k = \sum_{ij} \pi_{ij} p_{ki} p_{kj}$$

The average within-population diversity is defined as the mean of v_k in the general population, that is:

$$b_s = \sum_{ij} \pi_{ij} (p_i p_j + c_{ij})$$

The total diversity is defined as:

$$b_T = \sum_{ij} \pi_{ij} p_i p_j$$

The weights may be defined in any appropriate way, for example, the proportion of different nucleotides or restriction sites between the haplotypes, where $\pi_{ii} = 0$ and $\pi_{ij} = \pi_{ji}$. At its simplest, weights can be defined if the haplotypes are not identical, then $\pi_{ii} = 1$, and $\pi_{ii} = 0$ otherwise. In this case, b_s and b_T are simplified to:

$$b_s = 1 - \sum_i (p_i^2 + v_i) \text{ and } b_T = 1 - \sum_i (p_i^2), \text{ respectively.}$$

Genotypic diversity (D_G)

Simpson's index is often used as a measure of diversity, and has a maximum value approaching 1 and a minimum value of 0, when two samples are completely identical. The formula is:

$$1 - \frac{\sum [n_i(n_j - 1)]}{[N(N-1)]}$$

where n_i is the number of individuals of genotype I in a population of sample size N.

Shannon's index of diversity (Shannon 1948)

This is another diversity measure widely used in ecology but applied to genetics (Lewontin 1972), and is calculated as:

$$H = -\sum p_i \ln(p_i)$$

where p_i is the proportion of the ith allele in the population.

The effect of using Shannon's rather than Simpson's index is that the use of the natural logarithm reduces the importance of the most abundant values. Shannon's index has values that range from zero to infinity (the upper limit is determined by the number of loci sampled). However, values can be constrained by averaging estimates across loci. Shannon's index is generally preferred as the resultant values are normally distributed, which allows more refined statistical tests to be applied to the data.

Since Shannon's index uses binary data and the number of loci scored affects the maximum value of the measure it is very important that similar numbers of loci are used if valid comparisons are to be made. In particular, it is necessary to consider the effects of missing or equivocal data at a locus when comparisons are made between populations. Comparisons between studies should also check how individual locus have been combined.

3.4.4 Using dominant data to calculate diversity statistics

The main problem with using dominant data, derived from multilocus screening methods, to estimate diversity statistics is that the frequency of heterozygotes is unknown, as they are indistinguishable from homozygotes expressing the product under consideration. Therefore, it is not possible to assess directly whether a particular population is in Hardy–Weinberg equilibrium. There are two approaches to take with this problem. The first uses methods that ignore this problem and thus limitations must be borne in mind when results are considered (e.g., Nei's and Shannon's diversity indices). The second addresses the shortcomings of the data and uses specific statistics to offset these drawbacks, such as those developed by Lynch and Milligan (1994).

Nei's and Shannon's methods are not ideal for calculating diversity using dominant data. Nei's method was originally derived for use with (codominant) proteins, and is based on a measure of expected heterozygosity. For dominant data, the concept of heterozygosity is not applicable and the estimate becomes "gene diversity," simply a measure of genetic variability but still of statistical value (Nei 1987), and is defined as:

$$\text{Gene diversity, } h = 1 - \sum x_i^2$$

where (in the dominant case) x_i is the population frequency of each allele (1 and 0) at locus i. The average gene diversity is the average of this quantity across all loci. The value is adjusted for variation in sample sizes (Nei 1978) through multiplication by $2n/(2n-1)$, where n = sample size.

Shannon's index (H; Shannon 1948), originally developed as a measure of entropy in information theory, has become a widely used statistic for quantifying levels of diversity (Lewontin 1972). It may be applied to any linear array of symbols (Morowitz 1971), hence it is suitable given the problems of dominant data (Dawson et al. 1995, Gillies, Cornelius, and Newton 1997).

Both Nei's and Shannon's indices weight genetic diversity estimates, so that high diversity values are given to populations with intermediate frequency alleles (see Fig. 3.2). Therefore, populations with high numbers of high frequency alleles are estimated to have low diversity. For each locus Nei's index produces values from 0–0.5 and Shannon's produces values from 0–0.73, when the natural log is used. It is important to note, particularly when attempting interspecific comparisons, that there is variability in the published technique for calculating Shannon's index. Many authors use a base 2 logarithm, which whilst producing the same curve pattern, gives values on an entirely different scale (0–1.06). Approximate conversions may be attempted between indices, provided the authors have been explicit in their methods (Fig. 3.2; Cavers 2002). However, interspecific comparison should be treated with extreme caution, not only because of subtle variations in analytical method but also because of variations between studies in sample size, marker numbers, and primers. All of these factors contribute to variations between species and make comparative analyses difficult.

Lynch and Milligan (1994) recognized the shortcomings of dominant data and recommended that low-frequency markers are removed from the data set, as it is these alleles that cause most

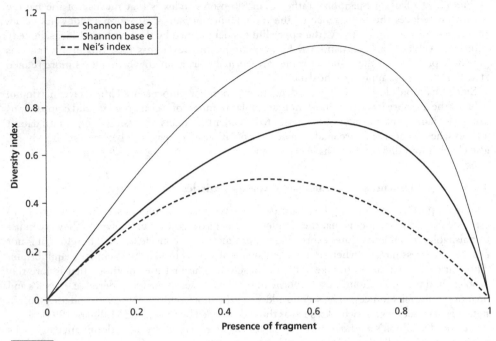

Fig. 3.2

Possible values of Nei's genetic diversity and Shannon's diversity index (calculated using both natural and base 2 logarithms). The model curves assumed dominant data, that is, two alleles, 1 and 0, with frequencies x and $(1 - x)$, respectively (from Cavers 2002).

variation between observed and expected heterozygosity-based parameters in simulation tests. As a guideline, only fragments with an observed frequency of less than $1 - 3/N$ (the total sample size) should be kept for analysis (e.g., for $N = 10$, those markers with an incidence of less than 0.7). Pruning of the data set in this fashion will mean that many more markers will have to be surveyed for complete analysis. This method does not allow correction for populations that are locally inbred, where the only solution would be to survey variation within offspring to assess Hardy–Weinberg equilibrium (Lynch and Milligan 1994) or to use an alternative codominant marker (e.g., SSRs), which allows direct assessment from the existing population.

Using the pruned allele frequencies, gene diversity measures can be derived from the conventional measure by adding a variance estimate:

$$H_j = 1/L \sum \{2q_j(i)[1 - q_j(i)] + 2var[q_j(i)]\}$$

where q is the frequency of the null allele, and the measure is averaged over L loci.

The sampling variance is then approximately:

$$var(H_j) = 1/L^2 \sum \{4[1 - 2q_i(i)]^2 \, var[q_i(i)]\}$$

where the within-locus sampling variance can be reduced to zero by increasing the number of individuals sampled and loci scored, the total sampling variance for n populations (H_w) can be partitioned between individuals (I), loci (L), and populations (P) and estimated as:

$$var_I(H_w) = 1/n^2 \sum var_I(H_j)$$

$$var_L(H_w) = 1/n^2 \sum var_L(H_j)$$

$$var_P(H_w) = var(H_w) - var_I(H_w) - var_L(H_w)$$

Alternative methods are suggested by Clark and Lanigan (1993) and Innan et al. (1999), which use nucleotide diversity, estimated from the average proportion of shared fragments between pairs of sampled genotypes, to measure genetic diversity.

3.5 Genetic differentiation

3.5.1 Nei's G_{ST}

Total gene diversity (H_T), measured in terms of the total expected heterozygosity, can be broken down in order to determine the proportion gene diversity of the species that is present within populations[1] (H_S) and among populations (D_{ST}; Nei 1973), indicating the differentiation of diversity between populations, such that:

$$H_T = D_{ST} + H_S$$

H_S is the mean of expected heterozygosities within each population:

$$H_S = 1 - \sum_{i=1}^{i=k} p^2$$

where p is the mean frequency of the ith allele at the kth locus in each population and the value is averaged over all populations.

These diversity indices (H_T, H_S, D_{ST}) can be used to calculate measures of genetic differentiation. G_{ST}, the gene diversity between populations relative to the combined populations, is termed the coefficient of gene differentiation (Nei 1973), where:

$$G_{ST} = D_{ST}/H_T$$

In contrast, R_{ST}, the between-population gene diversity relative to the within-population gene diversity, can be calculated as:

$$R_{ST} = D_m/H_S$$

where D_m is a measure of between-population diversity calculated from D_{ST} by excluding the comparisons of populations with themselves, and is estimated as $sD_{ST}/(s-1)$, where s is the number of populations surveyed. It should be noted that Nei's R_{ST} (renamed D_{ST}; Nei 1986) is different from Slatkin's (1995) R_{ST}.

An understanding of H_T and H_S and their relationship with differentiation estimates comes from a consideration of their maximum and minimum values. This is simplest in a diallelic system, with two alleles at equal frequency across all populations (maximum H_T is 0.5). If the two

---

[1] Nei (1973) originally termed H_T as the diversity within the population, which was then broken down into diversity within subpopulations (H_S). These are synonymous with the terms species and population, respectively, which are used here.

alleles are at equal frequency within a population, then H_S is also at its maximum (0.5). For more than two alleles, the maximum value of H_T or H_S will increase (Fig. 3.1). By comparison, G_{ST} varies between 0 and 1. G_{ST} is 0 when H_T is the same as H_S, that is, allele frequencies are identical across all populations. G_{ST} is 1 when H_S is 0, that is, there is no variation within populations, but populations are fixed for different alleles and therefore the populations are maximally differentiated, and all the surveyed variation is partitioned among populations. In practice G_{ST} is likely to lie between these two extremes. Sample calculations of Nei's gene diversity indices are included in Box 3.7.

3.5.2 Wright's F-statistics

Using the values for H_T, H_S, and deriving a new parameter, the mean observed heterozygosity per individual, H_I, the genetic structure of populations can be analyzed using F-statistics (Wright 1951). Wright described H_T and H_S as the total expected heterozygosity in the total population and the mean expected heterozygosity within populations, respectively (assuming Hardy–Weinberg equilibrium). Thus, his definitions of H_T and H_S differ from those of Nei (1973), although they are synonymous and have the same mathematical basis. Based on three levels of variation (the individual, the population, and the total population), Wright's approach distinguishes three levels of population structure.

F_{IS}, the inbreeding coefficient, describes the divergence of observed heterozygosity from the expected heterozygosity within populations assuming panmixia:

$$F_{IS} = (H_S - H_I)/H_S$$

F_{ST}, the fixation index, describes the reduction in heterozygosity within populations relative to the total population due to selection or drift:

$$F_{ST} = (H_T - H_S)/H_T$$

Box 3.7

Calculating diversity statistics (H_T, H_S, G_{ST}, and D_{ST}, Nei 1973) and Wright's (1951) F-statistics

The data shown in Table 3g are based upon three loci surveyed in three populations. Note that although the third loci is monomorphic it is still included in the calculations. Allele frequencies have been calculated assuming Hardy–Weinberg equilibrium. It is also assumed that sample sizes are consistent. All means calculated are arithmetic.

Gene diversity statistics
To calculate H_T:

H_T is the mean of the three expected population heterozygotes
$[(0.42 + 0.4982 + 0)/3] = 0.31$.

H_S is the mean of the three mean expected heterozygotes per population
$(0.33 + 0.44 + 0)/3 = 0.26$.

D_{ST} can be calculated by $H_T - H_S$ $(0.31 - 0.26)$ $= 0.05$.

G_{ST} is calculated using the equation $G_{ST} = D_{ST}/H_T$

D_{ST} and G_{ST} are then calculated for each locus and the means taken to give the overall value for the species (Table 3h).

So $D_{ST} = 0.0494$ $[(0.09 + 0.0582 + 0)/3]$

and $G_{ST} = 0.19$ $[0.21 + 0.1168 + 0)/3]$

Wright's F-statistics
With H_T and H_S calculated only H_I (the observed heterozygosity) is further required to calculate Wright's F-statistic,

Table 3g

Calculation of H_T, H_S, and H_I from genotype and allele frequencies at three loci in three populations in a single species. These values can be then be used to calculate D_{ST} and G_{ST} and Wright's F-statistics (see other text in this box).

Locus/Site	Phenotype frequency			Allele frequency		Expected number of heterozygotes per subpopulation
	a/a	a/b	b/b	a	b	
Pgi-1						
Pop. 1	0.3	0.4	0.3	0.5	0.5	0.5
Pop. 2	0.4	0.4	0.2	0.6	0.4	0.48
Pop. 3	1.0	0	0	1.0	0	0
Mean observed heterozygosity (H_I)		0.2667				
Mean population allele frequency				0.9	0.3	$0.33 = H_s(Pgi\text{-}1)$
Expected population heterozygosity (H_T for Pgi-1)				0.42		
Pgi-2						
Pop. 1	0.3	0.1	0.6	0.35	0.65	0.455
Pop. 2	0.25	0.5	0.25	0.5	0.5	0.5
Pop. 3	0.65	0.2	0.15	0.75	0.25	0.375
Mean observed heterozygosity (H_I)		0.2667				
Mean population allele frequency				0.33	0.47	$0.44 = H_s(Pgi\text{-}2)$
Expected population heterozygosity (H_T for Pgi-2)				0.4982		
Aat-1						
Pop. 1	1.0	0	0	1.0	0	0
Pop. 2	1.0	0	0	1.0	0	0
Pop. 3	1.0	0	0	1.0	0	0
Mean observed heterozygosity (H_I)		0				
Mean population allele frequency				1.0	0	$0 = H_s(Aat\text{-}1)$
Expected population heterozygosity (H_T for Aat-1)				0		

Table 3h

Measures of population differentiation for three allozyme loci calculated from data in Table 3g.

	D_{ST}	G_{ST}
Pgi-1	0.09	0.21
Pgi-2	0.0582	0.1168
Aat-1	0	0

and is derived by calculating the mean observed heterozygosity (phenotype a/b) for each locus, as follows for the data above:

$H_I = 0.18(0.2667 + 0.2667 + 0/3)$

With $H_T = 0.31$, $H_S = 0.25$, and $H_I = 0.18$

Wright's F-statistic can be derived as:

$F_{IT} = (H_T - H_I)/H_T = (0.31 - 0.18)/0.31 = 0.42$

$F_{ST} = (H_T - H_S)/H_T = (0.31 - 0.26)/0.31 = 0.16$

$F_{IS} = (H_S - H_I)/H_S = (0.26 - 0.18)/0.26 = 0.31$

For interpretation of these values refer to main text.

F_{IT}, the overall inbreeding coefficient, describes the reduction of heterozygosity within individuals relative to the total population due to non-random mating within subpopulations (F_{IS}) and population subdivision (F_{ST}):

$$F_{IT} = (H_T - H_I)/H_T$$

The three F-statistics are interrelated as $1 - F_{IT} = (1 - F_{IS})(1 - F_{ST})$.

F_{ST} should vary between 0 and 1 (although unequal population sizes can lead to slightly negative values). The values for F_{IS} and F_{IT} can be negative or positive. Considering a diallelic system, with the two alleles at equal frequencies, H_T and H_S have maximum values of 0.5. H_I can vary between 0 (no observed heterozygotes) to 1 (all individuals are heterozygous). The former case would make F_{IS} or F_{IT} positive, indicating a deficit of heterozygotes, with respect to Hardy–Weinberg expectations. An excess of heterozygotes would make F_{IS} or F_{IT} negative. The statistical significance of the deviation of these values from Hardy–Weinberg expectations (when F_{IT} or F_{IS} is 0) can be determined using a chi-squared test. Workman and Niswander (1970) provided the chi-squared formula $2NF_{ST}(k = 1)$ with $(k = 1)(s = 1)$ degrees of freedom, where N is the total sample size, k is the number of alleles at the locus, and s is the number of populations. Sample calculations of Wright's F-statistics are included in Box 3.7. F_{ST} is equivalent to G_{ST} (Nei 1975), although F_{ST} was developed for a diallelic loci and is a special case of the multiallelic approach of Nei's G_{ST}.

The hierarchical differentiation of diversity described by F_{ST} can be extended to other hierarchical levels by appropriate modifications of the formulae, for example, variation between populations within regions (Box 3.8).

3.5.3 Allele frequency variance methods

Other measures of population differentiation closely based on F_{ST} and G_{ST} (for example, Guries and Ledig 1982, Soltis and Soltis 1987), have used the variance of the allele frequencies among populations divided by the product of the mean allele frequencies. For a diallelic system, this is given by:

$$(\sigma p^2/p(1-p))$$

where σp^2 is the variance of the allele frequencies among populations and $p(1-p)$ is the product of the mean allele frequencies.

Perhaps the most widely used of these variance methods is that of Weir and Cockerham (1984), which is a method of estimation based on an ANOVA of the gene frequencies (Box 3.9).

Box 3.8

Hierarchical patterns of genetic diversity in *Persoonia mollis*

Persoonia mollis (Proteaceae) is a fire-sensitive shrub occurring in the fire-prone eucalypt-dominated dry sclerophyll forests and heaths of south-eastern Australia (Figure 3d). The species is morphologically variable and nine subspecies are largely contiguous. The species are completely outcrossing in natural populations, bees being responsible for pollen movement,

seed movement is initially via gravity followed by ingestion and subsequent dispersal by large birds.

A study by Krauss (1997) sampled two populations for seven of the subspecies, one population of subspecies *maxima* (the sole population), and three populations of subspecies *nectens* (due to small sample size). Subspecies and sample site are listed in the legend to Figure 3d. For each subspecies, 70–86 individuals were analyzed for allozyme variation at 11 variable loci.

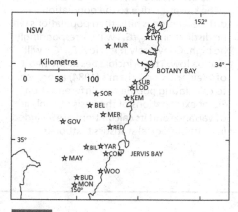

Australia

NSW   ☆ WAR
            ☆ MUR
                            152°
                          ☆LYR

Kilometres

0   58   100
                    BOTANY BAY   34°
          ☆SUB
          ☆LOD
    ☆SOR ☆KEM
    ☆BEL
            ☆MER
☆GOV   ☆RED

35°
          ☆BIL☆YAR
    ☆MAY ☆CON JERVIS BAY

          ☆BUD☆WOO
          ☆MON
150°

Fig. 3d

Collection sites of the various subspecies of
Persoonia mollis. Three-letter codes refer to sites.
The subspecies collected at each site is as
follows:
LYR, Lyrebird Gully, *maxima*; WAR, Waratah
Park, *mollis*; MUR, Murphies Glen, *mollis*; SUB,
Sublime Point, *nectens*; BEL, Belanglo State
Forest, *revoluta*; MER, Meryla State Forest,
ledifolia; RED, Red Rocks nature reserve,
ledifolia; YAR, Yarramummun Firetrail,
leptophylla; BIL, Bilys Hill, *leptophylla*; GOV,
Governers Hill, *liverns*; MAY, Mayfield, *liverns*;
CON, Conjola State Forest, *caleyi*; WOO,
Wood Burn State Forest, *caleyi*; BUD, Mt
Buddawang, *budawangensis*; MON, Mongstate
Forest, *budawangensis* (from Krauss 1997).

Instead of apportioning total gene
diversity (H_T) into H_S and D_{ST} (Nei 1973;
see main text), Krauss (1997) further
compartmentalized diversity, such that
total diversity (H_T):

$$H_T = H_P + D_{PS} + D_{ST}$$

where H_P is the gene diversity within
populations, D_{PS} is the gene diversity
among populations within subspecies
(such that $D_{PS} = H_S - H_P$), and D_{ST} is the gene
diversity among subspecies within the total

Table 3i

Levels and apportionment of genetic diversity
between 18 populations and nine subspecies
of *Persoonia mollis*. G_{PS}, G_{ST}, and G_{PT} were
calculated individually for each locus and
averaged.

H_T	0.139
H_S	0.114
H_P	0.109
G_{PS}	0.038
G_{ST}	0.179
G_{PT}	0.217

gene diversity ($D_{ST} = H_T - H_S$). With this
approach, Krauss aimed to establish the
contribution to the total gene diversity
of the difference within populations, the
difference among populations, and the
difference among populations within
subspecies.

Likewise, in the same way that G_{ST} can
be determined to identify the relative
apportionment of the genetic diversity
($G_{ST} = D_{ST}/H_T$), the ratios of the various
components of the total genetic diversity
in *P. mollis* can be considered.

Hence $G_{PS} = (H_S - H_P)/H_{ST}$, $G_{PT} = (H_T - H_P)/
H_T$, and G_{ST} is as defined in the main text
(section 3.4). The various components
of variation are listed in Table 3i. Of the
total gene diversity found in *P. mollis*
($H_T = 0.139$) 21.7% is distributed among
populations ($G_{PT} = 0.217$). Variation
within populations among subspecies
only accounts for a small proportion of this
population diversity ($G_{PS} = 0.038$).

The low allozymic differentiation
between subspecies possibly reflects the
substantially lower values of gene diversity
found in this species, compared with other
angiosperm species of similar breeding
systems.

The habitat and fire-sensitive nature of
the plant is likely to produce frequent
local extinctions and attendant
recolonizations, rather than broad-scale
burns. The haphazard nature and intensity
of these events and the subsequent
repopulation from other adjacent sources
will maintain genetic differentiation
among populations.

Analysis of variance – a brief explanation

Analysis of variance is used to examine the statistical significance of partitioning the variance of a data set into different groups or components. For instance, a mean allele frequency (X) can be calculated for a species that comprises several populations. Within a population allele frequencies at an individual locus (x_i) will vary from the species mean through sampling effects (d) and because that population is different from other populations (T) such that

$$x_i = X + d + T$$

ANOVA allows the sampling variation (d) to be separated from the population variation (T) and the relative magnitude of these effects over all loci can be considered. If the variation due to the population differentiation is high then populations will be distinct and F_{ST}/G_{ST} will correspondingly be high. Conversely if it is low F_{ST}/G_{ST} will also be low. This principle lies at the heart of Weir and Cockerham's (1984) approach to calculating population differentiation. Fuller explanations of the basis of analysis of variance and its calculations are included in most biological statistics textbooks.

Like the majority of genetic differentiation methods this powerful statistical method is based on the hierarchical structuring of diversity between populations, individuals within populations, and alleles within individuals. The variance can thus be described for each of these hierarchies as σ_a^2, σ_b^2, σ_w^2, respectively. Weir and Cockerham's (1984) approach estimates F_{ST}, F_{IS}, and F_{IT}, as the estimators θ, f, and F, respectively. These tests assume that the populations are in mutation-drift equilibrium and share the same time of divergence from a common ancestor, hence θ, f, and F are coancesty coefficients between their respective hierarchical levels.

$$\theta = \sigma_a^2/\sigma^2 (\approx F_{ST})$$

$$f = \sigma_b^2/(\sigma_b^2 + \sigma_w^2)(\approx F_{IS})$$

$$F = (\sigma_a^2 + \sigma_b^2)/\sigma^2 (\approx F_{IT})$$

Currently, Cockerham and Weir's method is one of the most widely used methods for estimating F_{ST}, F_{IS}, and F_{IT}, since it enables sophisticated *post hoc* statistical analysis to be undertaken to assess the validity of the results (Weir 1996). The method also differs substantially from that of Wright (1951) who developed his F-statistics to describe the genetic structure of species and their populations. F-statistics measure the fixation of alleles at three levels (individuals, population, and total population) relative to each other. Initially, they were developed for a single locus with two alleles. Where more than two alleles existed, Wright suggested using the most common allele and combining the frequencies of the other alleles to form a composite second allele. Nei (1977) subsequently developed statistics for multiple alleles, making no assumptions about the number of alleles per locus, and incorporated observed heterozygosity and expected heterozygosity at the population and total population levels to equate to the F-statistics.

The approach of Weir and Cockerham (1984), who were interested in relationship by descent, is philosophically different to that of Nei (1973). Nei's (1973) G_{ST} measure makes no assumptions about descent, simply describing how the total genetic variation is divided into within- and between-population components. By comparison, Weir and Cockerham (1984; θ) assume that a single population is the ancestor of all other populations. Sampling within each population identifies a suite of alleles that have been transferred to offspring, and are a subset of parental alleles. The assumptions underpinning this approach are that all populations are independent and constrained in size over time, and that there is no mutation, selection, or migration. Statistically, the method was developed to deal with small and unequal sample sizes, whilst Nei's approach is intended for large populations. Nei (1986) argues that many of the conditions necessary for Weir and Cockerham's method are not found in natural populations, for example, population sizes vary and are not constant over time. In addition, temporal variation in the numbers and relatedness of populations will also occur.

Two sources of error are present in the estimation of F-statistics: (i) the error associated with the estimation of a population allele frequency on the basis of a sample; and (ii) the error associated with the distribution of allele frequencies among the sampled populations. G_{ST} approaches this problem by calculating the bias associated with equating sample and population allele frequencies, and then using these to calculate unbiased estimates of H_S and H_T (Nei and Chesser 1983). In contrast, θ approaches the problem using analysis of variance of indicator variables. Under a particular model of population divergence, the mean squares associated with the ANOVA model are functions of coancestry coefficients, which are equivalent to F-statistics, assuming that the populations sampled are a random subset of all diverging populations.

The population differentiation estimates of G_{ST} and θ are different for two principal reasons: (i) θ includes a correction for the sampling error associated with the difference between the allele frequency distribution of the sample and the frequency distribution of the entire set of populations; and (ii) if θ is interpreted in terms of coancestry coefficients particular models of population divergence must be assumed.

With the differences in approach to the estimation of genetic differentiation, does it matter which method is used to calculate the various statistics? In terms of frequency of application, Culley et al. (2002) noted that 46% of studies surveyed used Nei's approach to calculating diversity. However, typically, when large numbers of populations are sampled, G_{ST} and θ differ relatively little. For example, Chakraborty and Danker-Hopfe (1991) compared G_{ST} and θ to sets of empirical data and found very little difference in the numerical values obtained by the two methods. Berg and Hamrick (1995) carried out a simulation study based on data from *Quercus cerris* and found that F_{ST} increased as sample sizes fell below 20, whilst samples of more than 30 individuals were sufficient to meet the assumptions of F_{ST} and G_{ST} approaches; θ was better if sample sizes were fewer than 30 individuals.

3.5.4 Calculating differentiation for special cases – multiallelic loci

The above differentiation statistics assume a Hardy–Weinberg equilibrium. However, one of these conditions, no mutation, may be inappropriate for microsatellite loci, which have a mutation rate of 10^{-3} to 10^{-4} per locus per replication, which is several orders of magnitude higher than the frequency of point mutations (10^{-9} to 10^{-10}; Hancock 1999). Acknowledging the role of mutation, two models have been developed to cope with the analysis of microsatellite data, the infinite alleles model (IAM, and the stepwise mutation model (SMM; Box 3.10)).

Under the the IAM, a mutation results in an allele not previously encountered, and is akin to the pattern of variation found in classical protein variants (allozymes). Under the SMM, the difference in size between alleles is proportional to the time elapsed since divergence from a common ancestor. The greater the duration of time since two particular lineages shared an ancestor, the more pronounced the difference between the two alleles of these lineages. The ability to infer a relationship between allele length difference and divergence time led Slatkin (1995) and Goldstein et al. (1995a) to develop similar methods of calculating genetic distance (R_{ST} and $\delta\mu$, respectively) as an alternative to F_{ST}, for loci which can be demonstrated to exhibit this behavior (usually microsatellites).

$$R_{ST} = \bar{S} - Sw/\bar{S}$$

where $\bar{S}$ is twice the variance in allele size over d populations and Sw is twice the average within-population variance in allele size (Goodman 1997).

Box
3.10

Microsatellites and the infinite alleles model and the stepwise mutation model

Microsatellites are repetitive sequences of short sections of DNA. These typically comprise 1–5 nucleotides repeated up to 100 times. Two models are suggested for the origin of new microsatellite alleles.

These are the infinite alleles model (IAM) and the stepwise mutation model (SMM).

Within IAM, any mutation converts an allele into any other allele by the instantaneous gain or loss of any number of tandem repeats. In contrast, within SMM, mutation results only in the gain or loss of a single repeat unit and thus variation is generated in a stepwise manner either up or down from the starting allelic state (Fig. 3e).

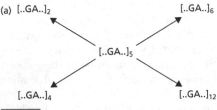

(a) [..GA..]$_2$ [..GA..]$_6$ (b) [..GA..]$_4$ ⟵⟶ [..GA..]$_5$ ⟶ [..GA..]$_6$

[..GA..]$_5$

[..GA..]$_4$ [..GA..]$_{12}$

Fig. 3e

Contrasting results of mutations (denoted by black arrows to indicate direction from starting state). Microsatellites in the sequence GA, under the infinite alleles model (a), where an allele can mutate into any other allele, and the stepwise mutation model (b), where an allele can only mutate into an allele one repeat unit or larger than the original allele.

This derivation assumes populations have equal sample sizes and that all loci have equal variances. However, these parameters are generally not met in natural populations and thus direct application of this formula will yield biased results. For estimates of R_{ST} based on multiple loci, a variance for each locus should be obtained before averaging over loci and calculating R_{ST} (Slatkin 1995). However, if a data set contains loci with very different variances, those loci with low variance will contribute little to the final value of R_{ST}, even if they exhibit a high degree of differentiation. Loci can be made comparable by transforming data before R_{ST} calculation so that alleles are expressed in terms of standard deviations from the global mean rather than repeat unit number. Each locus in the transformed data set will then have a global mean allele size of zero and a standard deviation of one.

Each allele at each locus is standardized as follows:

$$Y_S = (Y - G_M)/\text{std dev}$$

where Y_S is the standardized value of allele Y, Y is allele (n) at locus (l), and G_M is the mean allele size in repeat units for locus (l) over the whole data set. The std dev is the standard deviation in allele size in terms of repeat units for locus (l) over the whole data set.

Michalakis and Excoffier (1996) also developed an analog to R_{ST}, termed Rho_{ST} to account for differences in sample size between populations by obtaining variance components using conventional statistical approaches (the unbiased estimator; Sokal and Rohlf 1994); it is based on an analysis of molecular variance approach akin to Weir and Cockerham's (1984) estimate of F_{ST}:

$$Rho = Sb/(Sb + Sw)$$

where Sb is the component of variance between populations and Sw is the component of the variation within populations.

Initial modeling studies to assess the usefulness of this value (e.g., Shiver et al. 1995, Takezaki and Nei 1996), indicate that the R_{ST} statistic is an improvement on F_{ST}-based approaches for analyzing differentiation using microsatellite loci. These simulations were established with strict adherence to the SMM and no range constraints on allele size were imposed. However, there appears to be an upper limit to the number of tandem repeats within a microsatellite region, possibly due to the mutation process or natural selection. This property means that any genetic distance estimate reaches an asymptote. Despite any further passage in time no more differences between two separate lineages can accrue. Thus, the number of allelic states at a microsatellite locus is finite (Paetkau et al. 1997). A constraint on allele size makes it likely that once this upper limit is reached any further mutations will cause the recreation of an ancestral state. Incorporating these concerns, simulation models by Nauta and Weissing (1996) reached an upper limit of differentiation when population sample sizes were large or when the upper limit to microsatellite size was small.

The drawback to the utilization of R_{ST} and Rho_{ST} is that genetic distance measures rapidly approach a maximum when population size is large or the upper size limit for microsatellite allele is small. Thus R_{ST} and Rho_{ST} do not provide a more reliable estimator of differentiation than traditional approaches due to range constraints leading to an overestimation when population sample sizes are large. The statistics also have large variances and they ignore the influence of migration. When samples are small (e.g., <10) and/or few loci (number of loci <5) are sampled, R_{ST} may also result in biased estimation of Nm. Therefore the most conservative approach is to use θ (see Balloux and Lougon-Moulin 2002), alternatively if R_{ST} is to be applied then a large number of loci should be used (>10, Schlötterer et al. 1998). A comparison of R_{ST} (which utilizes microsatellite size differences) and chord distance (which does not incorporate microsatellite size differences, see section 3.6.2) in a natural study was undertaken by Estoup et al. (1998; Box 3.11).

3.5.5 Calculating differentiation for special cases – organelle genomes

Statistics of genetic differentiation (e.g., F_{ST} and G_{ST}) have traditionally been derived from diploid nuclear data. However, modifications to the standard formulae allow differentiation to be estimated based on haploid genomes (e.g., cpDNA and mtDNA). Thus, G_{ST} can be calculated based on haplotype frequencies.

Pons and Petit (1996) have also defined a measure N_{ST}, which takes into account the similarity between haplotypes. For this method similarity between haplotypes can be assessed by a phylogenetic method, or purely in terms of the number of substitutions/mutations separating haplotypes (network analysis, chapter 5). If v_s and v_T are introduced into the general formulation of Nei's differentiation parameter G_{ST}, then a parameter $N_{ST} = (v_T - v_s)/v_T$ is produced, that is:

$$N_{ST} = \frac{\sum_{ij} \pi_{ij} c_{ij}}{v_T}$$

π_{ij} is the genetic distance between haplotypes i and j, c_{ij} is the covariance between p_{ki} and p_{kj} in the general population, where p_i is the frequency of the ith haplotype in the general population and p_{ki} is the frequency of the ith haplotype within the kth population; and where p_{ki} are considered as random frequencies with mean p_i and variance v_T.

3.5.6 Calculating differentiation for special cases – dominant loci

Population differentiation can be assessed using dominant loci, although the same restrictions apply to calculating differentiation as for diversity measures, and a similar "pruning" of product

Box
3.11

Applying different genetic distance measures in the Brown Trout (*Salmo trutta*)

Difference in length of microsatellite alleles has led to the development of a genetic distance method incorporating this potential evolutionary divergence characteristic. This genetic distance is known as δµ (Goldstein et al. 1995). The advantage of this parameter compared with approaches based on allele similarity was investigated by Estoup et al. (1998) in Brown Trout (*Salmo trutta*). Eleven populations were surveyed in the Moselle and Ill drainages of the Vosges massif (Fig. 3f), none of which were thought to have been previously stocked. Therefore patterns of genetic variation in these populations are products of natural forces rather than artifacts of fishing

management. For comparison, samples from a local hatchery were also included.

Twelve microsatellite loci were surveyed on 15 individuals and the results analyzed using both chord distance (Cavalli-Sforza and Edwards 1967) and δµ (Goldstein et al. 1995). Pairwise distances were then used to construct two separate neighbor-joining trees, which were compared (details on constructing neighbor-joining trees are given in chapter 5, box 5.7). These trees are shown in Fig. 3g.

Both trees separated the Moselle from the Ill drainage basins (P2 and P3 on the trees represent the Ill drainage) but neither showed any pattern within the Moselle drainage itself. This absence of differentiation with either statistical technique is probably due to a combination of biological factors and sampling methods rather than any characteristics of the two statistics.

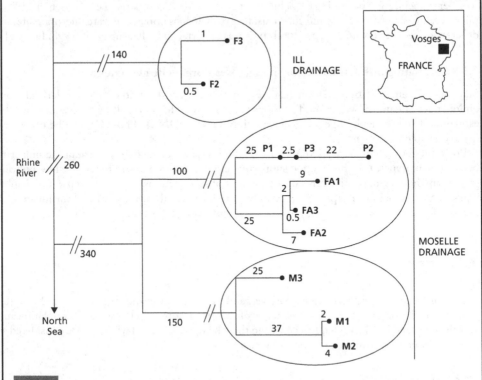

Fig. 3f

Sampling sites for microsatellite analysis of Brown Trout in Ill and Moselle drainage systems (from Estoup et al. 1998).

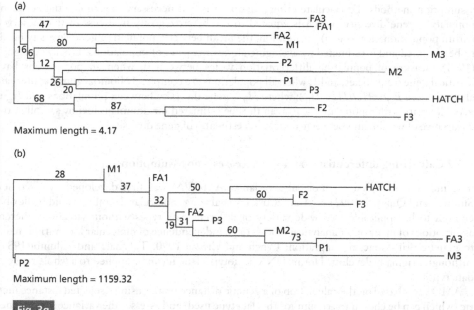

Fig. 3g

Neighbor-joining trees based on microsatellite data relating the natural population samples from the Moselle and Ill drainage plus samples from a trout hatchery (HATCH). Tree (a) was constructed using chord distance; (b) was constructed using δμ (from Estoup et al. 1998). In both cases the trout from the Ill drainage cluster together but neither show any pattern with those fish from the Moselle and tributaries.

The lack of meaningful resolution is likely to be a product of the small number of individuals and loci that show a low level of divergence between the populations. Sampling constrictions can be reduced by using a large number of individuals per population at loci with a high level of heterozygosity (Estoup et al. 1998).

However, an extended sampling regime may not have allowed separation of the various rivers of the Moselle. These rivers undergo great variation in water flow rate resulting in major population fluctuations. These bottlenecks in turn may have altered allele frequencies differently in the various separate tributaries, with these random effects effectively eliminating any geographic patterns.

frequencies is suggested (Lynch and Milligan 1994). However, some simulation studies to examine the behavior of dominant loci when calculating differentiation, indicate that these calculations were generally inflated relative to estimates from codominant data types (Isabel et al. 1999). Thus, one should be very careful when using population differentiation estimates derived from dominant data for comparative investigations. Dominant fragments can be used as allelic data, if the proportion of heterozygosity can be estimated (e.g., using codominant markers) or assumed (e.g., if Hardy–Weinberg equilibrium is thought to operate or individuals are known to self-fertilize, i.e., $F_{is} = 1$). Whilst these assumptions may compromise the analysis of data, it means that data from dominant markers can be analyzed using F-statistics or other differentiation measures.

Lynch and Milligan (1984) developed a specific method of calculating differentiation suitable for use with dominant data types, and this approach is recommended over other allelic assumption methods. To calculate differentiation it is first necessary to estimate the between-population gene diversity (which is equated to the heterozygosity in excess of that observed within populations across loci). In addition, the mean between-population gene diversity needs to be obtained and is estimated by averaging all distinct pairs of populations. Finally, Wright's (1951) measure of population differentiation varies between 0, when all populations have identical gene frequencies, and 1, when all populations are completely homozygous for alternate alleles, where $F_{ST} = H_B/(H_B + H_W)$, where H_B is gene diversity between populations and H_w is gene diversity within populations. F_{ST} can thus be estimated for dominant loci by sampling the covariance of within- and between-population estimates of gene diversity.

3.5.7 Calculating differentiation for special cases – no assumption

The method of analysis of molecular variance (AMOVA) was first developed by Excoffier, Smouse, and Quattro (1992) as a framework for analyzing molecular data that would be flexible enough to be applicable to a wide variety of different data types without violating inherent assumptions of many other analytical models (e.g., independence of molecular loci; variation due to genetic drift alone, no migration; Lynch and Crease 1990, Takahata and Palumbi 1985). Although originally developed for mtDNA haplotype data, it can be applied to a whole range of data types.

AMOVA is based on the calculation of a genetic distance matrix using a selected distance metric (which can be chosen to account for the data type used) and assesses the variance apportioned within and between predefined groups. Due to the absence of assumptions required, the test is both widely applicable and powerful, and the fact that allele frequencies are not calculated has led to the test being applied to analyze dominant data types (e.g., Huff, Peakall, and Smouse 1993). However, depending on which distance algorithm is applied (e.g., Jaccard's distance measure) the method may overestimate differentiation, and thus, an allele frequency pruning method is probably more appropriate for analyzing dominant data.

Significance levels for AMOVA are computed by non-parametric permutation of the data set with 1000 permutations. Φ-statistics are generated, which are correlation statistics directly analogous to F-statistics (Cockerham 1969, 1973), and derived from the variance components computed during AMOVA. They express the correlation of a pair of individuals drawn at random from a particular subgroup of the data set, relative to that of a pair of individuals drawn from a wider grouping, indicating the relative partitioning of diversity between the hierarchical levels being analyzed.

Correlation of samples within populations relative to samples from the whole data set:

$$\Phi_{ST} = \sigma_a^2 + \sigma_b^2/\sigma^2$$

Correlation of samples within a group of populations relative to samples from the whole dataset:

$$\Phi_{CT} = \sigma_a^2/\sigma^2$$

Correlation of samples within populations relative to samples from that subgroup of populations:

$$\Phi_{SC} = \sigma_b^2/(\sigma_b^2 + \sigma_c^2)$$

where σ_a^2 is the variance component between groups; σ_b^2 is the variance component between populations; σ_c^2 is the variance component within populations; and $\sigma^2 = \sigma_a^2 + \sigma_b^2 + \sigma_c^2$.

3.6 Genetic distance

Genetic diversity measures estimate the amount of variation that is found in a population, while genetic differentiation measures describe how this variation is partitioned among populations. Genetic distance quantifies the degree of similarity between two individuals, or groups of individuals (whether these are populations or even species). A genetic distance measure must be metric: (i) values must be positive; (ii) values must be symmetrical; (iii) values must be distinctive; and (iv) values must satisfy the triangular inequality. Conceptually, genetic distance measures how far apart two individuals or populations are in n-dimensional marker hyperspace, each axis corresponding to variation in a single marker. Ideally, a genetic distance method should produce values that vary between zero (when all markers are shared between two individuals or populations) and one (when no markers are shared between individuals or populations). For a genetic distance to be of evolutionary value, it is necessary that it is related to evolutionary divergence, that is, the more evolutionarily divergent two groups are the higher the genetic distance value. Numerous methods have been proposed to measure genetic distance between population and individuals (Appendix 1; Kalinowski 2002), while some of their sampling properties have been considered in chapter 2. Genetic distance is also considered in chapter 5 (Box 5.7). Here we have described two groups of methods: those which use unordered data and estimate genetic distance based on allele sharing criteria, and those which use ordered data. The assumptions of most of these methods fall under the infinite alleles model but some assume specific divergence rates depending on the number of alleles, common or different, between individuals or groups of individuals. Methods using ordered data allow the phylogenetic information content of the study loci to be applied to evolutionary divergence questions. Some are powerful tests which should only be applied under strict operational criteria. Chapter 5 provides a more detailed discussion of ordered methods of estimating evolutionary divergence rates and genetic distance.

3.6.1 Using unordered data

Using pairwise differentiation estimates

The differentiation measures described above in section 3.5 can be converted into distance estimates by calculating pairwise estimates between groups of individuals or populations. The most common treatment is to consider pairwise F_{ST} (Box 3.13), but the assumptions of these differentiation methods need to be met. The advantage of this method is that all data types (haploid, codominant, dominant) can be analyzed in this way using the appropriate differentiation measure.

Chord distance (Dc) is a measure that determines the distances between populations relative to their theoretical position on the surface of a hypersphere of unit radius (Cavalli-Sforza and Edwards 1967), where the position of populations is determined by the frequency of alleles at the surveyed loci. In a simple example, consider a single locus with three alleles in two populations. The three alleles ($p1, p2, p3$) correspond to three vectors $\sqrt{p1}, \sqrt{p2}, \sqrt{p3}$, the magnitude of which determine the position of the two populations on the surface of a sphere. Assuming the allele frequencies in the two populations are not identical, the two populations will take up different positions (Fig. 3.10). The distance between these two positions can be described as an arc $2\theta/\pi$ apart (where θ is the angle between the two radii that join each point on the surface of the sphere to the origin). However, as an arc is curved the shortest distance between the two populations in the chord of the arc, $[2\sqrt{2}/\pi][1 - \cos \theta]$, where $\cos \theta = \sum_{i=1}^{m} \sqrt{p^i p_i^!}$ and where p_i is the frequency of the ith allele in the first population and $p_i^!$ is the frequency of the same allele in the second population. An example of the use of chord distance is given in Box 3.11 and several distance measures using similar principles are outlined in Appendix B.

Nei's genetic distance (D)

The most widely used approach for codominant data is that of Nei (1972, 1978), and is based on the probability that two alleles chosen at random from two different populations, will be identical, relative to the proportion of two alleles chosen from the same population being identical. The method is based on an IAA model and in essence, D measures the similarity between pairs of populations based on allele frequencies, and varies between 0 (no similarity) and 1 (completely identical).

For two randomly mating diploid populations (X and Y), D can be described as:

$$D = -\ln \mathcal{J}_{XY} / \sqrt{\mathcal{J}_X \mathcal{J}_Y}$$

where $\mathcal{J}_{XY}, \mathcal{J}_X$, and $\mathcal{J}_Y$ are the arithmetic means of j_{XY}, j_X, and j_y, respectively. j_{XY} is the probability that an allele drawn from population X is the same as that from population Y. j_X is the probability that two alleles drawn randomly from population X are the same. j_Y is the probability that two alleles drawn randomly from population Y are the same.

Where multiple alleles segregate at a locus, X_i and Y_i are the ith alleles in populations X and Y, respectively.

$$\mathcal{J}_X = \sum X_i^2$$
$$\mathcal{J}_Y = \sum Y_i^2$$
$$\mathcal{J}_{XY} = \sum X_i Y_i$$

A sample calculation of D is given in Box 3.12. D has several advantages over other methods of genetic distance developed in the wake of easily available genetic variation data (e.g., Cavalli-Sforza and Edwards 1967, Hedrick 1971; Appendix): (i) it measures the accumulated number of gene substitutions per locus; (ii) if the rate of gene substitution per unit time is constant then it is linearly related to evolutionary time; and (iii) in some migration models it is related linearly to geographic distance or area. If evolutionary rate is non-constant, then D underestimates the number of gene substitutions per locus (Nei 1971a). When D is used for microsatellite loci the same consideration of uniformity of mutation rate across loci is required, if D is to be correlated with evolutionary time. An example of the use of D is given in Box 3.1.

Measures of distance for dominant markers

Estimating genetic distance for dominant data can be made using Jaccard's similarity measure (F) which relies on the number of shared bands between any two individuals, as a proportion of the total number of bands those two individuals display:

$$F = M_{xy} / (M_t - M_{xy0})$$

where M_{xy} is the total number of bands shared between accessions x and y; M_t is the total number of bands in the data set; and M_{xy0} is the total number of bands in the data set absent from either x or y.

Jaccard's measure is appropriate for analyzing dominant marker data since it relies only on shared presence of bands. Reasons for the absence of a dominant marker are difficult to clearly establish and therefore shared absence is not considered in estimates of similarity from dominant data (Rieseberg 1996). The reciprocal of the similarity measure $(1 - F)$ gives an estimate of genetic distance.

Other distance measures are also applicable for dominant data. For example, Huff, Peakall, and Smouse use a (1993) modified version of the Nei and Li (1979) distance estimate, which is a

Box
3.12

Nei's genetic distance sample calculation

Genetic diversity, D, can be calculated for one locus with three alleles common to two populations. Two populations X and Y have the following allele frequencies at a single locus (locus 1, Table 3j). The probability of choosing two identical alleles at random from population X (J_X) is:

$$J_X = (X_{1a}^2 + X_{1b}^2 + X_{1c}^2) = 0.3^2 + 0.4^2 + 0.3^2 = 0.34$$

Where X_{1a}^2 in population X is the squared value frequency of allele 1a.

Table 3j

Allele frequencies for locus 1 in populations X and Y.

	Population X	Population Y
Allele 1a	0.3	0.8
Allele 1b	0.4	0.1
Allele 1c	0.3	0.1

Similarly, the probability of choosing two identical alleles at random from population Y (J_Y) is:

$$J_Y = (Y_{1a}^2 + Y_{1b}^2 + Y_{1c}^2) = 0.8^2 + 0.1^2 + 0.1^2 = 0.66.$$

Where Y_{1a}^2 in population Y is the squared value frequency of allele 1a. (Note that J_Y is less than J_X (0.34 and 0.66, respectively) due to the more even spread of allelic variation in population X compared to population Y and which leads to a higher diversity estimation.)

The probability of taking the same allele at random from both population X and population Y (J_{XY}) is:

$$J_{XY} = (X_{1a} \times Y_{1a}) + (X_{1b} \times Y_{1b}) + (X_{1c} \times Y_{1c}) =$$
$$(0.3 \times 0.8) + (0.4 \times 0.1) + (0.3 \times 0.1) = 0.31$$

The identity of genes (I_j) between X and Y for this locus can then be calculated using the equation:

$$I_j = J_{XY} / \sqrt{J_X J_Y}$$

For the above example: $I = 0.6544$.

Considering more than one locus:
As defined by Nei, D is based on estimates from several loci (including monomorphic loci). Hence considering two more loci for populations X and Y (note: the third locus is monomorphic) as follows (Table 3k).

Table 3k

Allele frequencies at two loci in populations X and Y.

	Population X	Population Y
Locus 2		
Allele 2a	0.4	0.5
Allele 2b	0.6	0.5
Locus 3		
Allele 3a	1.0	1.0

Values of J_X, J_Y, and J_{XY} for the two additional loci are calculated using the methods described above. Along with the data for the first locus these are tabulated in Table 3l.

Table 3l

Mean values of J_X, J_Y, and J_{XY} calculated from three loci.

	J_X	J_Y	J_{XY}
Locus 1	0.34	0.66	0.31
Locus 2	1.00	1.00	0.5
Locus 3	0.80	0.5	1.00
Mean	0.71	0.72	0.60

Across all loci I is calculated using equation

$$I = J_{XY} / \sqrt{J_X J_Y} = 0.84$$

and $D = -\ln I = 0.174$.

Box
3.13

Genetic differentiation of polar bear populations (*Ursus maritimus*)

Polar bears (*Ursus maritimus*) are found on ice-covered waters throughout the circumpolar Arctic. An important question for the management of the bears is whether they exist as discrete populations or are effectively a single population. This issue has been approached using a variety of methods, including parasite loads, carbon isotopes, and skull morphometrics. In addition molecular approaches, using mitochondrial DNA and allozymes, have been utilized. Only tracking data has delineated populations and 19 polar bear populations are now recognized although there is normally movement of individuals between populations.

Paetkau et al. (1999) used 16 microsatellite loci to investigate the degree of genetic differentiation between these populations and hence to see if the currently recognized population boundaries are reflected in the genetic data. DNA was obtained largely from blood and tissue samples, with some samples being collected Inuk kills, from 17 of the world's 19 polar bear populations (Fig. 3h and Table 3m). Pairwise F_{ST} was calculated for each pair of populations and Fitch and Margoliash trees drawn utilizing the genetic distance values (Fig. 3i).

The genetic distances observed between populations were small although the tree revealed four clusters that reflected four geographic areas: Norwegian Bay, other Canadian Arctic populations, the southernmost populations, and polar basin populations.

The Norwegian Bay population is small (~100 individuals) and hence its separation may be due to drift. The absence of significant differentiation among populations in the polar basin is notable. This cluster covers a geographic area greater than that of the other three areas combined. Despite this, the largest genetic distances between samples within this cluster were similar to the smallest

Table 3m

Location, estimated population size, and numbers of samples in microsatellite analysis for the world's polar bear populations. Reliability of population numbers is indicated by (+) good; (±) fair; (−) poor; and (?) educated guess.

Population	Abbr.	Number	N
Western Hudson Bay	WH	1200 (+)	33
Southern Hudson Bay	SH	1000 (±)	0
Foxe Basin	FB	2300 (+)	30
Davis Strait-Labrador	DS	1400 (±)	30
Baffin Bay	BB	2200 (±)	31
Kane Basin	KB	200 (+)	30
Lancaster Sound	LS	1700 (+)	30
Gulf of Boothia	GB	900 (−)	30
M'Clintock Channel	MC	700 (−)	15
Viscount Melville Sound	VM	230 (+)	30
Norwegian Bay	NW	100 (±)	30
Queen Elizabeth Islands	QE	200 (?)	0
Northern Beaufort Sea	NB	1200 (+)	30
Southern Beaufort Sea	SB	1800 (+)	30
Chukchi Sea	CS	2000–5000 (?)	30
Laptev Sea	LV	800–1200 (?)	0
Franz Josef L. Novaja Zemlja	FN	2500–3500 (?)	32
Svalbard	SV	1700–2200 (?)	31
East Greenland	EG	2000–4000 (?)	31

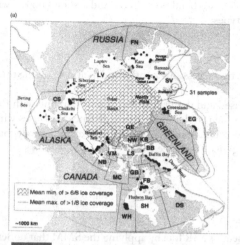
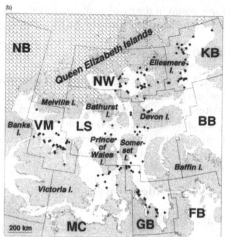

Fig. 3h

Polar bear populations and sampling locations (a) sampled around the North Pole; (b) the Canadian Central, and the High Arctic. Three areas were not sampled: Laptev Sea (LV); South Hudson Bay (SH); and Queen Elizabeth Island (QE).

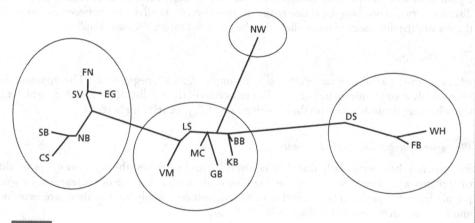

Fig. 3i

Fitch and Margoliash tree of genetic distance based on pairwise F_{ST} between polar bear populations (from Paethau et al. 1999). The four clusters represent the Norwegian Bay population (NW), polar basin (left), southernmost population (right), and the Canadian Arctic (center).

distances between a member of this cluster and a population immediately outside it. This low genetic separation possibly reflects the semicontinuous zone of polar bear habitat around the pole.

The difference between the clusters on the tree may reflect patterns of bear movements. For example in the polar basin radio tracking has shown mean home ranges to vary from 72,263–244,463 km^2 (Wiig 1995). In the Canadian central and high Arctic mean home range sizes were smaller (12,162–82,827 km^2). This in turn may reflect patchiness in food availability although there may also be other factors that limit movement within areas outside the polar basin: land features, human distribution, and seasonal ice variation.

simple count of the number of differences between two banding patterns and assumes no knowledge of the relationships between the different types:

$$D = 100\,[1 - (2n_{xy}/(n_x + n_y))]$$

where $2n_{xy}$ is the number of bands shared by the two accessions x and y; and n_x and n_y are the total numbers of bands in x and y, respectively.

Alternatively, Nei's genetic distance can be used and the fragments "pruned" in line with the Lynch and Milligan (1994) recommendations. One problem with using distance measures with dominant data is that homozygotes and heterozygotes are indistinguishable (except homozygous absent), which can lead to an overestimation of genetic distance.

3.6.2 Using ordered data

SMM estimator – δμ

A distance estimate of divergence can be obtained for SSR loci by applying the SMM that underpins the development of Slatkin's R_{ST} (1995) and its analogs to estimate evolutionary distance between alleles. It develops the average square distance (ASD) approach of Goldstein et al. (1995a), which depended upon the population size and had a high variance. The value is termed δμ (delta mu; Goldstein et al. 1995b) and assumes that the populations are in mutation-drift equilibrium and therefore the variances are on average constant over time. Theoretically, the distance between two samples is due to the mean difference in allele size between two samples; the greater the difference in mean allele size the more distant are the two samples.

$$(\delta\mu)^2 = (\mu_A - \mu_B)^2$$

where μ_A and μ_B are the means of allele size in samples A and B, respectively. This approach has been widely used for microsatellite loci. The criticisms of the validity of the SMM, on which this statistic is based and described in the consideration of R_{ST} are relevant here.

Phylogenetic methods and haploid genomes

For ordered data, particularly that derived from haploid genomes, there is also the possibility of applying a phylogenetic approach to examine evolutionary distance or coalescence between populations of a species. These principles and the most commonly used methods are examined and expanded in more detail in chapter 5.

3.7 Statistical approaches

3.7.1 Standardization of sample size

Allelic richness can be heavily influenced by variation in sample size. The level of influence varies with locus variability and highly polymorphic loci bias allelic richness to a greater degree than loci with low polymorphism (Spencer, Niegel, and Leberg 2000). This potential bias needs to be addressed when allelic richness is considered for studies using populations of unequal sample sizes. A simple approach is to standardize sample size with the lowest sample number and randomly select equal-sized subsamples from the other populations surveyed (single random reduction in N). However, this procedure leads to a loss of potentially useful data from the eliminated individuals (Leberg 2002). A refinement of this is to take repeat random samples of the lowest population size and estimate allelic richness from the values obtained (multiple random reductions in N). A third approach, originally developed in ecology, and applied to genetic investigations is called rarefaction (El Mousadik and Petit 1996, Petit, Mousadik, and Pons 1998). This utilizes the frequency of alleles at a locus to estimate the number of alleles that would occur in smaller samples of individuals. The results they produce are comparable to multiple random reductions in N (Leberg 2002).

3.7.2 Linkage disequilibrium

Linkage disequilibrium may occur due to drift, founder effects, or selection for loci linked to a neutral marker. It is therefore important that linkage disequilibrium is identified and accounted for in calculations of F_{ST} and related statistics (Box 3.2). The 2×2 contingency table, which results from consideration of linkage, lends itself to a chi-squared test, although it is important to ensure that the conditions of the test are met. Specifically, that all expected cells have values greater than one and that fewer than 20% of the cells have expected values of less than five (Berg and Hamrick 1997). Where these conditions are not met, rare alleles need to be summed until the minimum expected values are reached.

Linkage disequilibrium can be identified by analyzing the variance in the distribution of pairwise mismatch values (i.e., the number of loci at which pairs differ; Brown, Feldman, and Nevo 1980, Maynard-Smith et al. 1993). An index of association (I_A), a measure of linkage disequilibrium, can be calculated by comparing the observed variance (V_D) with that expected at equilibrium (V_E) for all pairwise combinations of loci:

$$I_A = 1/(L-1)((V_D/V_E) - 1)$$

where L is the number of loci analyzed.

Using this measure values of zero indicate complete linkage equilibrium, whereas values significantly different from zero indicate the presence of linkage disequilibrium.

3.7.3 Selection at loci

The neutrality of isozyme data can be tested using the Ewens–Watterson test (Ewens 1972, Watterson 1978) and the Ewens exact test (Ewens 1972, Slatkin 1994, 1996). Two special forms of natural selection can influence F_{ST}: overdominance (heterozygote advantage) and underdominance (heterozygote disadvantage), respectively. Overdominance will reduce the discrimination between populations by maintaining alleles in a population, underdominance will have the opposite effect.

3.7.4 Multiple comparisons and variance testing

In the case of more than one statistical test being performed there is the possibility of incorrectly detecting a significant result by accident due to conducting multiple simultaneous tests (i.e., Type 1 error, Bonferroni 1936). In the case of more than one statistical test, the chance of finding at least one test statistically significant due to chance fluctuation, and to incorrectly declaring a difference or relationship to be true, increases. Usually alpha (the value with which test statistics are compared) is set at 0.05 for a 95% confidence limit, but for each individual test alpha can be adjusted downward to ensure that the overall risk for all tests remains 0.05.

The Bonferroni correction is the simplest and most conservative approach and sets the alpha value for the entire set of n comparisons equal to α by taking the alpha value for each comparison equal to α/n (Shaffer 1995). Whilst Bonferroni's correction can be applied to a small number of simultaneous tests, if multiple pairwise comparisons are considered then an alternative statistic that considers multiple tests implicitly should be sought.

One such test is the analysis of variance (ANOVA) which estimates the likely range of variation of the averages given by the standard deviation of the estimated means ($\sigma/N^{1/2}$), where σ is the standard deviation of the measure and N the number of samples in a group. Thus, if the standard deviation of the group means is significantly larger than the standard deviation of the estimated means, there is evidence that the null hypothesis is not correct. Comparison between the actual variation of the group averages and that expected is expressed in terms of the F ratio, where:

$$F = \frac{\text{(observed variation of the group averages)}}{\text{(expected variation of the group averages)}}$$

Thus, if the null hypothesis is correct we expect F to be approximately equal to 1. However, if F is much larger than 1 then the null hypothesis is rejected.

ANOVA is used as a basis for several differentiation tests including Weir and Cockerham's F_{ST} estimator (1984), Rho_{ST}, and AMOVA. Thus, these estimates have a statistical confidence built into the test and do not require further statistical testing to establish the significance of differences between sample groups. ANOVA can of course be applied to measures to estimate variance components to establish if there is a significant difference between estimates (Sokal and Rohlf 1994). Alternatively, multivariate statistics can be used to present large numbers of multiple pairwise comparisons without committing a Type 1 error (chapter 6).

3.7.5 Resampling procedures

It may be difficult to estimate the accuracy of a particular statistic due to unknown variance components or the sampling distribution. In order to deal with this problem resampling procedures can be used, of which the most common are known as jackknifing and bootstrapping (Manly 1997). Both jackknifing and bootstrapping are non-parametric means of estimating variance parameters, and are based on repeated parameter estimation from a subsample of the data. The most important uses of these procedures are in the estimation of error limits, confidence intervals, and in the case of dendrograms, the statistical support for particular groupings of taxa. The distinction between the two procedures is based on how the subsamples are selected. In jackknife procedures, the parameter of interest is estimated for the total data set, and then the samples are removed one at a time and the parameter reestimated for each case. Using this procedure, standard errors can be easily calculated. However, confidence intervals cannot be estimated.

To estimate confidence intervals, the bootstrap method must be used. In this case, subsets of observations are selected at random with or without replacement (usually the former) and used to calculate the parameter. This procedure is often repeated many hundreds or even thousands of times, such that new sets are built up, containing some observations that can be repeated, but which comprise identical numbers of populations and observations. Thus, a distribution of parameter values is generated that can be used to estimate confidence intervals. To identify where the maximum variance lies within a data set it is possible to perform bootstrapping on both populations and loci, and if in doubt both tests should be performed.

3.7.6 Bayesian approaches

Recently, Bayes theorem has been used to provide a statistical test for many areas of population and evolutionary biology for which prior models or assumptions are present. Bayes theorem works out a conditional probability distribution and allows an assessment of the influence that the observed data have on the prior assumptions or model (Box 3.14). Bayesian methods have been applied to several areas including the extraction of allele frequencies and differentiation estimates from dominant data (Holsinger, Lewis, and Dey 2002, Zhivotovsky 1999), deriving gene flow parameters (chapter 4), and even for testing the likelihood of colonization pathways (e.g., Estoup and Clegg 2003). The method is particularly useful for testing the likelihood of complex scenarios where there are many potentially conflicting factors to be considered. The method applies a fully or non-fully likelihood method to rejection sampling. The principle of rejection-sampling methods consists in accepting (i.e., recording) only the values of the variable

Bayes theorem and Bayesian analysis

Bayesian analysis is based on the notion of posterior probabilities, that is, probabilities estimated based on a model (prior expectations), once part of the data is understood. Consider the case of a collection of coins, 95% of which are fair and 5% of which are biased to show tails 70% of the time when tossed. One the basis of this model, the chance that a randomly chosen coin is biased is 0.05. However, consider the case of the chosen coin being tossed 20 times and the probability of bias determined. In this case, the probability of bias will be modified in the light of the outcome of the tossing experiment and the knowledge of the initial model. This is the posterior probability and should be better than the estimate made with no knowledge.

If the run of tosses TTHTTHHTTTTTHTTHHHTTT is observed (the result R), the probability of this result given that the coin is fair is 0.5^{20}. The probability of this result given a biased coin is $(0.7)^{14}(0.3)^6$. Thus, Bayes theorem gives the posterior probability that the coin is biased:

$P[Biased \mid R] =$

$$\frac{P[R \mid Biased] \cdot P[Biased]}{(P[R \mid Biased] \cdot P[Biased]) + (P[R \mid Fair] \cdot P[Fair])}$$

$$= \frac{[(0.7)^{14}(0.3)^6] \cdot [0.05]}{\{[(0.7)^{14}(0.3)^6] \cdot [0.05]\} + \{[(0.5)^{20}] \cdot [0.95]\}}$$

$$= 0.21$$

Bayes theorem is used to revise the probability of a particular event happening based on the fact that a similar event has already happened.

$$P(A/B) = \frac{P(B/A) \cdot P(A)}{P(B)}$$

where $P(A)$ and $P(B)$ are the unconditional (or *a priori*) probabilities of the events.

Statistics are derived from probability theory through the use of Bayes theorem in the form:

$$P(H/d) = \frac{P(d/H) \cdot P(H)}{P(d)}$$

where H is the hypothesis and d is the experimental data (Bock 1998). The meaning of the different terms is:
- $P(H/d)$ is the degree of belief in the hypothesis H, after the experiment which produced data d.
- $P(H)$ is the prior probability of H being true.
- $P(d/H)$ is the likelihood function of H given d.
- $P(d)$ is the prior probability of obtaining data d.

parameters computed from simulated data sets, similar to the values of summary statistics computed from the observed data set. Hence, rejection-sampling methods give properties of the posterior distributions of the variable parameters of the model without explicit likelihood calculation (Estoup and Clegg 2003). It is thus a very flexible statistical test, applicable to a wide range of scenarios (Box 3.14).

3.8 Use of genetic diversity statistics

It is important to understand the maximum and minimum values that diversity and differentiation statistics can take and what they signify to allow data to be more fully interpreted. For example, within the plant kingdom G_{ST} values of zero (i.e., there is no differentiation among populations) have been found in some species (e.g., *Cucumis metuliferus*, Staub, Fredrick, and Marty 1987) while others have G_{ST} values of one (all variation is partitioned among populations, and all individual populations are monomorphic and fixed for different alleles), for instance *Emex spinosa* (Marshall and Weiss 1982). However, such extremes are rare.

Patterns of genetic variation summarized from allozyme data and arranged for comparison between geographic range and breeding system (from Hamrick and Godt 1989).

	Within species		Within populations		Among populations
	P_s	H_{es}	P_p	H_{ep}	G_{ST}
A. *Geographic range*					
Endemic	40.0c	0.096c	26.3c	0.063c	0.248a
Narrow	41.5bc	0.137b	30.6bc	0.105b	0.248a
Regional	52.9ab	0.150b	36.4ab	0.118b	0.216a
Widespread	58.9a	0.202a	43.0a	0.159a	0.210a
B. *Breeding system*					
Selfing	41.8b	0.124b	20.0c	0.074d	0.510a
Mixed – animal	40.0b	0.120b	29.2bc	0.090cd	0.216a
Mixed – wind	73.5a	0.194a	54.4a	0.198a	0.100c
Outcrossing – animal	50.1b	0.167ab	35.9b	0.124bc	0.197b
Outcrossing – wind	66.1b	0.162ab	49.7a	0.148b	0.099c

P is the proportion of polymorphic loci; H_e is the gene diversity at the species (s) and population level (p), respectively; and G_{ST} is the gene diversity among populations. Values within the same column that have different suffixes are significantly different ($p > 0.95$).

One might intuitively expect that populations of a species with high individual or propagule mobility will harbor different levels of diversity and differentiation compared with species that have restricted mobility. Indeed work summarizing the protein variation literature for plants has indicated that life history (e.g., breeding systems, pollination biology, and seed dispersal mechanisms) and ecological characters (e.g., habitat preference and population density) have a significant influence on the pattern of distribution of genetic diversity within populations (Hamrick and Godt 1989, Loveless and Hamrick 1984). Some of the other main findings of the Hamrick and coworker's meta-analyses are that wide-ranging species tend to possess more genetic variation than endemics, and that outcrossing species tend to maintain higher levels of diversity than those that are predominantly self-fertilizing (Hamrick and Loveless 1986, Loveless 1992, Loveless and Hamrick 1984). In addition, outcrossing species with highly mobile pollen or seeds exhibit lower differentiation between populations than selfing species with restricted pollen or seed dispersal mechanisms (Table 3.1).

There are, however, several problems when considering these types of meta-analyses. Firstly, they are restricted to allozyme studies, other methods are not included. Second, criticisms have focused on the statistical approach of treating taxa as independent samples when some may be phylogenetically related, and thus clearly not independent entities (Felsenstein 1985, Silvertown and Dodd 1996). Third, Culley et al. (2002) have noted other statistical difficulties, in particular Hamrick and Godt's (1989) table of G_{ST} is calculated differently from the measure originally defined by Nei (1973). Nei's estimate calculates means across all loci (whether they are polymorphic or monomorphic) whereas Hamrick and Godt (1989) only calculate means for polymorphic loci. Such a bias will inflate the diversity estimates of Hamrick and Godt, and of the two approaches Nei's original measure is probably more appropriate as it is more sensitive to variation in allele frequencies across populations, which is the factor promoting population differentiation. Fourth, many of the studies use only a limited number of loci and several studies combine loci with very different levels of diversity. Fifth, the sampling strategies of the combined studies are very different at the individual, population, and locus level and were not originally sampled to be combined as a meta-analysis. To rectify some of these difficulties, Culley et al. (2002) ask that

Box 3.15

Assessing the level of Bryophyte diversity

The perception of Bryophytes (mosses and liverworts) as relics of past geological eras with low genetic variation and hence low evolutionary potential was addressed by Stoneburner, Wyatt, and Odrzykoski (1991) by comparing previous Bryophyte isozyme work with the data for Tracheophytes (seed plants) (Table 3n).

This revealed that as a group mosses have almost identical levels of diversity to those found in outcrossing angiosperm and gymnosperm species. Liverworts have a lower level although there are significant levels of diversity within the group. This contradicts the traditional view of Bryophytes as possessing low diversity.

Table 3n

Measure of genetic diversity in Bryophytes and Tracheophytes. Values of diversity in Bryophytes and comparable with those in Tracheophytes.

	Mosses	Liverworts	Tracheophytes
Number of species surveyed	18	8	13
% polymorphic loci within population (P)	35.6	15.4	34.2
Mean number of alleles per polymorphic locus (A)	1.51	1.28	1.53
Gene diversity within subpopulations (H_S)	0.134	0.044	0.113

From Stoneburner, Wyatt, and Odrzykoski (1991).

for studies published in the future, methods of calculating G_{ST} are detailed along with H_T, H_S, D_{ST}, and G_{ST} values for individual loci and sample sizes for species and populations.

Despite the caveats, which must be borne in mind when considering meta-analysis data, and the variable nature of the combined studies, a large proportion of the total variation in genetic diversity (nearly 50%) was explicable by the life-history characters included in the analyses. Such analyses are clearly useful for understanding the wider context of a set of studies of genetic diversity and differentiation, and this is a view shared by many authors (Box 3.15). For example, Culley et al. (2002) note that 693 papers using G_{ST} compared their figures to those of Hamrick and Godt (1989).

The studies of Hamrick and coworkers are the most extensive conducted to date, but meta-analyses of the plant RAPD literature found similar trends of correlation between genetic diversity measures and life-history characters, although not as well supported as the findings of Hamrick et al. (Nybom and Bartish 2000). A study summarizing Bryophytes' genetic diversity found that comparatively high levels of diversity are maintained within populations of this taxonomic group and are potentially highly adaptive (Stoneburner, Wyatt, and Odrzykoski 1991; Box 3.15). Less detailed studies have also been carried out for groups of the animal kingdom. For example Ward, Skibinski, and Woodwark (1992) have compiled tables of H_T and G_{ST} for several major taxonomic groups and found some interesting trends in diversity and differentiation related to the relative vagility of members of the different groups. For example, differentiation between populations of highly mobile species like birds, fish, and insects is much lower than between species with a lower expected vagility, for example, amphibians, reptiles, and even mammals (Table 3.2). Echelle (1991) has listed genetic variation and its apportionment in endangered and threatened fishes of western North America. Packer et al. (1997) have listed the levels of heterozygosity in butterflies. However, neither of these two studies attempted extensive synthesis of data. The tabulation was undertaken to place the diversity values obtained by their own subject species into a wider context.

Table 3.2

Mean total heterozygosity (H_T) and proportion due to among-population differentiation (G_{ST}) in several major animal taxonomical groups (from Ward et al. 1992).

Taxon	H_T	No. of species	G_{ST}	No. of species
Vertebrates				
Fishes	51%	195	0.135	79
Amphibians	10.9%	116	0.315	33
Reptiles	7.8%	85	0.258	22
Mammals	6.7%	172	0.242	57
Birds	6.8%	80	0.076	16
Invertebrates				
Insects	13.7%	170	0.097	46
Crustaceans	5.2%	80	0.169	19
Molluscs	14.5%	105	0.263	44
Others	16%	15	0.060	5

Table 3.3

Ranges of Nei's genetic identities (D) for insects at various taxonomic levels (from Brussard et al. 1985). Note that there is considerable overlap in genetic identity between taxonomic levels.

	Genetic identity
Sibling species	0.56–0.94
Subspecies	0.77–0.99
Same species – different populations	0.88–1.00

One application of genetic diversity studies that has been heavily criticized is the comparison of rare with widespread species, and natives with introduced species, in attempts to address the genetic consequences of rarity. Kruckeberg and Rabinowitz (1985) when considering the biological characteristics of endemic species called for congeneric comparisons between widespread and endemic species. This targeted approach, is in contrast to the generalization provided by Hamrick and Godt (1989). In addition, Webb (1985) considered whether it is possible to distinguish introduced species from natives as the former should exhibit lower levels of diversity than the latter due to a bottleneck effect. The basis of the proposal is complicated by the interplay of other biological factors on genetic diversity (including breeding system and ecology) and a single figure cannot be used reliably to distinguish introduced species from natives, although general limits of diversity can be compared using the Hamrick and Godt tables. The effects of genetic bottlenecks are discussed in more detail in section 3.8.3.

Electrophoretic data have also been used in an attempt to delimit species. The genotypic cluster approach, used by Brussard et al. (1985), surveyed levels of genetic divergence values for insects (Table 3.3).

Considerable overlap between taxonomic levels is evident for genetic identity measures and indicates that this approach is not particularly good for delimiting species boundaries, although species may be non-integrating genotypic clusters within local areas (Mallett 1996). The concepts of species, their definition, and identification of boundaries is discussed in more detail in chapter 6.

3.8.1 Estimating gene flow

Gene flow, the movement of genes (via individuals or propagules) from one population to another, has long been of interest to population and evolutionary biologists as it is a central parameter offsetting the combined effects of mutation and genetic drift that prevents populations from differentiating over time. The different models, mechanisms, and parameters used to estimate gene flow are covered in detail in chapter 4, however, the inferred relationship between population genetic structure and indirect estimates of gene flow is introduced briefly here.

Using F_{ST} as a measure of population subdivision and noting that populations with high differentiation should have lower levels of gene flow between them than those with low differentation, Wright (1931) derived a parameter for gene flow, number of migrants per generation, Nm, which could be calculated directly from F_{ST} as:

$$Nm = (1 - F_{ST})/4F_{ST}$$

The relationship between Nm and F_{ST} is such that with Nm values of less than one ($F_{ST} = 0.2$), populations are expected to diverge genetically over time, but where Nm is greater than one populations are expected to retain genetic connectivity. The principle of one migrant per generation is also independent of population size. For large populations drift will be lower and so populations will be slower to diverge, thus one migrant per generation is effective at countering this slow drift. For small populations where drift is expected to be higher, a single migrant is a relatively high contribution and so often offsets the effect of drift.

Criticism has been levied for deriving such a powerful biological parameter (Nm) from a statistical property of allele frequency (F_{ST}) that has variance. In addition, Whitlock and McCauley (1999) document a series of instances when the relationship between Nm and F_{ST} does not hold, mostly because real populations often violate the assumptions of the model and that in most cases Nm cannot be accurately estimated from F_{ST}, although F_{ST} remains an excellent measure of population structure. The assumptions of the Nm model (and its criticisms) and alternative derivations of indirect and direct models and measures of gene flow are set out in chapter 4.

3.8.2 Spatial structuring of genetic diversity

In the situation where all populations of a species are not completely panmictic (and is probably the usual case), there will be genetic differentiation over some spatial scale due to lack of gene flow. If there is lower gene flow between more distantly separated populations, which consequently exhibit higher differentiation, this effect is termed isolation by distance (Wright 1943, 1946). The spatial scale at which isolation by distance can be observed varies considerably between species and can be observed between tightly grouped non-vagile individuals within a single population up to the regional distributional scale of a highly motile species. Depending on the scale of the structure several statistical procedures can be applied. At the regional scale where there is an effect suspected between widely spaced, discrete populations, a correlation between pairwise measures of geographic distance and genetic distance or differentiation can be plotted and the closeness of fit estimated using a Mantel's test (1967). Alternatively for continuously distributed individuals, usually within populations, pairwise distance estimates are grouped into categories and the observed and expected genetic distance or correlations are calculated between these groups. The combination of several individuals within a class allows the calculation of variance error and thus entities can be assessed for their significant positive or negative spatial structuring. These methods and the basis of the isolation by distance model and spatial autocorrelation analysis are described in more detail in chapter 4.

An alternative to divergence due to isolation by distance is due to complex separations of populations, when they will diverge due to a complete lack of gene flow (vicariance). To distinguish between isolation by distance and vicariance effects, a range of phylogeographic approaches can be adopted, including nested clade analysis (see chapter 5).

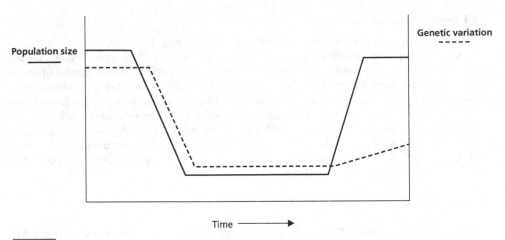

Fig. 3.3

Effect of population crash on genetic diversity. Note that following the population crash the genetic diversity does not rapidly return to its former level, unlike the population size.

3.8.3 Genetic bottlenecks and diversity statistics

The term genetic bottleneck refers to the process by which genetic variation is lost following a population crash. Whilst a population may rapidly recover its numbers following a crash, the level of genetic variation does not recover its previous value until restored by mutation or gene flow. A genetic bottleneck is effectively a sampling effect with the individuals surviving the crash often providing an unrepresentative and reduced sample of the variation present within the populations before the crash. The more severe the crash and the longer its duration the more extreme the sampling events are likely to be (Fig. 3.3).

Several measures of genetic distance (e.g., Nei 1973) assume that evolutionary divergence proceeds linearly at the same rate within populations. Hence distance is a product of time. However, if either or both populations have experienced a genetic bottleneck, then the genetic distance can increase much quicker than if population size remains constant. Under such circumstances, genetic distance is less an indication of time since divergence than a measure of the reduction in population size.

The degree of heterozygosity lost in each generation due to a genetic bottleneck is considered to be $(1 - 1/(2N_e))^t$, where N_e is the effective population size and t is the number of generations. Hence, the level of heterozygosity (H) in generation t can be expressed as:

$$H_t = (1 - 1/2N_e)^t \times H_o$$

where H_o is the initial observed heterozygosity.

The relationship between genetic distance and population size led Nei (1987) to modify D to incorporate bottleneck effects, as:

$$I = J_{XY}/J_X$$

where J_{XY} is defined as above and J_X is the homozygosity of the population that has not experienced a bottleneck. The modified approach typically gives a lower value for D than the original (1972) formula and assumes that the bottleneck has only occurred in one population, which can be easily identified.

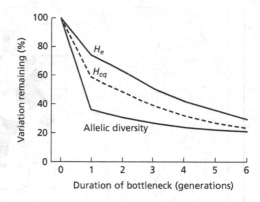

Fig. 3.4

Decline in measures of heterozygosity and allelic richness in generations following a demographic bottleneck. As the bottleneck continues allelic diversity declines faster than heterozygosity (from Luikart and Cornuet 1998).

Detecting bottlenecks

Genetic bottlenecks increase the possibility of population extinction through a combination of increasing demographic stochasticity and the deleterious effects of inbreeding. However, unless historical data on effective population size exist, a bottleneck is difficult to identify. Consequently measures of genetic diversity that require no consideration of population size are often used to identify genetic bottlenecks. Some studies incorporating historical data do exist (Box 3.16).

Comparisons between populations that have experienced bottlenecks with either samples collected prior to the bottleneck or populations that have not experienced a bottleneck have shown that both allelic richness (A) and heterozygosity decline with reduction in population size. However, allelic richness declines faster than heterozygosity, a finding confirmed by computer simulations (Cornuet and Luikart 1996; Fig. 3.4), and is the basis of the sign test of heterozygote excess.

The heterozygosity expected at a locus (H_{eq} – comparable to H_s) under Hardy–Weinberg equilibrium is compared to the number of heterozygotes (H_e) in a sample of the same size and number of alleles at mutation-drift equilibrium. The model calculates H_e with both SMM and IAM models of mutation. Populations under normal, non-bottleneck conditions, will be in mutation-drift equilibrium, and approximately 50% of loci will have $H_e > H_{eq}$ and 50% will have $H_e < H_{eq}$ as a

---

Box 3.16

Loss of genetic variation as a result of bottleneck in the Great Prairie Chicken

The Great Prairie Chicken (*Tympanuchus cupido pinnatus*) is a grassland prairie species that was distributed across the central plains of North America until the arrival of European settlers. Loss of habitat, coupled with competition and increased parasites due to the introduction of exotic birds, have caused a substantial decline in the numbers of the species.

In Illinois this has been considerably marked, with only 50 birds remaining in

1993. This is from an estimated population of a million in the 1860s. Even in 1933 an estimated 25,000 birds were present in the state. By comparison with the drastic population fall in Illinois, numbers in Minnesota, Kansas, and Nebraska have remained high, with 4000 in Minnesota to over 100,000 in Kansas and Nebraska.

Levels of genetic variation in the Greater Prairie Chicken were analyzed by Bouzut et al. (1998) using six polymorphic microsatellite loci. Samples were taken from Kansas, Minnesota, and Nebraska from live birds, with Illinois populations sampled with frozen tissue from incident mortalities (Fig 3j). To achieve a reasonable

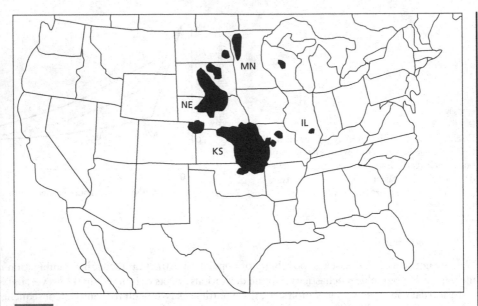

Geographic distribution of the Greater Prairie Chicken showing the distribution of sample areas (from Bouzut et al. 1998).

Sample size, observed heterozygosity, and mean number of alleles at six microsatellite loci in populations of Greater Prairie Chicken. Mean number of alleles per locus is significantly reduced in the Illinois samples compared with population from the other three states.

Population	Sample size	Observed heterozygosity	Mean no. of alleles per locus
Illinois	32	0.571[a]	3.67[a]
Kansas	37	0.597[a]	5.83[b]
Minnesota	38	0.654[a]	5.33[b]
Nebraska	20	0.626[a]	5.83[b]

Superscripts denote which values were significantly different from others of the same measurement.

sample size (n = 32) some of the Illinois samples were from much older samples than those from the other three states.

The mean number of alleles per locus were calculated plus the observed heterozygosity as allele frequencies in all populations were at Hardy–Weinberg equilibrium. These are listed in Table 3o.

There was no difference in the observed heterozygosity at the loci surveyed, however, the Illinois population had a significantly lower level of mean number of alleles per locus than the other three populations. Illinois has a small subset of the total alleles found elsewhere in the survey, with no unique alleles. Similarly few unique alleles are found at the Kansas and Nebraska populations suggesting that all populations were originally a single population with genetic variability similar to the levels currently recorded in the larger populations. Thus Illinois has lost a significant part of its diversity as a result of the demographic bottleneck.

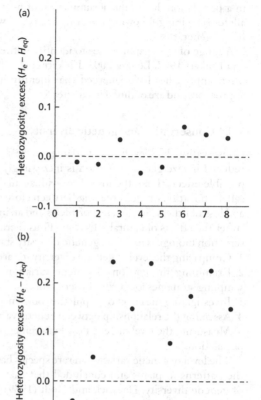

Fig. 3.5

Pattern of heterozygosity excess observed at each of eight polymorphic microsatellite loci in (a) a non-bottlenecked population (brown bears) and (b) a bottlenecked population (wombats). The horizontal dashed line represents the boundary between heterozygote excess (above the line) and heterozygote deficiency (below the line). Note that the non-bottlenecked population has as many loci above the dashed line as below it while the bottlenecked population has all but one locus with an excess of heterozygotes (from Luikart and Cornuet 1998).

result of chance fluctuations. By comparison the mutation-drift equilibrium of bottlenecked populations is temporarily broken and a significant proportion of loci will have an excess of heterozygotes ($H_e > H_{eq}$; Fig. 3.5). The bottleneck will only be detectable for a short time (0.2–4.0 N_e generations) and assumes an immediate and permanent bottleneck in population size. However, the method fails to identify bottlenecks that are not severe or recent (Luikart and Cornuet 1998).

The other conditions of the test are that loci should be in Hardy–Weinberg equilibrium. In particular, loci where heterozygote excess may be due to overdominance (heterozygotic advantage), should be removed from the analysis. Such loci are easily identified due to their idiosyncratic heterozygosity behavior relative to other loci as all loci should be equally affected by a bottleneck. It is recommended that 10–20 polymorphic loci are used to conduct the test due to the possibility of redundancy. During analysis, results may be sensitive to the mutation model applied and a significant excess of heterozygotes is more likely under IAM than SMM. However, given that most microsatellites probably have an intermediate mutation pattern between these two models, calculation of both parameters provides boundaries within which to compare results. By contrast allozyme variation profiles most closely fit the IAM and hence SMM may be unreasonably conservative and cause a bottleneck to be overlooked. Following the bottleneck, recovery of heterozygote proportions, which are in equilibrium ratio, will be quicker than the recovery of allelic richness. The increase in rare allele numbers will take time to achieve in a population due to the accumulation of new mutation events (and gene flow), whereas the alleles remaining following the sampling effect will still contribute to overall heterozygosity in future generations.

A range of allelic approaches to identify bottlenecks is available (Bouzut et al. 1998, Cornuet and Luikart 1997, Leberg 2002, Luikart et al. 1998, Richards and Leberg 1996). More recently other approaches have emerged that identify bottlenecks using phylogenetic and coalescent approaches and are outlined in chapter 5.

3.8.4 Conservation and genetic diversity

Conservation biology has a fundamental basis in genetic diversity. As populations become reduced in size genetic variation is increasingly likely to be lost, and is likely to result in two possible effects. First, the loss of individuals may make inbreeding more likely (either through self-fertilization or interbreeding between close relatives) with attendant inbreeding depression and loss of fitness. Second, loss of alleles and an increasingly uniform population is more likely to suffer the effects of natural selection. Thus there are several important issues that have led conservation biologists to utilize genetic diversity statistics. These include:

1 Comparing the level of genetic diversity in rare species with that in more widespread ones.
2 Examining the portion of genetic variation within and among populations as a guide to sampling strategies for *ex situ* conservation.
3 Investigating the effect of population bottlenecks on genetic variation.
4 Assessing the relationship between genetic variation and fitness components.
5 Measuring the level of gene flow between populations and identifying unique allele units in a population.

The level of genetic variation in rare species has been widely studied. Stebbins (1980) reviewed the patterns in plants and concluded that there was no relationship between rarity and level of genetic diversity. Hamrick and Godt (1989) reported reduced levels of genetic diversity in rare species. A summary of 25 plant species (Gitendanner and Soltis 2000) comparing rare and widespread congenic species (as a suggested approach by Karron 1987) revealed a slight reduction in the former group compared to the latter. However, rare species did not have consistently lower levels of diversity and the diversity ranges of the two abundance classes overlapped considerably. The last word on rarity and genetic diversity in both plants and animals has not been written and it will continue to be an area of active research. In addition, whether low diversity is a cause of rarity or an effect will continue to be debated.

A frequently cited aim of both *in situ* and *ex situ* conservation projects is to preserve a representative sample of the genetic diversity present in a species. Measures of genetic differentiation are therefore extremely important to understanding the distribution of diversity in a population. Species with low G_{ST} may reasonably be sampled over a few populations, and the number of populations required to sample a reasonable proportion (at least in conservation genetic terms) of genetic variation has been expressed mathematically as:

$$1 - G_{ST}{}^n$$

where n is the number of populations proposed for sampling (Ceska, Affolter, and Hamrick 1997). This formula has been used to sample and preserve germplasm of endangered and threatened species (Box 3.17; Ceska, Affolter, and Hamrick 1997).

There are drawbacks to using this approach. Schoen and Brown (1991) have shown that as G_{ST} increases, the variation in genetic diversity between populations also increases. Species with high G_{ST} will have some populations where H_s (within-population diversity) is high and some where H_s is low. For those species identifying the populations with high H_s is critical to ensure effective sampling.

The sampling intensity required to collect a representative amount of genetic diversity has been the subject of considerable debate, particularly concerning the significance of allelic fre-

Box
3.17

Using allozymes to develop an *ex situ* sampling strategy for *Baptisia arachnifera*

An increasing priority for *ex situ* conservation is to develop collections that maximize the genetic diversity found within a species. To achieve this an understanding of the level and apportionment of genetic diversity within the species to be conserved is required.

In plants such knowledge informs seed sampling for subsequent collections and housing in a gene bank. Cesska, Affolter, and Hamrick (1997) measured genetic diversity in the rare plant *Baptisia arachnifera* (Hairy Rattleweed, Fabaceae).

B. arachnifera is a rare, small (up to 30 cm), semi-woody plant found in a 16 km strip in South Georgia, USA. The species is restricted to artificial habitats, most notably pine plantations. Historically it was found in longleaf pine/saw palmetto flats where spring burning is common. In its current habitat it cannot survive the cyclic harvesting of pine with the associated tilling and bulldozing. In addition it is outshaded in dense pine plantations or by dense undergrowth and it is therefore threatened.

Genetic diversity of *B. arachnifera* was measured using 37 allozyme loci, nine of which are polymorphic in 10 populations. The among-population variation was calculated and gave a G_{ST} value of 0.096, therefore just under 10% of the variation found in the species is among populations. Using the value of G_{ST} and the equation $1 - (G_{ST})^n$, two populations would suffice to sample 99% of the genetic variation present in the species ($1 - (0.096)^2 = 0.99$). This is consistent with the empirical data. Pairwise comparison of samples showed that the two populations with the highest degree of variation, when considered together, possessed 48 of the 49 alleles found in the 10 populations surveyed.

quency. Hamrick et al. (1991) recommend that 50 individuals from each population should be sampled to sufficiently represent allele frequencies in a population and to ensure that rare alleles are sampled. Brown and Briggs (1991) consider that the extra effort in maintaining rare alleles through larger sample sizes is not justified, arguing that rare alleles are not important in species preservation. Finally, Holsinger and Gottlieb (1991) recommend maintaining alleles if they occur above a frequency of 0.05. Consideration of the species allelic distribution can also help identify populations with unique suites of alleles. Such an assemblage of alleles may indicate the importance of according the population higher conservation status (e.g., Echelle et al. 1989; *Pecos gambusia*).

Of the other areas of interest within conservation genetics bottlenecks have previously been considered, fitness components are largely beyond the scope of this book but are discussed briefly below (section 3.8.6), while gene flow is the subject of the following chapter.

3.8.5 Historical processes

Using markers which can be interpreted phylogenetically allows application of coalescent approaches to assess the historical dynamics of populations. The approximate age of populations, their historical size, whether they have been expanding or contracting and even the influence of selection at linked loci, can be determined using such techniques. The basis of the coalescent and its potential applications to population and evolutionary questions are presented in chapter 5.

3.8.6 Adaptation

Selection biases allele frequency always from Hardy–Weinberg equilibrium and can clearly cause incorrect interpretation of population level processes if undetected. However, it is only those loci that are under selection, or closely linked to adaptive genes, which will display such behavior and so comparison of allele frequencies across all loci should allow removal of those that are behaving idiosyncratically. Selection is only one of several possible causes of non-concordant allele frequency behavior, other reasons include null alleles (where the observed number of heterozygotes will be reduced) and linkage. Whilst presenting a problem for the interpretation of neutral gene diversity and differentiation statistics, markers for particular adaptive genes are available. The suite of markers is increasing rapidly and will soon allow powerful examinations of the distribution of adaptive genes in natural populations. Comparing the distribution of adaptive gene variation with neutral locus markers within the same individuals and populations allows testing for the action of selection (Ford 2002), and should prove to be a developing area of ecological genetics study in the future.

3.9 Concluding remarks

The choice of appropriate genetic diversity, differentiation and distance statistics rests upon a careful consideration of the question that the researcher seeks to answer. This is particularly important to consider when the various computer packages that are available can quickly calculate a large number of values. The most straightforward statistics are the measures of allelic richness. Assuming that sampling is of sufficient breadth and any vagaries in sample size have been addressed by approaches such as rarefaction, the measures are easy to calculate and understand. Whilst allelic richness measures have very little statistical power, they are important for conservation purposes and are a more reliable measure of a species' ability to respond to selection than change in allele frequency.

More sophisticated analysis can be undertaken with the measures of genetic diversity and population differentiation. In addition to providing powerful insights into population structure at the spatial and temporal scales, including identifying historical population bottlenecks, most recent common ancestor, and isolation by distance, they can also be utilized to identify gene flow parameters. It is worth noting though that the increased statistical power is based upon an increasing set of assumptions. These assumptions and hence a wider understanding of the species' biology must be borne in mind when interpreting results.

3.10 Further reading

An excellent introduction to population genetics and the role of the Hardy–Weinberg equilibrium is found in both Hartl (1988) and Hartl and Clark (1997). Nei (1987) provides a comprehensive overview of the calculation of genetic variation statistics, whilst Nei and Kumar (2000) present a good introduction to measures of DNA variation. Weir (1996) provides a comprehensive introduction to genetic data analysis and this text also has a good chapter on the analysis of genetic variation measures.

Frankham, Ballou, and Briscoe (2002) and Frankel, Brown, and Burdon (1995) are excellent introductions to the importance of genetic variation in conservation biology while Allendorf and Ryman (2002) focus on a more specialized area of conservation genetics, considering the role of genetics in population viability analysis (PVA).

REFERENCES

Allendorf, F.W. and Ryman, N. 2002. The role of genetics in population viability analysis. In S.R. Beissinger and D.R. McCullough, eds. *Population Viability Analysis*. Chicage: Chicago University Press.
Balloux, F. and Lugon Moulin, N. 2002. The estimation of population differentiation with microsatellite markers. *Molecular Ecology*, **11**: 155–65.
Berg, E.E. and Hamrick, J.L. 1995. Fine-scale genetic structure of a Turkey Oak forest. *Evolution*, **49**: 110–20.

Berg, E.E. and Hamrick, J.L. 1997. Quantification of genetic diversity at allozyme loci. *Canadian Journal of Forestry Research*, **27**: 415–24.

Bonferroni, C.E. 1936. Teoria statistica delle classi e calcolo delle probabilità. *Pubblicazioni del R Istituto Superiore di Scienze Economiche e Commerciali di Firenze*, **8**: 3–62.

Bouzut, J.L., Cheng, H.H., Lewin, H.A., Westemeir, R.I., Brawn, J.D., and Paige, K.N. 1998. Genetic evaluation of a demographic bottleneck in the Greater Prairie Chicken. *Conservation Biology*, **12**: 836–43.

Brown A.H.D. and Briggs, J.D. 1991. Sampling strategies for genetic variation in ex situ collections of endangered plant species. In D.A. Falk and K.E. Holsinger, eds. *Genetics and Conservation of Rare Plants*. New York: Oxford University Press.

Brown, A.H.D., Feldman, M.W., and Nevo, E. 1980. Multilocus structure of natural populations of *Hordeum spontaneum*. *Genetics*, **96**: 523–36.

Brusard, P.F., Ehrlich, P.R., Murphy, D.D., Wilcox, B.A., and Wright, J. 1985. Genetic distances and the taxonomy of checker-spot butterflies (Nymphalidae; Nymphalinae). *Journal of Kansas Entomological Society*, **58**: 403–12.

Byrne, K. and Nichols, R.A. 1999. *Culex pipiens* in London underground tunnels: differentiation between surface and subterranean populations. *Heredity*, **82**: 7–15.

Cavalli-Sforza, L.L. and Edwards, W.F. 1967. Phylogenetic analysis: models and procedures. *Evolution*, **21**: 550–70.

Cavers, S. 2002. *Population Genetic Structure and Phylogeography of Two Commercially Important Neotropical Tree Species*: Vochysia ferruginea *Mart. and* Cedrela odorata L. PhD thesis, University of Edinburgh.

Ceska, J.E., Affolter, J.M., and Hamrick, J.L. 1997. Developing a sampling strategy for *Baptisia arachnifera* based on allozyme diversity. *Conservation Biology*, **11**: 1133–9.

Chakraborty, R. and Danker-Hopfe, H. 1991. Analysis of population structure: a comparative study of different estimators of Wright's fixation indices. In C.L. Rao and R. Chakraborty, eds. *Handbook of Statistics, Vol. 8, Statistical Methods in Biological and Medical Sciences*. New York: Elsevier Science, pp. 203–54.

Charlesworth, B., Nordberg, M., and Charlesworth, D. 1997. The effects of local selection, balanced polymorphism and background selection on equilibrium patterns of genetic diversity in subdivided populations. *Genetic Research*, **70**: 155–74.

Clark, A.G. and Lanigan, C.M.S. 1993. Prospects for estimating nucleotide divergence with RAPDs. *Molecular Biology and Evolution*, **10**: 1096–111.

Cockerham, C.C. 1969. Variance of gene frequencies. *Evolution*, **23**: 72–84.

Cockerham, C.C. 1973. Analyses of gene frequencies. *Genetics*, **74**: 679–700.

Cornuet, J.M. and Luickart, G. 1996. Description and evaluation of two tests for detecting recent population bottlenecks from allele frequency data. *Genetics*, **144**: 2001–14.

Culley, T.M., Wallace, L.E., Gengler-Nowak, K.A., and Crawford, D.J. 2002. A comparison of calculating G_{ST}, a genetic measure of population differentiation. *American Journal of Botany*, **89**: 460–5.

Darwin, C. 1859. *The Origin of Species*. London: Murray.

Davis, J.L. and Nixon, K.C. 1992. Populations, genetic variation, and the delimitation of phylogenetic species. *Systematic Biology*, **41**: 421–35.

Dawson, I.K., Simons, A.J., Waugh, R., and Powell, W. 1995. Diversity and genetic differentiation among subpopulations of *Gliricidia sepium* revealed by PCR-based assays. *Heredity*, **74**: 10–18.

Echelle, A.A. 1991. Conservation genetics and genetic diversity in fresh water fishes of western North America. In W.L. Minckley and J.E. Deacon, eds. *Battle against Extinction: Native Fish Management in the American West*. Tuscon, AZ: University of Arizona Press, pp. 141–53.

Echelle, A.A., Echelle, A.F., and Edds, D.R. 1989. Conservation genetics of a spring dwelling desert fish, the Pecos gambusia, *Gambusia nobilis* (Poeciliidae). *Conservation Biology*, **3**: 159–69.

El Mousadik, A. and Petit, R.J. 1996. High level of genetic differentiation for allelic richness among populations of the argan tree [*Argania spinosa* (L.) Skeels] endemic to Morocco. *Theoretical and Applied Genetics*, **92**: 832–9.

Ennos, R.A., Worrell, R., and Malcolm, D.C. 1998. The genetic management of native species in Scotland. *Forestry*, **71**: 1–23.

Estoup, A. and Clegg, S.M. 2003. Bayesian inferences on the recent island colonization history by the bird *Zosterops lateralis lateralis*. *Molecular Ecology*, **12**: 657–74.

Estoup, A., Rousset, F., Michalakis, J.-M., Cornuet, M., Adriamanga, M., and Guyomard, R. 1998. Comparative analysis of microsatellite and allozyme markers: a case study investigating microgeographic differentiation in brown trout (*Salmo trutta*). *Molecular Ecology*, **7**: 339–53.

Excoffier, L., Smouse, P.E., and Quattro, J.M. 1992. Analysis of molecular variance inferred from metric distances among DNA haplotypes: application to human mitochondrial DNA restriction data. *Genetics*, **131**: 479–91.

Ewens, W.J. 1972. The sampling theory of neutral alleles. *Theoretical Population Biology*, **3**: 87–112.

Felsenstein, J. 1985. Phylogenies and the comparative method. *American Naturalist*, **125**: 1–15.

Ford, M.J. 2002. Applications of selective neutrality tests to molecular ecology. *Molecular Ecology*, **11**: 1245–62.

Frankel, O.H., Brown, A.H., and Burdon, J.J. 1995. *The Conservation of Plant Biodiversity*. Cambridge: Cambridge University Press.

Frankham, R., Ballou, J.D., and Briscoe, D.A. 2002. *Introduction to Conservation Genetics*. Cambridge: Cambridge University Press.

Franklin, I.R. 1980. Evolutionary change in small populations. In M.E. Soule and B.A. Wilcox, eds. *Conservation Biology: An Evolutionary-Ecological perspective*. Sunderland, MA: Sinauer, pp. 135–9.

Gillies, A.C.M., Cornelius, J.P., and Newton, A.C. 1997. Genetic variation in Costa Rican populations of the tropical timber species *Cedrela odorata* L., assessed using RAPDs. *Molecular Ecology*, **6**: 1133–45.

Gitzendanner, M.A. and Soltis, P.S. 2000. Patterns of genetic variation in rare and widespread plant congeners. *American Journal of Biology*, **87**: 783–92.

Goldstein, D.B., Linares, A.R., Feldman, M.W., and Cavalli-Sforza, L.L. 1995a. An evaluation of genetic distances for use with microsatellite loci. *Genetics*, **139**: 463–71.

Goldstein, D.B., Linares, A.R., Cavalli-Sforza, L.L., and Feldman, M.W. 1995b. Genetic absolute dating based on microsatellites and the origin of modern humans. *Proceedings of the National Academy of Sciences USA*, **92**: 6723–7.

Goldstein, D.B. and Schlotterer, C. 1999. *Microsatellites, Evolution and Applications*. Oxford: Oxford University Press.

Goodman, S.J. 1997. RST CALC: a collection of computer programs for calculating unbiased estimates of genetic differentiation and determining their significance for microsatellite data. *Molecular Ecology*, **6**: 881–5.

Grant, V. 1981. *Plant Speciation*, second edition. New York: Columbia University Press.

Guries, R.P. and Ledig, F.T. 1982. Genetic diversity and population structure in Pitch Pine (*Pinus rigida* Mill). *Evolution*, **36**: 387–402.

Hamrick, J.L. and Godt, M.J.W. 1989. Isozyme diversity in plant species. In A.H.D. Brown, M.T. Clegg, A.L. Kahler, and B.S. Weir, eds. *Plant Population Genetics, Breeding, and Genetic Resources*. Sunderland, MA: Sinauer.

Hamrick, J.L., Godt, M.J.W., Murawski, D.A., and Loveless, M.D. 1991. Correlations between species traits and allozyme diversity. In D.A. Falk and K.E. Holsinger, eds. *Genetics and Conservation of Rare Plants*. New York: Oxford University Press, pp. 75–86.

Hamrick, J.L. and Loveless, M.D. 1986. The influence of seed dispersal mechanisms on the genetic structure of plant populations. In A. Estrada and T.H. Fleming, eds. *Frugivores and Plant Dispersal*. The Hague: Junk.

Hancock, J. 1999. Microsatellites and other simple sequences: genomic context and mutational mechanisms, In D.B. Goldstein and C. Schlötterer, eds. *Microsatellites, Evolutuion and Applications*. Oxford: Oxford University Press.

Hardy, G. H. 1908. Mendelian proportions in a mixed population. *Science*, **28**: 49–50.

Hartl, D.L. 1988. *A Primer of Population Genetics*, 2nd edn. Sunderland, MA: Sinauer.

Hartl, D.L. and Clark, A.G. 1989. *Principles of Population Genetics*, 2nd edn. Sunderland, MA: Sinauer.

Hedrick, P.W. 1971. A new approach to measuring genetic similarity. *Evolution*, **25**: 276–80.

Holsinger, K.E., Lewis, P.O., and Dey, D.K. 2002. A Bayesian approach to inferring population structure from dominant markers. *Molecular Ecology*, **11**: 1157–64.

Holsinger, K. and Gottlieb, L. 1991. Conservation of rare and endangered plants: principles and prospects. In D.A. Falk and K.E. Holsinger, eds. *Genetics and Conservation of Rare Plants*. New York: Oxford University Press.

Huff, D.R., Peakall, R., and Smouse, P.E. 1993. RAPD variation within and among natural populations of outcrossing buffalo grass [Buchlo' dactyloides (Nutt.) Engelm.]. *Theoretical and Applied Genetics*, **86**: 927–34.

Innan, H., Terauchi, R., Kahl, G., and Tajima, F. 1999. A method for estimating nucleotide diversity from AFLP data. *Genetics*, **151**: 1157–64.

Isabel, N., Beaulieu, J., Theriault, P., and Bousquet, J. 1999. Direct evidence for biased gene diversity estimates from dominant random amplified polymorphic DNA (RAPD) fingerprints. *Molecular Ecology*, **8**: 477–83.

Kalinowski, S.T. 2002. Evolutionary and statistical properties of three genetic distances. *Molecular Ecology*, **11**: 1263–73.

Karl, S.A. and Avise, J.C. 1992. Balancing selection at allozyme loci in oysters: implication from nuclear RFLPs. *Science*, **256**: 100–2.

Karron, J.D. 1987. A comparison of levels of genetic polymorphism and self-compatibility in geographically restricted and widespread plant congeners. *Evolutionary Ecology*, **1**: 47–58.

Kimura, M. 1953. "Stepping-stone" model of population. *Annual Report of the National Insitute of Genetics, Japan*, **3**: 62–3.

Koehn, R.K. and Hilbish, T.J. 1987. The adaptive importance of genetic variation. *American Scientist*, **75**: 134–41

Kruckeberg, A.R. and Rabinowitz, D. 1985. Biological aspects of endemism in higher plants. *Annual Review of Ecology and Systematics*, **16**: 447–79.

Krauss, S.L. 1997. Low genetic diversity in *Persoonia mollis* (Proteaceae), a fire sensitive shrub occurring in a fire prone habitat, *Heredity*, **78**: 41–9.

Lande, R. 1995. Mutation and conservation. *Conservation Biology*, **9**: 782–91.

Leberg, P.L. 2002. Estimating allelic richness: effects of sample size and bottlenecks. *Molecular Ecology*, **11**: 2445–9.

Levin, D.A. 2000. *The Origin, Expansion and Demise of Plant Species*. Oxford: Oxford University Press.

Lewontin, R. 1972. The apportionment of human diversity. *Evolutionary Biology*, **6**: 381–98.

Loveless, M.D. 1992. Isozyme variation in tropical trees: patterns of genetic organisation. *New Forests*, **6**: 67–94.

Loveless, M.D. and Hamrick, J.L. 1984. Ecological determinants of genetic structure in plant populations. *Annual Review of Ecology and Systematics*, **15**: 65–95.

Luikart, G. and Cornuet, J. 1998. Empirical evaluation of a test for identifying recently bottlenecked populations from allele frequency data. *Conservation Biology*, **12**: 228–37.

Luikart, G., Sherwin, W.B., Steele, B., and Allendorf, F.W. 1998. Usefulness of molecular markers for detecting population bottlenecks and monitoring genetic change. *Molecular Ecology*, **7**: 963–74.

Lynch, M. and Crease, T.J. 1990. The analysis of population survey data on DNA-sequence variation. *Molecular Biology and Evolution*, **7**: 377–94.

Lynch, M. and Milligan, B.G. 1994. Analysis of population genetic structure with RAPD markers. *Molecular Ecology*, **3**: 91–9.

Mallett, J. 1996. The genetics of biological diversity. In K.J. Gatson, ed. *Biodiversity: A Biology of Numbers and Difference*. Oxford: Blackwell Science.

Manly, B.F.J. 1997. *Randomization, Bootstrap and Monte Carlo Methods in Biology*. London: Chapman and Hall.

Mantel, N. 1967. The detection of disease clustering and a generalized regression approach. *Cancer Research*, **27**: 209–20.

Marshall, D.R. and Allard, R.W. 1970. Isozyme polymorphisms in natural populations of *Aveba fatua* and *Avena barbata*. *Heredity*, **25**: 373–82.

Marshall, D.R. and Weiss, B.S. 1982. Isozyme variation within and among Australian populations of *Emex spinosa* (L), *Campd*. *Australian Journal of Biological Science*, **35**: 327–32.

Maynard-Smith, J., Smith, N.H., O'Rourke, M., Spratt, B.G. 1993. How clonal are bacteria? *Proceedings of the National Academy of Sciences USA*, **90**: 4384–8.

Michalakis, Y. and Excoffier, L. 1996. A generic estimation of population subdivision using distances between alleles with special reference to microsatellite loci. *Genetics*, **142**: 1061–4.

Morowitz, H.J. 1971. *Entropy for Biologists: An Introduction to Thermodynamics*. New York: Academic Press.

Nauta, M.J. and Weissing, F.J. 1996. Constraints on allele size at microsatellite loci: implications for genetic differentiation. *Genetics*, **143**: 1021–32.

Nei, M. 1971. Interspecific gene differences and evolutionary time estimated from electrophoretic data on protein identity. *American Naturalist*, **105**: 385–98.

Nei, M. 1972. Genetic distance between populations. *American Naturalist*, **106**: 283–92.

Nei, M. 1973. Analysis of gene diversity in subdivided populations. *Proceedings of the National Academy of Sciences USA*, **70**: 3321–3.

Nei, M. 1975. *Molecular Population Genetics and Evolution*. Amsterdam: North Holland.

Nei, M. 1977. F-statistics and analysis of gene diversity in subdivided populations. *Annals of Human Genetics*, **41**: 225–33.

Nei, M. 1978. Estimation of average heterozygosity and genetic distance from a small number of individuals. *Genetics*, **89**: 583–90.

Nei, M. 1986. Definition and estimation of fixation indices. *Evolution*, **40**: 643–5.

Nei, M. 1987. *Molecular Evolutionary Genetics*. New York: Columbia University Press.

Nei, M. and Chesser, R.K. 1983. Estimation of fixation indices and gene diversities. *Annals of Human Genetics*, **47**: 253–9.

Nei, M. and Kumar, S. 2000. *Molecular Evolution and Phylogenetics*. Oxford: Oxford University Press.

Nei, M. and Li, W.-H. 1979. Mathematical model for studying genetic variation in terms of restriction endonucleases. *Proceedings of the National Academy of Sciences USA*, **76**: 5269–73.

Nyakaana, S. and Arctander, P. 1999. Population genetic structure of the African elephant in Uganda based on variation at mitochondrial and nuclear loci: evidence for male-biased gene flow. *Molecular Ecology*, **8**: 1105–15.

Nybom, H. and Bartish, I.V. 2000. Effects of life history traits and sampling strategies on genetic diversity estimates obtained with RAPD markers. *Perspectives in Plant Ecology, Evolution and Systematics*, **3**: 93–114.

Packer, L., Taylor, J.S., Savignano, D.A., Bleser, C.A., Lane, C.P., and Sommers, L.A. 1997. Population biology of an endangered butterfly, Lycaeides, *Melissa samuelis* (Lepidotera; Lycaenidae): genetic variation, gene flow and taxonomic status. *Canadian Journal of Zoology*, **76**: 320–9.

Paetkau, D., Waits, L.P., Clarkson, P.L., Craighead, L., and Strobeck, C. 1997. An empirical evaluation of genetic distance statistics using microsatellite data from bear (Ursidae) populations. *Genetics*, **147**: 1943–57.

Paetkau, D., Amstraup, S.C., Bourne, E.W., Calvert, W., Derocher, A.E., Garner, G.W., Messier, F., Sterling, I., Taylor, M.K., Wiig, Ø., and Strobeck, C. 1999. Genetic structure of the world's polar bear populations. *Molecular Ecology*, **8**: 1571–84.

Petit, R.J., El. Mousadik, A. and Pons, O. 1998. Identifying populations for conservation on the basis of genetic markers. *Conservation Biology*, **12**: 339–48.

Pons, O. and Petit, R.J. 1996. Measuring and testing genetic differentiation with ordered versus unordered alleles. *Genetics*, **144**: 1237–45.

Provan, J., Powell, W., and Hollingworth, P.M. 2001. Chloroplast microsatellites: new tests for studies in plant ecology and evolution. *Trends in Ecology and Evolution*, **16**: 142–7.

Richards, C. and Leberg, P.L. 1996. Temporal changes in allele frequencies and a population's history of severe bottlenecks. *Conservation Biology*, **10**: 832–9.

Rieseberg, L.H. 1996. Homology among RAPD fragments in interspecific comparisons. *Molecular Ecology*, **5**: 99–105.

Schlötterer, C., Ritter, R., Harr, B., and Brem, G. 1998. High mutation rate of a long microsatellite allele in *Drosophila melanogaster* provides evidence for allele-specific mutation rates. *Molecular Biology and Evolution*, **15**: 1269–74.

Schoen, D.J. and Brown, A.H.D. 1991. Intraspecific variation in population gene diversity and effective population size correlates with the mating system in plants. *Proceedings of the National Academy of Sciences USA*, **88**: 4494–7.

Shaffer, J.P. 1995. Multiple hypothesis testing. *Annual Review, Psychology*, **46**: 561–84.

Shannon, C.E. 1948. A mathematical theory of communication. *Bell Systems Technical Journal*, **27**: 379–423, 623–56.

Shiver, M.D., Jin, L., Boerwinkle, E., Deka, R., and Ferrel, R.E. 1995. A novel measure of genetic distance for highly polymorphic tandem repeat loci. *Molecular Biology and Evolution*, **12**: 914–20.

Skelton, P. 1993. *Evolution: A Biological and Palaeontological Approach*. Wokingham, UK: Addison-Wesley.

Silvertown, J. and Dodd, M. 1996. Comparing plants and connecting traits. *Philosophical Transactions of The Royal Society of London Series B*, **351**: 1233–9.

Slatkin, M. 1985. Rare alleles as indicators of gene flow. *Evolution*, **39**: 53–65.

Slatkin, M. 1994. An exact test of neutrality based on the Ewens sampling distribution. *Genetic Research*, **64**: 71–4.

Slatkin, M. 1995. A measure of population sub-division based on microsatellite allele frequency. *Genetics*, **139**: 457–62.

Slatkin, M. 1996. A correction to the exact test based on the Ewens sampling distribution. *Genetic Research*, **68**: 259–60.

Sokal, R.A. and Rohlf, T.J. 1994. *Biometry*, 3rd edn. London: W.H. Freeman.

Soltis, P.S. and Soltis, D.E. 1987. Population structure and estimates of gene flow in the momosporous fern *Polystichum munitum*. *Evoution*, **41**: 620–9.

Soltis, P.S. and Soltis, D.E. 1993. Molecular data and the dynamic nature of polyploidy. *Critical Reviews in Plant Sciences*, **12**: 243–73.

Soltis, P.S. and Soltis, D.E. 1999. Polyploidy: origins of species and genome evolution. *Trends in Ecology and Evolution*, **14**: 348–52.

Soule, M.E. 1980. Thresholds for survival: maintaining fitness and evolutionary potential. In M.E. Soule and B.A. Wilcox, eds. *Conservation Biology: An Ecological–Evolutionary Perspective*. Sunderland, MA: Sinauer, pp. 151–69.

Spencer, C.C., Niegel, J.E., and Leberg, P.L. 2000. Experimental evaluation of the usefulness of microsatellite DNA for detecting demographic bottlenecks. *Molecular Ecology*, **9**: 1517–28.

Staub, J.E., Fredrick, L., and Marty, T.L. 1987. Electrophoretic variation in cross-compatible wild diploid species of *Cucumis*. *Canadian Journal of Botany*, **65**: 792–8.

Stebbins, G.L. 1980. Rarity of plant species: a synthetic viewpoint. *Rhodora*, **82**: 77–86.

Stoneburner, A., Wyatt, R., and Odrzykoski, I.J. 1991. Application of enzyme electrophoresis to bryophyte systematics and population biology. *Advances in Bryology*, **4**: 1–27.

Takahata, N. and Palumbi, S.R. 1985. Extranuclear differentiation and gene flow in the infinite island model. *Genetics*, **109**: 441–57.

Takezaki, N. and Nei, M. 1996. Genetic distances and reconstruction of phylogenetic trees from microsatllite DNA. *Genetics*, **144**: 389–99.

Wahlund, S. 1928. Zusammersetung von Populationen und Korrelation-sercheinungen von Standpunkt der Verebungslehre aus betrachtet. *Hereditas*, **11**: 65–106.

Ward, R.H., Skibinski, D.O.F., and Woodwark, M. 1992. Protein heterozygosity, protein structure and taxonomic differentiation. *Evolutionary Biology*, **26**: 73–159.

Watt, W.B. 1977. Adaption at specific loci. 1. Natural selection on phosphoglucose isomerase of *Colias* butterflies: biochemical and population aspects. *Genetics*, **87**: 177–94.

Watt, W.B. 1983. Adaptation at specific loci. 2. Demographic and biochemical elements in the maintenance of the *Colias* PGI polymorphism. *Genetics*, **103**: 691–724.

Watterson, G. 1978. The homozygosity test of neutrality. *Genetics*, **88**: 405–17.

Webb, D. 1985. What are the criteria for presuming native status? *Watsonia*, **15**: 231–6.

Weinberg, W. 1908. Über den Nachweis der Verebung beim Menschen. *Jahreshefte d. Ver. f. vaterlandische Naturkunde in Wurttemberg*, **64**: 368–82.

Weir, B.S. 1990. *Genetic Data Analysis: Methods for Discrete Population Genetic Data*. Sunderland, MA: Sinauer.

Weir, B.S. 1996. *Genetic Data Analysis*, 2nd edition. Sunderland, MA: Sinauer.

Weir, B.S. and Cockerham, C.C. 1984. Estimating *F*-statistics for the analysis of population structure. *Evolution*, **38**: 1358–70.

Whitehouse, A.M. and Harley, E.H. 2001. Post-bottleneck genetic diversity of elephant populations in South Africa, revealed using microsatellite analysis. *Molecular Ecology*, **10**: 2139–49.

Whitlock, M.C. and McCauley, D.E. 1999. Indirect measures of gene flow and migration: FST ≠ 1/(4*Nm* + 1). *Heredity*, **82**: 117–25.

Wiig, Ø. 1995. Distribution of polar bears (*Ursus maritimus*) in the Svalbard area. *Journal of Zoology*, **237**: 515–29.

Wolfe, K.H., Li, W.-H., and Sharp, P.M. 1987. Rates of nucleotide substitution vary greatly among plant mitochondrial chloroplast and nuclear DNAs. *Proceedings of the National Academy of Sciences USA*, **84**: 9054–8.

Workman, P.L. and Niswander, J.D. 1970. Population differentiation in the Papago. *American Journal of Human Genetics*, **22**: 24–9.

Wright, S. 1931. Evolution in Mendelian populations. *Genetics*, **16**: 97–159.

Wright, S. 1943. Isolation by distance. *Genetics*, **28**: 114–38.

Wright, S. 1946. Isolation by distance under diverse systems of mating. *Genetics*, **31**: 39–59.

Wright, S. 1951. The genetical structure of populations. *Annals of Eugenics*, **15**: 323–54.

Zhang, D.-X. and Hewitt, G. 2003. Nuclear DNA analyses in genetic studies of populations: practice, problems and prospects. *Molecular Ecology*, **12**: 563–84.

Zhivotovsky, L.A. 1999. Estimating population structure in diploids with multilocus dominant DNA markers. *Molecular Ecology*, **8**: 907–13.

Zink, R.M. 1986. Patterns and evolutionary significance of geographic variation in the Schistacea group of the Fox Sparrow (*Passerella iliaca*). *Ornithological Monographs*, **40**.

Zink, R.M. 1994. The geography of mitochondrial DNA variation, population structure, hybridisation and species limits in the Fox Sparrow (*Passerella iliaca*). *Evolution*, **48**: 96–111.

4

Gene flow and mating system

"Be fruitful, and multiply, and replenish the earth"

Genesis 1: 28

Summary

1 Gene flow is defined as the proportion of newly immigrant genes moving into a population. The definition implicitly separates gene dispersal from successful establishment, but the two processes constitute gene flow.

2 Biological characteristics of an organism, particularly vagility of individuals and their propagules, and reproductive system (asexual or sexual, selfing or outcrossing) are important determinants of the magnitude of gene flow both between and within populations. The physical and heterogeneous environment in which organisms live also determines gene flow parameters, including physical barriers, predominant wind or water direction, and habitat.

3 Indirect estimates of gene flow are derived from genetic structure estimates for adult populations. Whilst poor at measuring changes in contemporary gene flow, they hold a wealth of historical and evolutionary information accessible using population and phylogenetic statistics.

4 Direct estimates of gene flow are derived from genetic analysis of adult and progeny cohorts in a population. Direct measures trace particular gene dispersal events, are excellent for studying contemporary changes in dispersal, and are easily integrated into a spatial context.

5 Advances in understanding the spatial context and evolutionary history of gene flow are expected, and should further enlighten studies of population genetics structure. Measuring gene flow of adaptive markers relative to neutral gene models will be an important future step.

4.1 Introduction

Several definitions of gene flow have been offered but perhaps the most widely adopted is that of Endler (1977), who described gene flow as the proportion of newly immigrant genes moving into a given population, as opposed to gene movement which refers to intrapopulation phenomena (Devlin, Roeder, and Ellstrand 1988, Schnabel and Hamrick 1995).

Gene flow is an essential aspect of the proliferation and spread of species, and migration of genes may be mediated by adult or juvenile individuals or their reproductive propagules in the form of sexual/asexual gametes, zygotes, and/or clones. It is only through gene flow that populations of a species can maintain genetic connectivity. Without gene flow populations will diverge and differentiate over time, the ultimate consequence of which may well be speciation.

The nature and extent of gene flow is largely dependent on two intrinsic biological characteristics of the organism: mode of reproduction (asexual versus sexual; hermaphroditism versus biparental outcrossing) and the mobility of individuals (vagility) and their propagules (gametic and zygotic dispersal mechanisms and behavior). However, in natural populations, there are other biological characteristics that dramatically influence partitioning of genetic variation, including:

• Timing of reproduction can result in temporal isolation of populations and affect potential gene flow patterns.
• Individual density can also affect the size of breeding neighborhoods (Franceschinelli and Bawa 2000).
• Behavior of animals or plant pollinators and seed dispersers.
• Progeny survival and establishment will be dependent on their adaptation to microsite conditions (Nathan and Muller-Landau 2000).

In addition to these intrinsic characteristics, it is becoming increasing clear that the magnitude and rate of gene flow are strongly affected by the environment an organism inhabits. Topographic and hydrological features of an organism's environment strongly influence the scale and direction of genetic connectivity between individuals and populations of a species. Environmental selection pressures can strongly influence the distribution of taxa, and gene flow across steep adaptive clines may be limiting (Loveless and Hamrick 1984). However, only once we understand how organisms disperse their genes and the ecological requirements for propagule establishment, will we be able to predict the likely effects of contemporary environmental change on genetic diversity.

A suite of methods and models has been developed to study gene flow, and can be classified into two approaches. Indirect methods use the distribution of genetic variation within contemporary adult populations to infer the magnitude of gene flow between them. Using indirect methods, only successful gene flow events are considered, that is, those that have led to the establishment of a successful migrant/propagule. However, the problem with using an indirect approach is that genetic structure is also influenced by other factors. For example, current patterns of gene flow that have only recently been contributing to the genetic structure of a population may be masked by the influence of historical gene flow or colonization processes that have little relevance within a contemporary landscape.

Direct methods use genetic variation in progeny arrays to identify parental contribution or variability and to calculate gene movement parameters of gametes or propagules. Direct measures offer a much more accurate way of assessing gene dispersal. However, an unknown proportion of propagules/gametes will be successful and therefore it is unclear whether this parameter is a good approximation of successful gene flow. Gene movement studies can be easily integrated into survival and ecological studies to provide information on successful establishment/migration. Indeed the study of contemporary patterns of gene flow has been critically underinvestigated, yet can reveal important observations of fundamental importance to ecological and conservation genetic studies (Adams 1992, Ellstrand and Elam 1993).

In this chapter the term gene flow will be used to refer to the successful movement of genes (via permanent migrants, fertilized gametes, and established propagules) between and within populations, as this process can affect genetic composition at large and small scales, and the definition of scale of a population is in itself contentious. The movement of genes as measured directly from gametes or propagules will be referred to as gene dispersal, and is only translated as gene flow following successful establishment and proliferation. Using these terms, it is the purpose of this chapter: to review those intrinsic (biological, particularly reproductive system) and extrinsic (environmental) factors which can determine the magnitude and direction of gene flow; to review some of the complications and offer some guidelines for undertaking gene flow studies, including choice of marker and sampling strategy; and to present summaries of the predominant statistical methods to analyze gene flow and dispersal using both indirect and direct methodologies.

4.2 Factors governing gene flow

Both intrinsic (i.e., biological; reproductive system, summarized in Table 4.1, vagility and dispersal behaviors) and extrinsic (i.e., environmental; physical barriers and selection gradients) factors are expected to influence gene flow parameters, and are expanded below.

Table 4.1

Details of sexual system, zygote composition, and taxonomic distribution of the most common asexual and sexual reproductive strategies of plants and animals.

	Asexual Clonal reproduction	Sexual Autogamy (selfing)	Mixed mating	Outcrossing
Sexual system	Vegetative or apomixis	Hermaphrodite or monecy in plants (reproductive organs on same individual)	Hermaphrodite or monecy in plants (reproductive organs on same individual)	Dioecy (reproductive organs on different individuals)
Composition of zygote	Exact genetic copy of parent	Combination of alleles derived from single parent	Mixture of progeny some combining alleles from a single parent some from two parents	Combination of alleles derived from two parents
Taxonomic distribution	Lower animals and plants	Lower animals and plants	Lower animals and plants	Higher animals and plants

4.2.1 Gene transfer and reproductive system

Individuals and asexual reproduction

The movement of individuals between populations may constitute gene flow (migrants). Migrations may, however, have a range of outcomes from no influence if the individual moves back to its source population, to a long-lasting influence if a group of individuals with high longevity permanently relocate to a new population (immigration). Under these migration scenarios, the period of influence is dependent solely on the life span of the individual(s) involved. No lasting consequences will result after the death of the migrant(s), unless additional gene transfer via asexual or sexual reproduction/propagation has occurred.

Most plant species and some lower animals (e.g., corals) do not move of their own accord once established, but vegetative propagation (runners, stolons, rhizomes, bulbs, root or stem suckers, tissue fragments, fission, and derived polyps) and asexual modes of egg/seed production (apomixis, somatic fragmentation, parthenogenesis, gynogenesis, and fissiparity) may allow an exact genetic copy (clone) of the individual to be dispersed far and wide. Asexual methods of propagation may result in many free-living and separate individuals (ramets) that, as long as they have the exact genetic composition of the founder(s), constitute a single genetic individual (genet).

The size of such genetic individuals may be significant. Recent studies using genetic fingerprinting methods (RAPD and iSSRs) have shown that a single clone of the Japanese knotweed (*Fallopia japonica*), which was introduced to Britain in the mid-nineteenth century, has effectively dispersed and established across most river systems in the British Isles. Due to its widespread ramets, it has earned the title of the largest vascular plant in the world (Hollingsworth and Bailey 2000).

Some clones can be very long-lived, although their evolutionary viability is theoretically limited, due to the lack of recombination that would allow adaptation to new environments and the purging of deleterious recessive mutations (their genetic load), which may accumulate. One complicating factor in identifying clones is that some long-lived individuals may accumulate additional mutations, resulting in individuals that have been derived asexually from the same progenitor but are genetically differentiated.

Autogamy

Autogamy or self-fertilization (selfing) is a reproductive system that produces a zygote following the fusion of two gametes derived from the same individual. Self-fertilization is common in many plant and lower animal groups and species that possess male and female reproductive organs on the same individual are described as hermaphrodites. In plants a distinction is made between species which have both male and female reproductive organs within the same flower (herma-phrodite), or different flowers (monecious). Whilst selfing allows the fixation of preferential gene combinations, a lack of recombination can lead to an accumulation of deleterious mutations, which are expressed as a reduction in an individual's fitness, an effect known as inbreeding depression. The only way to purge such genetic loads from selfing populations is through selection.

Whilst having some inherent disadvantages, under certain circumstances selfing may be a preferential form of reproduction. For example, selfing may be the only way to achieve reproduct-ive assurance at the edge of a species' range when only a few or single individuals are present (a principle known as Baker's law, Baker 1955, 1967, Pannell and Barrett 1998). Selfing is also a mechanism that limits the influx of genes from another portion of a species' range, which may disrupt adaptive gene complexes adapted to new local environments (a problem known as out-breeding depression).

Outcrossing and dioecy

Dioecy is a reproductive system in which an individual has only a single sex, and ensures that zygotes are the product of gametes originating from two different individuals (referred to as out-crossing). Outcrossing is the normal mode of reproduction for most higher animals. Outcrossing can result from other reproductive systems in plants and lower animals (e.g., hermaphroditism and monoecy) if there are mechanisms to limit the fusion of gametes derived from the same individual (e.g., self-incompatibility). In fact there are a large number of mechanisms that have evolved to prevent self-fertilization and promote outcrossing but these are too complex to cover in detail here and should be investigated in detail for the particular study organisms selected (e.g., for plants see Richards 1996).

One consequence of having truly random outcrossing is that alleles should be distributed amongst individuals of a population according to Hardy–Weinberg equilibrium principles (see chapter 3). For diploid organisms, the proportion of homozygous and heterozygous genotypes in a population is derived according to the relative frequency of alleles in the population, calculated according to the equation, $p^2 + 2pq + q^2$, where p and q are allele frequencies. Deviation of the population from these principles can thus be used to test whether a population is randomly mating or not. If not then further investigation of potential inbreeding, assortative mating, or reproductive cliques/clusters may be required (see chapter 3 for derivation of Hardy–Weinberg equilibrium, inbreeding coefficient, test of assortative mating, and test of spatial autocorrelation).

Ploidy considerations for reproduction and family groups

Under normal conditions, gametes produced by animals and plants contain half the chromosome complement of the parent (usually referred to as haploid, even for polyploid taxa). Such ploidy conditions are rarely important, and of little consequence, as these gametes are short-lived and easily identifiable as haploid reproductive cells.

However, within some plants (e.g., coniferous gymnosperms) the haploid female gamete is supported by a megagametophyte, a tissue which is haploid and incorporated into the seed structure after the zygote has been fertilized. The megagametophyte can be excised and since it represents a haploid derivative of the maternal genotype, genetic fingerprinting methods can be used to identify the originating seed mother. In a similar way, the pericarp tissue of many plant seeds is derived from the diploid maternal tissue complement and therefore carries a somatic copy of the diploid maternal complement. For ferns, the situation is more complex as the gametophyte (haploid structure that produces the gametes) is a free-living entity and separate from the sporophyte, which produces the fertilized spores. Such tissues and life stages should be clearly distinguished from zygotic/sporophytic or gametic material, if genetic analyses are to be applied.

For animals there are also some interesting and complex ploidal interactions within family reproductive systems. Eusocial insects within the Hymenoptera produce males from unfertilized eggs and therefore contain a haploid combination of the queen's chromosome complement. Worker females on the other hand are diploid, produced when a haploid male fertilizes a haploid egg. Under conditions where the queen is mated by a single male, and only one queen lays eggs within a nest, this ploidal condition means that worker females share 75% of their genes with each other but are only related to their mother by 50%. It is this close genetic relationship that has been used to explain the ultimate altruistic behavior exhibited within eusocial colonies, where all females, except the queen, forgo the opportunity to reproduce and instead tend to the rearing of their closely related siblings (Avise 1994). Whilst theoretically this situation appears to offer a suitable explanation for the eusocial behavior of these insects, in reality queens can be multiply inseminated (polyandry) or several queens can lay eggs within a single nest (polygyny). There is evidence of evolutionary mechanisms that may allow the assessment and recognition of the relatedness of individuals and allow the existence of closely related cliques within diverse nests.

4.2.2 Vagility and dispersal

In addition to the reproductive system, which determines the composition of transmitted genetic material, the vagility of individuals and dispersal of propagules (including gametes and zygotes) and behavior of progeny are also very important factors in considering the extent of gene flow between populations of a species.

Some animals may range over extremely large distances and therefore have reproductive neighborhoods spread over very large areas. Marine animals (e.g., whales) and birds in particular may migrate thousands of kilometers in a season, potentially maintaining genetic connectivity of extremely dispersed populations (Avise 1994). Whereas animals with low vagility and most plants may not move very far at all and therefore it is their gametes and/or zygotes, which must be dispersed to allow sexual reproduction, genetic connectivity, and colonization to occur. Male gametes can be dispersed over hundreds of kilometers by wind (e.g., tree pollen) and marine currents (e.g., sperm of marine animals), or have very limited dispersal distance (e.g., many animals need direct contact between individuals to pass sperms from a male to female partner, and some plants are pollinated by insects which have very limited flight distances, e.g., colony or forest bees; Dick 2002). Female gametes are usually confined within the female individual/reproductive structure, and therefore only dispersed as a fertilized zygote.

Zygote dispersal can also be asynchronous. In plants, seed dispersal varies from light, wind-dispersed to heavy, gravity-dispersed propagules. For animals, pelagic larvae may float freely in the oceanic current for hundreds of kilometers before establishment or offspring may live within tightly knit family units and be in lifetime contact with their parents (as is the case for many primate species).

For species with low vagility the effect of restricted dispersal will be most evident at the fine spatial scale of genetic structuring. Advances in the analysis of fine-scale structure have resulted from the availability of highly polymorphic markers and concurrent development of statistical analyses for dealing with specific requirements and assumptions of the data and have helped to infer gene dispersal dynamics at field sites (Sork et al. 1999). Recent studies have involved direct, spatially and temporally explicit methods of analysis (Sork et al. 1999). For example, in a study of spatial genetic structure, much stronger spatial structure was evident in the subpopulation of pedunculate than sessile oak (*Quercus robur* and *Quercus petraea* respectively; Streiff et al. 1998). The authors speculate that the difference is probably due to seed dispersal differences of the species, as *Q. petraea* has acorns borne on short seed stems (peduncles), whilst *Q. robur* has long peduncles, and so acorns are more likely to be conspicuous to and removed by birds and other potential dispersers (Bossema 1979, Streiff et al. 1998). *Q. robur* acorns are also able to develop within a broader range of microsites than *Q. petraea* and are interspersed with other species, which probably contributes to the low spatial genetic structuring of pedunculate oak (Cottrell et al. 2003, see chapter 7, oaks case study).

Differences in gene dispersal (both gametic and zygotic) can have such a profound impact on the magnitude of gene flow and the structuring of genetic diversity within populations (see chapter 3), that it is important to know the relative vagility and dispersal mechanisms of a study organism if one is to interpret gene flow parameters accurately.

Gender differences

Gender bias in animal vagility may be strongly determined by size or strength, resulting in different neighborhood sizes for the two sexes, which will affect the distribution of genes on the sex chromosomes (e.g., Y chromosome) or uniparentally inherited genomes (e.g., mtDNA).

For example, a genetic analysis of island populations of the Galapagos Marine iguana, *Amblyrhynchus cristatus*, found that the mtDNA gene pools are significantly differentiated and related to the sequence in which the islands were historically colonized by the iguanas (Rassmann et al. 1997). Conversely, analysis of nuclear DNA indicated substantial interpopulation gene flow, and the geographical distribution of markers fit an isolation-by-distance model. The contrasting nuclear and mitochondrial DNA patterns are explained by asymmetric migration behavior of the two sexes of iguana, with higher (active and passive) interisland dispersal for males than females. The smaller size of males and their consequent weaker swimming capacity makes them more likely to be swept between islands by channel currents, which is not the case for the larger, strong-swimming females (Rassmann et al. 1997).

The relative dispersal potentials of gametes and progeny can also have strong influences on the genetic connectivity of uniparentally inherited genomes relative to biparentally inherited genomes (nuclear DNA), and is an important consideration when choosing markers to examine gene flow parameters. For example, in plants differences in dispersal between the male gamete (pollen) and maternally produced propagule (seed) may establish different spatial patterns of variation across different genomes. In addition, gene flow via male gametes may antagonize the spatially limited gene flow of progeny in the spatial structuring of nuclear genes (Berg and Hamrick 1995, Cottrell et al. 2003, Streiff et al. 1998).

Breeding behavior

High vagility of individuals may offer high potential gene flow across the whole range of a species, however, if individuals return to their birth site to reproduce (philopatry) then gene flow will be severely reduced and a strong differentiation between populations based on natal site is expected.

For a range of highly mobile animals (e.g., whales, Baker et al. 1990; salmon, Bermingham et al. 1991; and birds, Avise and Nelson 1989), significant genetic differentiation between populations or species ranges has been demonstrated due to philopatric effects. Perhaps the most significant effect discovered to date is that within the green turtles (*Chelonia mydas*). Using mtDNA markers, Bowen et al. (1992) clearly differentiated turtles sampled from rookeries in the Indian and Pacific Oceans from those from the Atlantic Ocean and Mediterranean Basin. In addition, significant differentiation was detected between sites within these groupings. In contrast to the picture of gene flow painted using mtDNA markers (which are maternally inherited in the organisms), markers from the nuclear genome indicated much lower differentiation and therefore indicated significant inter-rookery gene flow mediated by male turtles (Karl, Bowen, and Avise 1992).

Mating between closely related individuals (assortative mating) may also increase spatial structure for species with low vagility, particularly if closely related individuals have synchronous reproductive timings (Bacilieri, Labbe, and Kremer 1994). However, studies of temperate forest trees have found little correlation between phenological groups and genetic similarity (Bacilieri-Labbe, and Kremer 1994, Gram and Sork 2001). Such observations indicate that spatial clumping of phenological groups may be due to common environmental conditions or that phenological differences are not great enough to isolate groups reproductively (Gram and Sork 2001).

4.2.3 Historical processes

Many early studies of gene flow focused on summary statistics inferred from population genetic structure between populations. These estimates reflect gene flow that is averaged over time and space. Conclusions from these estimates therefore reflect historical effects as well as contemporary events. Thus, indirect estimates of gene flow may not accurately reflect contemporary processes, as the population genetic structure may carry a predominant signal from a drastic but historical event (e.g., major bottleneck or colonization).

One way to establish contemporary gene flow parameters within the present landscape is to use a direct estimate of gene flow. For example, one may wish to examine gene flow between populations that have recently had a drastic habitat fragmentation. Such studies are best conducted if field samples can be replicated and controlled either by sampling several different sites with the same treatment or temporally through time as the population landscape changes (ideally both).

It is also possible to use a phylogenetic approach to population genetics to infer historical events within populations. Such studies apply coalescent theory to estimate the historical size, connectivity, and relative colonization history of populations (see chapter 5 for an expansion of these theories and methods).

4.2.4 Extrinsic factors

In addition to innate biological differences in breeding system and dispersal ability, there are extrinsic factors that may lead to gene flow differences across the range of a single species. The physical environment within which an organism lives and grows will naturally limit its vagility and potential for dispersing propagules.

Physical environment

Physical barriers (e.g., mountains, large bodies of water, predominant oceanic currents) will have a tremendous effect on the genetic connectivity of individuals occurring on either side of such barriers. Thus, gene flow between populations of a species is a complex interaction between the innate vagility/dispersal ability of a species and its physical environment. Physical environmental factors need to be considered when interpreting physical distance estimates of gene flow events (e.g., see Box 4.1).

Environmental selection factors

In addition to large physical barriers, the local environmental conditions can be a determinant to the survival of a particular species, and therefore limit its distribution. Habitat change may be as strong as the physical barriers mentioned above, and in certain circumstances is part of the same effect (the high altitudinal effect of mountains causes significant environmental changes, in addition to increasing the distance between populations by adding a vertical dimension).

Physical and environmental barriers to gene flow in Spanish cedar

Spanish cedar, *Cedrela odorata* (Meliaceae), is a widespread neotropical tree. It is monoecious, predominantly outcrossing, pollinated by insects, and has wind-dispersed seed. It is highly valued for its timber (being as it is a close relative of mahogany), which has resulted in significant overexploitation for more than two centuries. Due to the effects of forest fragmentation and exploitation, Spanish cedar is now a priority for conservation in the Costa Rican Biodiversity Action Plan and it is rare to find individuals within forests, but is more often seen as remnant trees on farmland or regenerating on disturbed land. The species shows significant adaptability, manifest primarily as tolerance for both dry and moist habitat.

To help develop a management strategy for the available genetic resources of Spanish cedar, Cavers, Navarro, and Lowe (2003a) used AFLP variation to define management units within Costa Rica. Up to 20 individuals from each of 10 Costa Rican populations (Fig. 4a(a)) were screened for 145 AFLP fragments and the total species diversity, as expected for an outcrossing, long-lived, woody species, was high ($H_T = 0.27$). However, this concealed a very deep and highly significant divergence ($\Phi_{ST} = 0.83$, Fig. 4a(b)), also noted by previous analyses of RAPD (Gillies et al. 1996) and cpDNA variation (Fig. 4a(a), Cavers, Navarro, and Lowe 2003b), which splits the species into two groups with different habitat preferences, and two ecotypes, wet and dry, can be distinguished. The two ecotypes, besides being ecologically different, are morphologically differentiated and appear to be phenological isolated, the latter probably being environmentally influenced. The available data suggest that no contemporary gene flow exists between these two ecotypes even when they occur in sympatry, and a phylogeographic analysis of the Meso-American range of the species suggests they have divergent origins (Cavers, Navarro, and Lowe 2003b).

Even between populations of a single ecotype there appear to be environmental factors restricting gene flow. Within the wet ecotype group, Mantel's test demonstrated that pairwise genetic

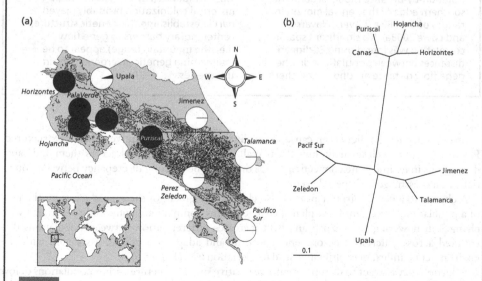

(a) Relief map of Costa Rica with locations of Spanish cedar populations and the distribution of one of two cpDNA haplotypes identified as black or white. (b) Neighbor-joining tree constructed using pairwise Φ_{ST} estimates.

(c)

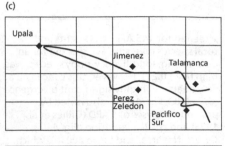

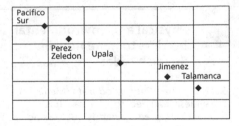

Fig. 4a

(c) Distribution of populations used for distance matrices in Mantel tests. Top: Euclidean geographic distribution, line shows approximate position of mountain ranges (Mantel's correlation, $r = 0.15$, $p = 0.39$). Middle: distribution after "north" linearization through a point in the center of Upala population (Mantel's correlation, $r = 0.35$, $p = 0.17$). Bottom: distribution after "south" linearization through 8.5°N, 82.0°W, representing the southern end of the mountains (Mantel's correlation, $r = 0.63$, $p = 0.04$).

distance between populations fitted an isolation-by-distance model around the southern edge of the central mountain ranges of Cost Rica (Cavers, Navarro, and Lowe 2003a). No significant spatial correlation was found using Euclidean distances between populations or when a gene flow distance around the northern limit of the mountain ranges of Costa Rica was considered (Fig. 4a(c)). Although historical colonization probably played its part in establishing this genetic structure, contemporary barriers to gene flow (i.e., the mountain range) appear to be maintaining genetic divergence within this species.

Strong selection gradients can cause sharp clines across which adaptive genes are structured. However, adaptive clines may limit the overall level of gene dispersal across them and cause disruptions in neutral gene connectivity, thus there may be a strong discrepancy between potential and successful gene dispersal events.

A change to a new environment may also change an element of the reproductive mechanism of a population. For example, in plants, changes in environment are commonly associated with changes in flowering time. Such an effect increases differentiation between populations distributed across different environments and drift and adaptation effects, associated with the environmental influx, may ultimately lead to speciation (see chapter 6).

Selection may also act to disrupt or enforce positive spatial structure within populations of low vagility species (Epperson and Allard 1989). Much of the spatial genetic correlation caused by familial clumping of juveniles may subsequently be removed by intensive competition, leaving only a few adults to survive to reproductive maturity (Dow and Ashley 1996). Jensen et al. (2003) found that spatial autocorrelation significantly decreased in the adult compared to juvenile cohorts of oaks at Hald Wood in Denmark. They ascribed the high spatial autocorrelation within the juvenile cohort to limited seed dispersal within the *Q. petraea* dominated wood and which is disrupted in the older generation. However, the mechanism controlling this phenomenon is less clear. Selective elimination of inbreds has been suggested (Streiff et al. 1998), or a simple increase in competition between the more densely clustered sibs may operate. However, direct evidence is scant. Reports that inbred seedlings have poorer germination than outbred seedlings suggest a selective role at this life stage (Jensen et al. 2003). Sampling for parentage of both the seed and seedlings of early and late stages would allow detection of any inbreeding depression across these stages, and has been recommended for such cases by Schnabel (1998).

4.3 Considerations for measuring gene flow

4.3.1 Observational versus genetic methods

Observational methods have traditionally been used to quantify gene flow between populations and have taken the form of directly marking individuals, gametes, and/or progagules and either recapturing a sample to estimate vagility or to follow dispersal directly from a *source*. For example, in plants, pollen can be tracked from source to recipient using fluorescent dyes (Campbell 1985, Campbell and Waser 1989), pollinator behavior and paths can be followed and mapped (e.g., Roubik 2002), and metal-tagged acorns have been used to estimate oak seed dispersal (Sork 1984).

Process-based methods provide good direct estimates of the vagility and propagule dispersal abilities of a species and can provide some useful insights into behavioral or asynchronous aspects of gene flow mediated by organisms or through a heterogeneous environment. However, there are at least four reasons why taking a pattern-based approach, using molecular markers, provides a superior method for assessing vagility, dispersal, and ultimately gene flow:

• It is more efficient to use molecular markers than direct observation. There may be a very high mortality rate for propagules, which would mean that the total number of events that would need to be examined is logistically high. For example, in an observation-based study, Sork (1984) initially used 3000 metal-tagged acorns to document effective dispersal of seed. As well as a low recovery rate of tracked acorns, the 0.9% survival rate meant that only a small sample size remained from which information about "successful" seed could be made. Unlike studying the fate of seed itself, questions relating to parentage of successful seed (and seedlings) require a more efficient pattern-based approach.

• Observation-based approaches often miss rare events or events which occur over large distances, despite such events being clearly important, or even essential, for understanding a species' overall gene flow or colonization dynamics (Cain et al. 2000). In particular, concentrating on pollen, in terms of the source plant, is likely to underestimate pollen flow since very long distance pollen dispersal events are likely to be overlooked (Campbell 1991, Snow and Lewis 1993, Streiff et al. 1999).

• Tracing methods that reflect gamete movement rarely take into account post-dispersal factors, such as viability, sexual compatibility, gamete competition, or selective embryo abortion. Therefore observational methods are only measuring dispersal not successful gene flow (Cain, Milligan, and Strand 2000, Snow and Lewis 1993). A pattern-based approach considers only the successful fertilization events and identifies the source individual. Gene flow is therefore traced *a posteriori*. However, further monitoring of successfully established progeny or their genetic contribution to recipient populations may have to be followed to show a significant gene flow event.

• Process-based approaches tend to concentrate on a few individuals taken from a source point and hence, do not provide a population estimate of dispersal. Using genetic structure to infer gene flow on the other hand directly provides a population average.

4.3.2 Indirect versus direct methods

Gene flow estimates can be made using genetic markers in two fundamentally different ways. Indirect methods derive gene flow parameters from analyzing the genetic structure of adult populations whereas direct methods analyze gametes or progeny arrays and utilize identified gene dispersal events to calculate gene flow parameters.

Indirect methods

Indirect estimates of gene flow can be obtained using F-statistics under the assumptions of the island model (Wright 1951; see chapter 3). Whilst estimates are essentially a cumulative combination of gene flow contributions across the generations that compose populations, it is also important to note that only successful gene flow events are reflected. However, population structure-based estimates also reflect evolutionary phenomena (Slatkin and Barton 1989), and therefore, it is questionable whether they should be used at all to estimate current patterns of gene flow as this violates the island model assumptions both on temporal and spatial scales (Whitlock and McCauley 1999).

Indirect estimates of gene flow have been useful for assessing the relative contribution of historical gametic and zygotic dispersal (Hu and Ennos 1997). For example, Ennos (1994) used existing data, describing the differences in partitioning of variance between the maternally inherited chloroplast genome and the biparentally inherited nuclear genome, to estimate the asymmetry of gene flow due to seed and pollen in European oaks (a ratio of 1:196, respectively).

An alternative to using population genetic differentiation to make indirect estimates of gene flow is to use genealogy reconstruction methods based on the inference of a relationship between parents and offspring. The rare marker approach tracks a gene dispersal event from an individual parent which exhibits a particular rare marker allele, whilst the model approach determines the probable genetic structure within progeny and estimates the most likely generating dispersal parameters (Adams 1992, Roeder, Devlin, and Lindsay 1989). Genealogical methods can also use a coalescent approach (chapter 5) to calculate the level of sharing of common ancestors between populations, and therefore act as a measure of historical gene flow (Neigel 1997).

Despite the limitations of indirect gene flow methods, they have their place and can be extremely effective as an estimate of the historical genetic connectivity between populations, or for assessing the lifetime contribution of gene flow to the genetic structure across a range of populations. However, the following more serious problems should be considered before the techniques are applied:

• Data can only be used to give estimates of the relative contribution of gene flow between different gametes or between zygotes and gametes. There are often cases where absolute distances and frequencies of gene flow are required (e.g., for ecological modeling of the impact of gene escape, or for devising genetic resource management strategies that consider genetic erosion by gene flow).

• Many indirect estimations of gene flow require several assumptions regarding genetic structure in the population which are often not met in reality (e.g., genetic equilibrium between gene flow and drift, and a randomly mating metapopulation of equal population sizes that are linked by constant gene flow; Ennos 1994, Sork et al. 1999), and illustrate how indirect gene flow estimates are so easily influenced by historical factors that perturb the equilibria (Cottrell et al. 2003, Gram and Sork 2001, Sork et al. 1999). Although the fact that indirect methods obtain a cumulative result of genetic partitioning has been used as justification for using such data to assess changes in historical land use (Sork et al. 1999).

• The analysis of indirect gene flow parameters is actually a complex parameter that also includes selection. Therefore it is impossible to separate the difference in genomic diversity that is attributable to actual dispersal or to factors such as post-dispersal selection.

Direct methods

For assessing contemporary gene flow dynamics it is recommended that a direct method be used (Cain, Milligan, and Strand 2000, Schnabel 1998). Genetic identification of the parents of gametes and/or progeny provides a direct estimate of realized short-term gene dispersal at the individual or population level (Schnabel 1998), and requires few assumptions. However, direct methods also have limitations:

• Direct estimates are not influenced by historical events, as gene dispersal is measured for a given period of time only. However, due to the limited temporal dimension of the dispersal estimate, it is vulnerable to variability between recording periods.

• The effort required to sample and analyze the genotypes of many parents and gametes/offspring is significant and if only small numbers of progeny arrays are considered then there may be considerable interindividual variation effects. Ideally to mitigate against the variability factors of points 1 and 2, an estimate of direct gamete/progeny dispersal would be obtained by repeated samplings over time for many individuals, but resource limitation clearly restricts the scope of such replication.

• Direct estimates of gene flow may only actually be estimating dispersal if the survival or success of a gamete/progeny is not assessed. Whilst not suffering to the same extent as observational methods (section 4.3.1), the magnitude of the problem depends on the stage at which parental contribution of propagules was assessed. If measured directly on pollen or sperms then there is no estimation of success and this is purely a dispersal measure. However, if juvenile individuals are followed and their reproductive contribution to the next generation assessed then this can be a good indication of successful gene flow. Selection pressures can of course be assessed for each stage, if time and resources permit. However, a combination of indirect and direct approaches may also provide similar resolution of dispersal, stochastic, and selection impacts on population genetic structure.

The recent proliferation of highly variable markers for a range of species (which allows calculation of the high exclusion probabilities), alongside statistical advances, has seen a sharp increase in the number of studies of contemporary gene flow (Carvalho 1998). A summary of the pros and cons of direct and indirect approaches for assessing gene flow is provided in Table 4.2.

4.3.3 Suitable genetic markers for gene flow analysis

Depending on the approach taken a variety of molecular markers can, and perhaps should, be applied to study gene flow (see Table 4.3).

Isozymes

Isozyme markers have been the mainstay of indirect estimates of gene flow. Indeed, many models based on population differentiation were developed with this marker in mind (Kimura 1953, Latter 1973, Slatkin 1985). Whilst isozymes do not highlight particularly high levels of variation, which is unordered, their codominant nature, known genomic origin (usually nuclear), and the fact that they can be applied to almost any species (chapter 2), means that they are very suitable for use with indirect measures of gene flow (both differentiation and private allele-based), and if enough polymorphic loci are used (approximately 10), sufficiently high exclusion probabilities can be achieved to calculate direct estimates of gene flow. However, some loci may not be neutral and so should be chosen with caution (see Box 4.2).

Microsatellites

Microsatellite markers are highly polymorphic, codominant loci and are abundant in most species' nuclear genomes. However, this type of marker needs to be specifically developed for the

Table 4.2

Type of information that can be derived using the available direct and indirect estimates of gene flow.

Information type	Individual/clone mapping	Nm	Phylogenetic	Spatial	Mating system	Exclusion analysis
Measure of contemporary gene flow?	Recent (extension of clone due to historical and contemporary processes)	No	No	Recent (analysis of juveniles within population will allow recent-scale gene flow estimate)	Yes (progeny array analysis directly indicates mating system)	Yes (direct estimate of dispersal parameters leading to gene flow at chosen spatial or temporal scales)
Indication of mating system?	Yes (asexually produced ramets will be genetically identical)	Yes (inbreeding coefficient, F_{is}, can be calculated)	No	Possibly (inbreeding coefficient can be calculated for population compartments)	Yes (direct measure)	Yes (easily extracted from propagule/parent identification)
Measure of historical gene flow?	Recent (extension of clone due to historical and contemporary processes)	Yes (gene flow accumulated over life time of populations)	Yes (gene flow accumulated over life time of populations)	Recent (gene flow accumulated over life time of populations)	No	No
Evolutionary timing?	Yes (phylogenetic analysis of long-lived clones with de novo mutations)	No	Yes (coalescent approach can be applied)	No	No	No
Spatial scale?	Yes (map clonal extent)	Yes (pairwise differentiation can be tested or use spatial autocorrelation between populations)	Yes (spatial position of populations can be included in analysis)	Yes (explicit use of spatial criteria to estimate gene flow spatial scale)	Yes (spatial location of individuals allows neighborhood breeding size estimate)	Yes (spatial location of individuals allows calculation of neighborhood parameters)
Estimation of dispersal distances?	No	Yes (spatial autocorrelation can be used to estimate gene flow parameters)	Yes (colonization or dispersal distance/speed can be examined using random walk method)	Yes (explicit use of spatial criteria to estimate gene flow dispersal distances)	Possibly (if spatial position of individuals is recorded then average dispersal distances can be calculated)	Yes (direct estimate of frequency of dispersal distance from propagule/parent identification)

Table 4.3

Suitability of markers for gene flow analysis (indirect and direct).

| Marker | Mutation rate | Ordered? | Inheritance | Indirect methods | | | | Direct methods | | |
				Individual ID	Nm	Phylogenetic	Spatial	Mating system	Exclusion analysis
Isozymes	Slow	No	Biparental	**	****	–	**	***	***
Microsatellites	Fast to v. fast	No (yes SMM)	Biparental	****	****	**	****	****	****
Dominant multilocus markers	Slow to v. fast	No	Biparental	****	**	–	****	***	***
MtDNA animals	V. fast	Yes	Maternal	**	***	****	***	**	*
CpDNA plants	Slow	Yes	Maternal or paternal	*	***	**	**	**	*
MtDNA plants	V. slow	No	Maternal	*	**	–	*	*	
Specific/non-specific nuclear loci	Slow to v. fast	Yes	Biparental	***	***	***	***	***	***

The number of stars rates the use of the method for a particular application, where **** is very useful, * is of limited use, and – is of no use.

Box
4.2

Gene flow and selection in the pine beauty moth

The pine beauty moth, *Panolis flammea* (Lepidoptera: Noctuidae) is widely distributed in Europe and occurs naturally at low density in forest populations of Scots, *Pinus sylvestris*, or Corsican pine, *P. nigra* var. *maritime*. However, since the extensive commercial planting in the 1970s of the North American lodgepole pine, *P. contorta*, the pine beauty moth has made a successful host switch and become a significant pest of lodgepole plantations over the last 40 years. To examine geographic variation of the pest in more detail, Wainhouse and Jukes (1997) undertook a population genetic analysis of samples collected from eight British pine forests (Fig. 4b(a)) using four polymorphic allozyme systems. Overall genetic differentiation between populations was significant ($F_{ST} = 0.109$), and suggests limited dispersal of these moths between sites. Although a calculation of indirect gene flow (*Nm*) indicates that migration or dispersal across Britain is sufficient to counter the effects of drift between sites. Chi-squared tests found that two loci were not in Hardy–Weinberg equilibrium, and for one of these loci (*6-Pgd*), there was a significant correlation between the population frequency of an allele (*fast*) and latitude (Fig. 4b(b)). The authors suggest that this locus may be subject to selection pressures across the range of the species and note that clinal variation has been observed at this locus in several other species, particularly *Drosophila* (Wainhouse and Jukes 1997).

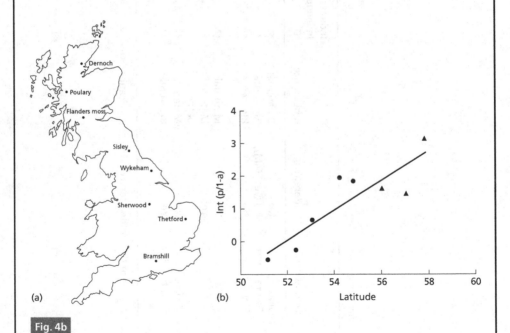

(a)   (b)

Fig. 4b

(a) Location of sample sites of *Panolis flammea* for allozyme study of Wainhouse and Jukes (1997).
(b) Proportion of the most abundant allele (*fast*) of *6-Pgd* (after logit transformation) in relation to latitude of sample origin of *P. flammea* populations.

focal species of the study requiring additional time and resources. The high number of alleles per locus, particularly low frequency alleles, allows calculation of high exclusion probabilities, and offers much potential for identifying private alleles (Estoup and Angers 1998, Schnabel 1998). A stepwise mutation model can also be applied to repeat number variation to infer evolutionary relationships between alleles, although such estimates suffer from homoplasy problems (Schlötterer 1998) and a large number of loci (more than 10) should be applied if such estimates are to be used. These properties make SSRs highly suitable for almost all gene flow estimates, and are often the marker of choice (Parker et al. 1998). However, one problem with this marker type might be high mutation rate, which would lead to inconsistencies between parental and offspring genotypes. The mutation rate of highly polymorphic loci may have to be estimated and progeny arrays carefully screened for evidence of *de novo* mutation.

Dominant markers

The class of marker represented by AFLPs, RAPDs, and iSSRs are dominantly expressed multilocus variation, which represents unordered alleles derived from all cellular genomes. The application of dominant markers to differentiation-based indirect estimates of gene flow is problematic due to the unknown proportion of heterozygotes in populations. However, several estimates of F_{ST} have been proposed for use with dominant data, and their careful application can allow some inference of gene flow parameters using a differentiation basis (Lynch and Milligan's (1994) method and AMOVA). Codominant data are still clearly superior to dominant data for such estimates. Dominant multilocus markers are an excellent source of private alleles, and can be used without problem in direct estimates of gene flow. Comparisons between AFLP and microsatellite markers suggest that although the latter had higher exclusion probabilities, AFLP analysis can be used successfully for parentage analysis (Gerber et al. 2000). The authors note that low-frequency AFLP fragments produce higher exclusion probabilities, but are unlikely to be found consistently in natural populations. Therefore a large number of fragments (more than 200) may have to be screened to derive enough markers for calculation of satisfactory exclusion probabilities.

MtDNA – animals

Variation in the animal mitochondrial genome is high and ordered. It is therefore an excellent marker to apply to all indirect estimates of gene flow, including differentiation and private allele models. However, the main advantage of this marker comes with applying phylogenetic and co-alescent models to gene flow estimation. Being predominantly maternally inherited in animals it can only be used to assess female-mediated gene flow. In addition, high intrapopulation mtDNA variation should also allow this marker to be used to assist maternity analysis, although exclusive identification is not usual unless a unique haplotype is identified.

Organelle markers – plants (cpDNA and mtDNA)

Chloroplast genomes tend to harbor low levels of variation although mutations can be ordered. Indirect estimates of gene flow (differentiation, private allele phylogenetic) are possible using this marker as long as enough variation can be found. Within populations cpDNA diversity tends to be low, but some species harbor population-level polymorphism, which can be used to assist in parental identification. Microsatellite loci in the chloroplast genome appear to be more variable than indel mutations (Provan et al. 1999), and have been used to make direct estimates of gene flow (Ziegenhagen et al. 1998). Chloroplast genomes are predominantly inherited maternally in angiosperms and paternally in conifers and can thus only be used for uniparental estimates of gene flow.

Plant mitochondrial DNA has low levels of unordered variation and is maternally inherited in angiosperms and gymnosperms. Variation in this genome may help provide markers to assess female-mediated gene flow. However, the low level of variation limits application of this genomic marker, particularly within populations (Wolfe and Liston 1998).

nDNA variation

Variation at neutral loci within the nuclear genome offers great potential for gene flow studies. Nuclear encoded variation is codominant and can be ordered (although there may be paralogy issues, see chapter 2). These properties make nuclear loci suitable for estimating a range of indirect gene flow parameters using differentiation, private allele or phylogenetic models. If loci harbor high levels of polymorphism then they can be used to calculate direct estimates of gene flow. The utility of nuclear markers (other than SSRs) is presently limited due to relatively slow screening procedures, which are not particularly suitable for interpreting codominant expression of alleles (i.e., RFLP and sequencing). However, the screening of single nucleotide polymorphisms (SNPs) at nuclear loci does offer the potential for developing a highly effective and rapid technique for screening codominant, polymorphic nuclear loci.

4.4 Measuring gene flow – indirect estimates

Indirect methods of assessing gene flow use the observed genetic structure within adult populations (not propagules) to compare with models of the genetic structure expected under a range of models. Some of these tests can be very powerful but several have restrictive conditions of population composition and relations before they can be correctly applied.

4.4.1 Clonality, inbreeding, and family structure

Clones

The distribution of a clonal individual (genet) in terms of number of ramets and size of area covered can be assessed easily using genetic fingerprinting approaches. The problem is, at what level is genetic identity accepted as indicating two ramets of a clone, rather than two individuals which are genetically very similar but are not identical, yet are indistinguishable using the chosen marker system? Unless the entire genomes of both individuals are sequenced this cannot be truly proven. However, in practice, it is possible to apply a reasonable number (usually defined by available resources) of highly polymorphic markers (e.g., RAPD, AFLP, SSRs, iSSRs, RFLPs), to demonstrate that individuals are identical and to show the distribution of a single clone. The confidence of correct clonal identity can be shown by examining the variation present within individuals of non-clonal origin (if they exist) and by calculating an exclusion probability. Gerber et al. (2000) define an exclusion probability as "the average capability of any marker system to exclude any given relationship" and it is conditional on the genotypes of the reported relatives, frequency of alleles at loci, and the number of independent marker systems (loci) tested (Jamieson and Taylor 1997). An exclusion probability basically allows a probability to be assigned to the possibility of identifying two non-identical individuals as identical-by-accident, and a common derivation is outlined in Box 4.3. As long as the exclusion probability is close to 1 then identity of clones can be made fairly confidently.

An example of using genetic fingerprinting techniques to identify a wide-ranging clone is presented in Box 4.4, and describes the distribution of a single genet of the recently introduced Japanese knotweed across many British waterways. One complication that occurs for long-lived clonal lineages is that separate mutations may have accumulated in different ramets. A phylogenetic reconstruction could be used to trace the evolutionary mutation history of separate ramets (see chapter 5).

Box 4.3

The exclusion probability

The power of any individual or parental discrimination method depends on the exclusion probability calculable within a given population using the loci under study (Devlin, Roeder, and Ellstrand 1988). The exclusion probability is defined as "the average capability of any marker system to exclude any given relationship" (Gerber et al. 2000) and it is conditioned by the genotypes of the reported relatives, frequency of alleles at loci, and the number of independent marker systems (loci) tested (Gerber et al. 2000, Jamieson and Taylor 1997). A high exclusion probability equates to high resolution of individuals and their gametes/propagules (Schnabel 1998). The importance of the exclusion probability to parental assignment is that an increase in exclusion probability increases the probability of correct identification among a set of non-excluded parents (Devlin, Roeder, and Ellstrand 1988). Whilst the exclusion probability measures the power of a data set, complementary analysis is necessary to make ecological inferences about population or individual characteristics.

The probability that a male randomly sampled in the population is excluded as father, knowing the genotypes of the offspring and its mother (i.e., paternity analysis), is derived as follows (after Chakraborty, Meagher, and Smouse 1988):

Let $p_1, p_2 \ldots p_k$ and $q_1, q_2 \ldots q_k$ equal the allele frequencies of k codominant alleles at an autosomal locus (l) in males and females respectively.

The exclusion probability at locus l (PE_l) is:

$(1 - p_i)^2$ with probability $p_i[1 - q_i + q_i^2]$ for $i = 1, 2, \ldots, k$

$PE_l =$

$(1 - p_i - p_j)^2$ with probability $q_i q_j (p_i + p_j)$ for $j > i = 1, 2, \ldots, k$.

$$E(PE_l) = \sum p_i (1 - p_i)^2 (1 - q_i + q_i^2)$$
$$+ \sum \sum q_i q_j (p_i + p_j)(1 - p_i - p_j)^2$$
$$= 1 - 2a_{20} + a_{30} + 3(a_{11}a_{21} - a_{32}) - 2(a_{11}^2 - a_{22})$$

where $a_{rs} = \sum p_{ir} q_{is}$ for $r, s = 0, 1, 2, 3$.

Considering multilocus genotypes, a combined exclusion probability $PE(C)$ can be obtained as follows:

$$PE(C) = 1 - \prod (1 - PE_l)$$

where PE_l is the exclusion probability afforded by the lth system.

PE_l is maximized when alleles are in equal frequencies and $PE(C)$ increases as the number of loci and the number of alleles per locus increase. Calculation assumes random mating and equal frequency of alleles.

An exclusion probability can also be empirically estimated (Adams 1992) as the sum of the number of male parents in the population that can be excluded for each observed female and progeny pair divided by the total number of observed pairs (Devlin, Roeder, and Ellstrand 1988).

Inbreeding

Under the Hardy–Weinberg principle, differences in the proportion of heterozygotes observed (h_{obs}) and expected (h_{exp}) within a population, can be used to estimate the proportion of inbreeding present, and an inbreeding coefficient F_{IS} can be calculated as follows (but see chapter 3 for full derivation and discussion of method):

$$F_{IS} = 1 - (h_{obs}/h_{exp})$$

If one assumes that all the deficit in heterozygosity is due to selfing (s) then the relationship between selfing and the inbreeding coefficient, according to Wright (1965), is:

$$F_{IS} = s/(2 - s)$$

Box 4.4

Mapping the extent of a Japanese Knotweed clone

Since its introduction in the 1850s, the Japanese Knotweed (*Fallopia japonica*, Polygonaceae) has spread extremely successfully to establish large colonies along the banks of most of the waterways of the British Isles. Once established this plant is exceedingly difficult to eradicate. It is regarded as the most pernicious alien weed in the British Isles, and since 1981 it has been a criminal offence to knowingly introduce Japanese Knotweed into the wild. To test whether populations from the British Isles in fact represent dispersal of a single clonal individual by vegetative propagation, Hollingsworth and Bailey (2002) undertook a genetic fingerprinting screen of 150 British samples (see Fig. 4c). Ten arbitrary decamer RAPD primers were used and all samples produced an identical multiprimer RAPD profile. A subset of these individuals had previously been studied with a larger set of markers including inter-SSRs (Hollingsworth et al. 1998), the genetic identity of samples across both screens reassured the authors that all samples analyzed represent ramets of a single, exceptionally widespread clonal individual. No male-fertile individuals of this species have ever been found in Britain, and so the clone probably represents one of the world's largest vascular plants.

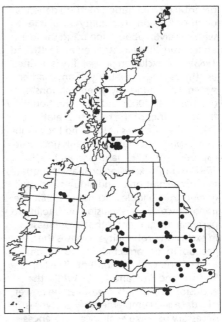

Fig. 4c

Distribution of Japanese Knotweed sample used in the survey of genetic variation in the British Isles by Hollingsworth and Bailey (2002).

However, there are two reasons why F_{IS} may not be a good indicator of selfing. First, mating between close relatives may be mistaken for selfing as it causes an increase in homozygosity using this parameter. Second, homozygotes resulting from selfing are likely to suffer from inbreeding depression, and may be under selective pressure and removed from a population, causing an upward bias in the proportion of observed heterozygosity.

Overall, F_{IS} remains a reasonable indicator of the level of inbreeding within a population or species, and should be estimated before applying estimates of gene flow based on population genetic structure if any level of autogamy or family clustering is suspected.

Coefficients of relatedness and family relationships

Genetic relatedness can be expressed as a single measure, r, between individuals (Avise 1994). For full sibs and parent–offspring pairs $r = \frac{1}{2}$, for half-sibs and between aunts/uncles and nieces/nephews or between grandparents and grandchildren $r = \frac{1}{4}$, for unrelated individuals $r = 0$. These principles can be used to calculated relatedness where the pedigree of relationship between individuals is known (Michod and Anderson 1979). However, where the pedigree is unknown, which is the usual case in natural populations, a mean coefficient of relatedness can be calculated among group/population members using polymorphic markers according to comparisons between observed and expected levels of heterozygosities within family groups compared to a wider population (Queller and Goodnight 1989). One coefficient of relatedness, interpreted as a genotypic correlation among group members in a subdivided population is defined by Pamilo and Crozier (1982) as:

$$r = \frac{H_{exp} - (1/c)\sum h_{exp,m} - (1/c)\sum [1/N_m - 1)] [h_{exp,m} - (1/2)h_{obs,m}]}{H_{exp} - (1/2)H_{obs}}$$

where $h_{obs,m}$ is the observed heterozygosity at a locus and $h_{exp,m}$ that expected under Hardy–Weinberg within a group m with N_m individuals, and H_{obs} and H_{exp} the observed and expected heterozygosities in the wider population composed of c colonies.

Where conditions exist that allow higher relatedness scores between individuals and the offspring of a colony than would be possible to an individual's own offspring, then individuals that make a reproductive sacrifice and rear the colony offspring will be favored by selection (known as inclusive fitness), and the evolution of such systems will be favored under the right relatedness conditions (known as Hamilton's rule; Hamilton 1964). This principle is applied to explain the existence of eusocial colonies of insects and some mammals (Box 4.5). The level of gene flow between colonies is expected to be very low to maintain high relatedness coefficients.

4.4.2 Using population genetic structure

Gene flow can be inferred by examining the partitioning of genetic variation between populations and calculating expectations of genetic structure according to theoretical models (see chapter 3). One such model is the infinite island model (Slatkin 1985, Wright 1931, 1940; Figure 4.1), which assumes that a species is subdivided into an infinite number of populations (or islands) of finite and equal size N, which all exchange alleles (i.e., produce and receive migrants) with equal probability. The number and relative location of island populations is not specified but as the source population is infinite it is not subject to genetic drift. This model has been altered to the finite island model (Latter 1973), which has n finite populations, and the migration rate m represents the fraction of each population derived from migration. Another model, the stepping-stone model (Kimura 1953), assumes that gene flow occurs only between neighboring populations (or demes). Predictions from the stepping-stone model are made with reference to the number of steps between pairs of populations, and thus allow gene flow to occur between non-neighboring populations by using intermediary populations as "stepping stones."

Besides gene flow, drift can also change allelic frequencies within populations and its effect is increased with decreasing effective population size (chapter 3). As these two processes are difficult to discern, population genetic approaches only allow the inference of the number of individuals exchanged between populations per generation, Nm, where according to Wright (1951):

$$F_{ST} \cong 1/(1 + 4Nm) \qquad \text{or} \qquad Nm \cong (1 - F_{ST})/4F_{ST}$$

F_{ST} was originally described by Wright (1951) to partition departures from random mating into components due to non-random mating within and between populations, and chapter 3 outlines the basis for its calculation using a range of models, estimators, and markers. However, several other population subdivision statistics can also be used and have a similar or approximated relationship to Nm (see chapter 3), including gene diversity (i.e., G_{ST}, Nei 1973), organelle partitioning (G_{ST} for haploid genomes, Pons and Petit 1992, Takahata and Palumbi 1985), nucleotide diversity (N_{ST}, Lynch and Crease 1990), and dominant variation (a ϕ_{ST} derived from an analysis of molecular variation; Excoffier, Smouse, and Quattro 1992; and a corrected F_{ST} using mid-frequency markers; Lynch and Milligan 1994; Box 4.1).

Box
4.5

Inbreeding and family structure within naked mole-rat colonies

When eusociality was first discovered in some Hymenoptera species, it was thought to be a unique and peculiar consequence of the haploid/diploid relationship between males, females, and the queen of the colonies exhibiting this behavior. However, eusociality is predicted for any communal group of animals within which a genetic fitness advantage is gained by females due to caring for the dominant female's offspring rather than having her own (inclusive fitness; Hamilton 1964). Amazingly a eusocial mammal has been discovered, the naked mole-rat, *Heterocephalus glaber* (Jarvis 1981), and female workers inhabiting colonies of this species in southern Africa forgo reproduction to raise and tend the offspring of the dominant and only reproductively fertile female of the colony (the queen). Whilst no haploid/diploid relationship exists between females and males of the colony, multilocus fingerprints were used to examine band-sharing probabilities (0.88–0.99) and found that genetic relatedness values amongst individuals of the colonies were extremely high ($r = 0.88 \pm 0.10$; Faulkes et al. 1997), and which satisfy the criteria of relatedness for eusociality to evolve.

Screening mtDNA in individuals of 15 colonies from Ethiopia, Somalia, and Kenya, demonstrated considerable genetic divergence between colonies (Faulkes et al. 1997), and the sequence divergence between Ethiopian and southern Kenyan populations was 5.8% for cytochrome-b, which is approaching the interspecific values seen between other Bathyergids. Such high levels of differentiation suggest very restricted or non-existent intercolony gene flow. However, geographically local colonies in Kenya often shared the same mtDNA haplotype, suggesting recent colonization and establishment of a common maternal ancestor over short distances (Faulkes et al. 1997).

The mole-rat colonies themselves are burrowed into extremely hard earth, and the collective effort of the entire colony is required to extend a nest underground, thus old sites tend to be restructured rather than new ones dug. Colonies have a communal toilet area and it is suspected that the dominant queen releases a hormone in her urine that suppresses the reproductive maturity of the colony's female helpers, keeping them non-reproductive (Jarvis 1981). The behavioral adaptations required for the naked mole-rats to exist in such harsh climatic conditions have caused ecologists to speculate whether eusociality could evolve as a system without the high relatedness ratios of the inclusive fitness model. Recently support has been found for this argument. Burland et al. (2002) used microsatellite markers to investigate 15 colonies across the range of the eusocial Damaraland mole-rat (*Cryptomys damarensis*). Most breeding pairs in wild colonies were unrelated ($r = 0.02 \pm 0.04$) and mean colony relatedness was little more than half that previously reported for naked mole-rats ($r = 0.46 \pm 0.01$), and below the level predicted by Hamilton to drive a shift towards eusociality. These results demonstrate that even with normal familial levels of relatedness, eusociality can still occur in mammals. Burland et al. (2002) speculate that it is ecological constraints and extrinsic factors that play the central role in driving mole-rat population adaptation to cooperative breeding, and that inbreeding in naked mole-rats is a response to extreme constraints on dispersal.

The theoretical relationship between Nm and drift is such that, on average, the movement of one individual per generation between populations ($Nm = 1$) is sufficient to prevent genetic differentiation by genetic drift alone (Allendorf 1983). This Nm value equates to a mean F_{ST} of 0.20 and is used as a criterion to classify populations or species as having high (above this value) or low (below this value) gene flow. Several animal groups with high vagility also have low popu-

(a)

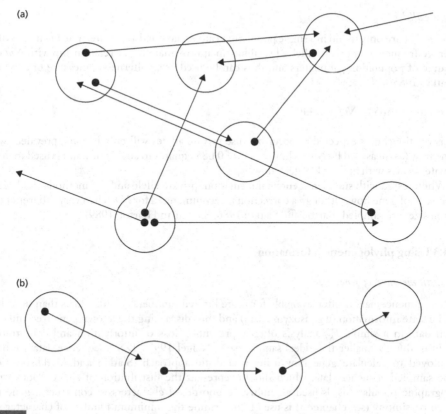

(b)

Gene flow models. (a) The island model where migrants move amongst populations of equal size; the number of populations can be finite, infinite, or there can be an infinitely large source population (Latter 1973, Slatkin 1985, Wright 1931, 1940). Migrants move at random between source and receiver populations and frequency of transfer is independent of distance. (b) The stepping-stone model (Kimura 1953) asserts a spatial dimension to gene flow. Gene flow only occurs between neighboring populations and gene flow between non-neighboring populations must use intermediary populations as "stepping stones."

lation differentiation and consequently high Nm values, for example, the pine beauty moth (Box 4.2). Where genetic differentiation estimates are available for markers that have possible different inheritance characteristics (e.g., maternal vs. biparental), it is possible to estimate the relative ratio of gene flow occurring due to male gametes and maternally produced propagules. Although most commonly used to derive pollen vs. seed gene flow ratios for plants, this measure can also be used for animal cases (e.g., the Galapagos iguanas, section 4.2.2). For hermaphrodite species, male to propagule gene flow ratio has been estimated by Ennos (1994) as:

$$\{[(1/F_{ST(B)} - 1)(1 + F_{IS(B)}] - 2(1/F_{ST(m)} - 1)\}/(1/F_{ST(m)} - 1)$$

where $F_{ST(B)} = F_{ST}$ for biparentally inherited nuclear markers. $F_{IS(B)} = F_{IS}$ for biparentally inherited nuclear markers. $F_{ST(m)} = F_{ST}$ for maternally inherited markers.

Using this method, estimates of pollen to seed flow ratio in plants have a tremendous range of variation from 1.43:1 for the broad-leaved Helleborine to close to 1:650 for *Fagus crenata* (Squirrell et al. 2001).

Private alleles

Alleles that are only found in a single population can be classified as private. The logarithm of the average frequency of private alleles [$p(1)$] has an approximate linear relationship with Nm over a range of population parameters and thus can be used as an alternative measure of gene flow (Slatkin 1985), where:

$$\ln[p(1)] = -0.505 \ln(Nm) - 2.440$$

This relationship is expected to occur because private alleles will only become prevalent when gene flow (as measured by Nm) is low. A $p(1)$ of 0.085 equates to an Nm of 1 and is used to differentiate species with high or low gene flow.

When tested with sufficient genetic information, private allele and F_{ST} methods yield similar estimates of gene flow, although a correction is recommended for Nm when very different population sizes are sampled (Barton and Slatkin 1986, Slatkin and Barton 1989).

4.4.3 Using phylogenetic information

Cladistic measures of gene flow

DNA sequence data is often available for more limited numbers of individuals than unordered allelic state information (e.g., isozyme data) and thus discarding the known gene tree and using such data in a simple F_{ST} analysis often represents a loss of information and compromised analysis, due to smaller than ideal sample sizes (Neigel 1997). DNA sequence data are better employed to calculate gene flow using a cladistic approach (Slatkin and Maddison 1989). The sampled sequence data, which should represent the distribution of allelic clades within geographic populations, is used as multistate unordered characters to construct a gene tree. Then parsimony (see chapter 5) is used to determine the minimum number of character state transitions (i.e., migration events) required and to construct a robust phylogenetic tree of sequence relationships. The distribution of this minimum number of migration events was shown through simulation studies to be a simple function of Nm (Slatkin and Maddison 1989). Simulations also showed that a linear relationship occurred between Nm and the distance between samples in a stepping-stone model (Slatkin and Maddison 1990). Thus a powerful indirect estimator of gene flow can be obtained using the genealogical information of sequence data.

Unfortunately, this method cannot be applied evenly to all genomic regions for which phylogenetic information can be obtained. Additional simulations by Hudson, Slatkin, and Madison (1992) indicate that the best correlation between F_{ST} and the cladistic method is obtained for loci with low recombination and moderate to high migration, and thus this method is probably best applied to animal mtDNA variation, which is not subject to recombination (unlike nDNA) and can provide fully resolved genealogies (unlike cpDNA).

Timing dispersal distance

An evolutionary clock for molecular divergence can be used to calculate the expected distribution, due to dispersal, of various aged, gene tree lineages by applying a multigeneration "random-walk" process from the specified center of origin for each clade (Neigel, Ball, and Avise 1991). This approach is most useful when considering low-dispersal species and using a rapidly mutating, non-recombining genetic marker, such as animal mtDNA. In these circumstances, mutations that delineate new descendant lineages may be dispersed at rates sufficiently low to prevent the attainment of equilibrium between genetic drift and gene flow that many earlier models assume.

This method was used to provide a single generation dispersal distance for the deer mouse (200 m), which was very similar to that obtained from mark–recapture experiments (Lansman et al. 1983).

A coalescent approach

As phylogenetically interpretable data have been introduced to population genetic analyses then genealogical approaches have been developed for testing evolutionary hypotheses and population genetics parameters (Felsenstein 1992). Genealogical approaches exploit two properties of gene trees, and can be used to make inference about gene flow (Neigel 1997). First, branch lengths that connect sequences correspond to a time of coalescence to the most recent common ancestor. Second, the relative order of branches in a tree corresponds to the order of coalescence events and therefore constitutes a cladistic relationship amongst the sequences (see chapter 5 for an expansion of these principles and assumptions).

Slatkin (1991) demonstrated a relationship between coalescence time and F_{ST} for loci where mutation rate is too low to affect F_{ST}:

$$F_{ST} = (t - t_0)/t$$

where t_0 is the average coalescence time of two genes drawn from the same subpopulation and t is the average coalescence time of two genes drawn from the same total population. This F_{ST} estimator can then be used in the Nm calculation and allows the demographic processes of migration and genetic drift to be separated from mutation (Neigel 1997). The power of this analysis means that further advances in calculating gene flow parameters utilizing coalescent approaches are expected. The use of related techniques for demonstrating ancient gene flow and colonization events is shown in Box 4.6.

4.4.4 Using a spatial context

Gene flow in organisms with limited vagility may be restricted by distance alone. This effect is known as isolation-by-distance (Wright 1943, 1946), and may operate within or between populations. Several models have been developed to consider estimates of gene flow in a spatial context. Continuum models define the position of individuals by the distribution of dispersal movements, births, and deaths, rather than by using a location array (Neigel 1997). Where individuals produce only a single propagule to replace themselves, then the distribution of individuals will be random and similar to the movement of particles under Brownian motion (Skellam 1951). Where multiple offspring are produced, siblings will produce clusters, the size of which is determined by dispersal distances (Sawyer 1967). In this model, clusters will increase over multiple generations, becoming infinitely large, and complicating the analysis of population genetic consequences (Felsenstein 1975, 1976). For simulations of natural populations, lattice models, which use a cellular automaton approach defining the exact position, density, and size of populations, have been most useful in examining the dynamics of isolation-by-distance (Slatkin and Barton 1989).

When using an indirect approach to calculate gene flow, the isolation-by-distance concept has been used to infer the extent of genetic connectivity within and/or between populations. In a simple form, pairwise F_{ST} estimates can be calculated between populations. These pairwise differentiation estimates can be plotted as a dendrogram (see chapter 5) or used to examine an isolation-by-distance effect using Mantel's test of spatial correlation (see chapter 3), and have been used to examine range-wide and fine-scale genetic structure and infer gene flow dynamics. For example, an examination of isolation-by-distance within a population of sessile oaks, *Quercus petraea*, found significant clustering of genetically similar individuals (using SSRs and isozymes) and was explained by the limited seed dispersal strategy of the species, in spite of evidence for long-distance pollen dispersal (Streiff et al. 1998, 1999; chapter 7). An examination of genetic connectivity between Costa Rican populations of Spanish cedar (*Cedrela odorata*; Cavers, Navarro, and Lowe 2003a), found a significant isolation-by-distance effect when the geographic distance between populations included the physical barrier of the central mountain range (rising above 4000 m in many places). This result suggested that the physical topography of the country affects the contemporary genetic connectivity of populations (see Box 4.1).

Box
4.6

Phylogeny and gene flow

The European grayling, *Thymallus thymallus*, is widespread throughout most of Europe's major water systems and rivers. However, declining population sizes have prompted management strategies that involve translocation of brood stocks between water basins for rearing and release, especially in southern European basins. To examine the genetic differentiation between water basins and to examine if this policy is "best practice" for existing genetic resources of the species, Weiss et al. (2002) undertook to examine the colonization and gene flow history of the species across Europe using mtDNA variation.

The complete mitochondrial DNA (mtDNA) control region (1043 base pairs) and 162 base pairs of flanking transfer RNA genes were sequenced in 316 European grayling, from 44 populations throughout the western European range of the species. In total, 58 haplotypes were found and pairwise divergence estimates ranged from 0.001 to 0.038. In addition to finding well-supported phylogenetic clades within the Danube basin, one highly divergent clade in the Adriatic basin, and one large, diverse group representing most other European populations (Fig. 4d(a)). The available data reveal a complex pattern of interglacial and postglacial expansions originating from disjunct refugia throughout central Europe.

A parsimony network-based nested-clade analysis (NCA) was used in support of specific hypotheses of postglacial expansion (Figs. 4d(b) and (c)), including expansion into the Baltic and the upper Rhine and Danube from sources in the lower Rhine (Moselle) and Elbe systems (Danish populations); and colonization of the

(a)

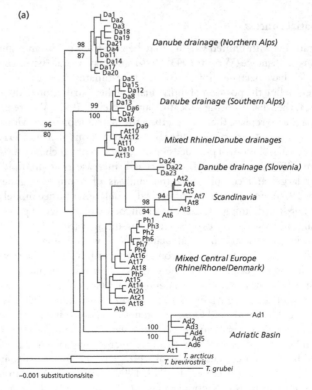

Fig. 4d

(a) Neighbor-joining phylogram of mtDNA haplotypes; % node support for the tree is shown above. Predominant regional location of haplotypes is indicated.

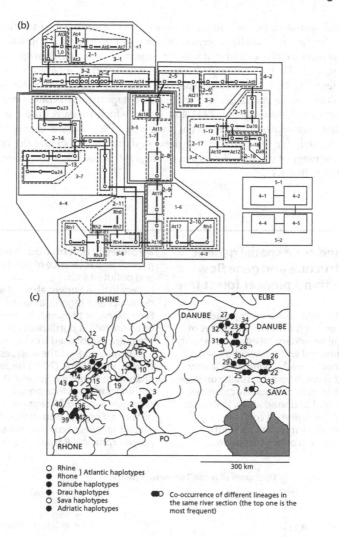

Fig. 4d

(b) Nested clade analysis of mtDNA variation, predominantly representing Rhone and Atlantic populations. Clades are built sequentially from the smallest one-step up to four-step clades. The final nesting level, two five-step clades is depicted separately in the lower right-hand corner.
(c) Distribution of major mtDNA lineages, corresponding to the five disjunct networks or subnetworks based on the 95% parsimony criteria.

Rhone from identified Rhine populations (Doller, Orbe, and Reuss). The divergent clades of the Danube and a deep divergence between Adriatic and Loire basin haplotypes (Fig. 4d(a)), support the theory that European graylings have had a long history in western Europe that pre-dates the Pleistocene glacial cycles. The patterns of mtDNA divergence demonstrate high genetic diversity within basins and differentiation between basins (Fig. 4d(c)) that has been established due to historical colonization and subsequent isolation. Weiss et al. (2002) recommend that the practice of translocation of brood stocks for rearing and release be stopped to maintain this genetic divergence and probable local adaptation.

More formal spatial autocorrelation analysis can also be undertaken to allow a comparison between the correlation in allelic or genotypic state of portions of a population grouped into defined spatial distance classes and that which is expected under random spatial distribution of variation (Ennos 2001). Positive autocorrelation indicates that genetically similar individuals cluster together spatially, for which one explanation is limited gene flow. Several types of analysis can be conducted to show spatial genetic structure. Perhaps the most commonly applied method is Moran's autocorrelation statistic, I (Moran 1950), which measures the correlation in gene frequencies between samples of genotypes within specific distance classes. Analysis can be conducted separately for individual loci or data combined over loci and the combined I value depicted for the range of distance classes as a single graph (known as a correlogram, Smouse and Peakall 1999; e.g., Box 4.7).

Box 4.7

Fine-scale spatial genetic structure and gene flow within a pioneer forest tree

Vochysia ferruginea is a widespread, long-lived pioneer, and a dominant species of neotropical secondary forest. It is primarily insect-pollinated and has wind-dispersed seed. It is occasionally found as a canopy tree in old-growth forests, usually on slopes or less fertile soils, and quickly colonizes disturbed or abandoned areas. In Central America *V. ferruginea* is commercially important where it is used for light construction, and offers tremendous

agroforestry potential due to its fast growth and tolerance of poor quality and polluted soils.

Very little is known about the population dynamics and regeneration of *V. ferruginea*, or the genetic consequences of colonization bottlenecks when new areas are established from old forest sources. To address these issues, Lowe, Cavers, and Wilson (2001) undertook an analysis of fine-scale genetic structure within four Costa Rica populations of *V. ferruginea* with different colonization histories. For each population, 50 individuals were screened for 61 polymorphic AFLP fragments. Diversity

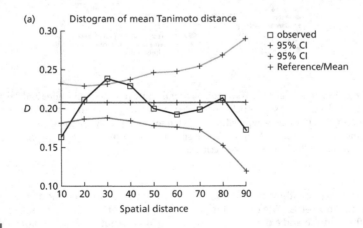

Fig. 4e

(a) Correlogram of genetic distance plotted by spatial distance class in m for *V. ferruginea* at Penjamo site. A significant (95%) observed correlation between genetic distance and distance class is indicated if points are plotted outside of the limits of the confidence interval plots. If the observed correlation is below the lower 95% CI line then individuals within that distance class have lower genetic distances between one another (i.e., are more genetically similar) than expected at random and a significant positive spatial structure is observed.

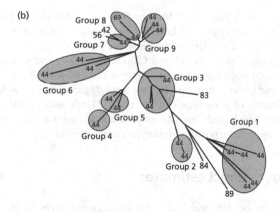

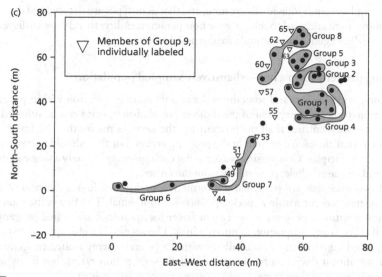

Fig. 4e

(b) Neighbor-joining dendrogram of Jaccard's distance estimate calculated from AFLP profiles. Clusters of genetically similar individuals are indicated by the grey shaded groups and the locations of the individuals are plotted in (c). (c) Plot of tree positions within an intensively sampled plot at Penjamo. Shaded Groups 1 to 7 indicate individuals that are clustered together based on genetic distance analysis (b). Members of Group 9 also clustered together genetically but are dispersed across the sampled plot.

was found to be significantly lower and fine-scale genetic structure significantly higher for recently colonized populations compared to old growth forests (see Fig. 4e). A further examination of the spatial distribution and the genetic similarity of individuals within a population, which was only established 30 years ago (Penjamo), found significant spatial clustering of genetically similar individuals. It appears that colonization of cleared areas proceeds via limited seed migrants (causing low diversity) and that half or full sib sapling communities are established within recently colonized sites by predominant local seed dispersal around the early colonist. Older colonist populations recover diversity and spatial genetic structuring decreases markedly, and is probably due to additional gene flow from divergent source populations (via pollen and seed), selection and thinning pressures within the populations (Lowe, Cavers, and Wilson 2001).

If each sample is composed of a single individual then the frequency of alleles within an individual can be used (reflecting the homozygous or heterozygous state at each locus surveyed). Under these conditions, there is a direct relationship between I and Wright's coefficient of relatedness $\rho_{(x)}$ (Pamilo and Crozier 1982, Queller and Goodnight 1989) and so a direct link between spatial autocorrelation and population genetic analysis can be made (Ennos 2001). If populations are at equilibrium and genetic structure has been solely generated by isolation-by-distance, then quantitative estimates of gene flow can be made (Hardy and Vekemans 1999). However, the assumptions under which such analysis may be conducted are easily violated in natural populations, and greatly reduce the power of interpretation of such a test.

4.5 Measuring gene flow – direct estimates

Direct measures of gene flow can be distinguished from indirect measures by the fact that propagules and not adult individuals are used, and that genotypes are determined directly following gene movement and used to retrace gene flow parameters directly relating to dispersal events. A number of direct gene flow methods are discussed below.

4.5.1 Using progeny from a non-exhaustively sampled population

By screening codominant (and sometimes dominant) marker variation within progeny arrays, the predominant breeding system of the mother/population/species can be inferred. Clonality (apomixis) can be identified if all the progeny are the same as the mother. Selfing is assumed if only the maternal alleles are present in the progeny arrays, but they should segregate according to Mendelian principles. Outcrossing is assumed if each progeny has only a single maternal allele and a second "foreign" allele, presumably from the father.

Indeed two measures are particularly informative, and give a first indication of the type of breeding system present within a species/population/individual. The first is the total number of alleles present within a progeny array (two or fewer for apomictic and selfed progeny but more than two if there is any proportion of outcrossing). The second is a chi-squared test of observed versus expected segregation ratios of alleles within a progeny array (selfed progeny should not show any significant deviation from expected allelic segregation ratios, but if any apomixis or outcrossing is involved then there should be a significant relationship).

Mating system parameters

Tests of mating system parameters have been developed into formal models by Ritland and colleagues (Ritland 1986, 2002, Ritland and Jain 1981). Genetic variation from progeny arrays (where both the maternal genotype is known and unknown) at codominant or dominant loci can be used to calculate the proportion of inbreeding to outcrossing. To do this two parameters are estimated. A single locus estimate of outcrossing (where information from single loci are used and then averaged) is made based on the number of non-maternal alleles, the total number of alleles, and their segregation ratios. Similarly a multilocus estimate of outcrossing is made, where multiple or all loci are used to calculate this parameter.

Differences in these two statistics are also applied to calculate two further parameters. The first is the proportion of biparental inbreeding (mating between relatives, which causes increased homozygosity) and is calculated as the difference between the single and the multilocus estimates. For truly selfed progeny arrays, the single locus and multilocus estimates will be similar/identical, whereas for matings involving closely related individuals, the single locus outcrossing estimate will be significantly lower than the multilocus outcrossing rate. The second parameter is correlated matings, the degree to which siblings share the same male parent, which is calculated as the relationship between the probability that two progeny taken from a single array are outcrossed full sibs (known as the "correlation of paternity") or selfed full sibs (known as the "correlation of selfing"). This estimate is made once an overall estimate of selfing rate has been made across families and variances are normalized. These statistics can be made spatially explicit if the location of the mother is known, although in general they are used as a population average.

This suite of measures provides a very powerful test of the mating system of a species and gives an indication of the number of males in a population contributing to the family of a particular female.

Number of males and effective reproductive neighborhood size

Differences in genetic differentiation between the parent and progeny generation can be used to great effect to infer gene flow parameters, and with further development, these methods offer the potential to dissociate historical gene flow and contemporary dispersal events.

Smouse et al. (2001) have proposed a new measure of male dispersal, which only needs genotype information from a non-exhaustive sample of females and their progeny (Austerlitz and Smouse 2001). The method defines an F_{ST} equivalent for male gametes in a population. This allows a quantification of the heterogeneity of the male gamete pool sampled by females scattered across a landscape, and is used to estimate mean pollination distance and effective neighborhood size. The method appears to be sensitive to marker systems with low exclusion probabilities, but the model can be confidently applied using a relatively small number of regularly segregating, highly polymorphic SSR markers (e.g., four loci). Similar principles have been applied by Burczyk et al. (2002), who have adapted earlier work by Adams and Birks (1991) to develop a model that can be applied to angiosperm and gymnosperm plants. The model apportions offspring to three categories: selfed; outcrossing with males in a circumscribed area around the mother (neighborhood); and outcrossing to males outside the neighborhood. The model allows the derivation of average and specific mating system, number of males, neighborhood size for outcrossed events, and the proportion of matings within and outside this area. An example of using this differential differentiation approach to determine gene flow parameters is presented for *Dinizia excelsa* (Box 4.11).

4.5.2 Assignment of parental contribution within an exhaustively sampled area

Surveying the genetic variation of progeny or juveniles of a population also offers the potential to directly identify the parents of these progeny. However, to undertake such analyses with any hope of obtaining a meaningful result requires two conditions. First, the parent population must be sampled sufficiently that potential parents are included in the sample (usually an exhaustive sample within a particular area is chosen when not working with small, closed systems). Second, a suitable molecular marker system (preferably codominant and highly polymorphic) must be chosen so that exclusion probabilities are sufficiently high to allow confident assignment of parental identity (Box 4.3).

Successful identification of one or both parents of progeny within or between populations may allow calculation of several parameters, including:
• Fertility or mating success of an individual or population.
• Propagule dispersal frequency between populations or amongst regions/family groups within a population.
• For low-vagility species, the distance and frequency of propagule dispersal or a dispersal distance/frequency curve can be calculated.

Calculation of such parameters for a range of propagules (gamete, dispersed progeny, established progeny, juvenile, recently matured adult) will also provide an insight into selection pressures acting at the different life stages and the total proportion of dispersal that translates as effective gene flow.

Box
4.8

Simple exclusion and extra-pair mating in snow goose families

Pairs of snow goose, *Chen caerulescens*, rear their brood of several goslings in nests. However, to test whether the male and female nest attendants were actually the biological parents of the gosling broods Quinn et al. (1987) undertook a simple exclusion analysis following genetic typing of putative parents and their offspring for 14 single-copy nuclear RFLP loci (a subset of the data are presented in Table 4a and are adapted from Avise 1994). There are several exclusion possibilities with this type of analysis. (a) One of the parents must be excluded. However, both parents share one allele with the offspring but neither have the other offspring allele, therefore there is

uncertainty which parent to exclude, e.g., gosling 4 locus A and G and gosling 12, locus A. (b) The mother is excluded as the offspring possess an allele found in the male but the other allele is not present in the female, e.g., gosling 11 and 12, locus E. A maternal exclusion suggests intraspecific brood parasitism. (c) The father is excluded as the offspring possess an allele found in the female but the other allele is not present in the male, e.g., gosling 12, locus G. A paternal exclusion suggests an instance of extra-pair mating. (d) Both parents are excluded if the offspring possess alleles that are not present in either putative male or female, e.g., gosling 4, locus C.

Overall the genotyping results of Quinn et al. (1987) corroborate field observations documenting frequent intraspecific brood parasitism and extra-pair mating within snow geese populations.

Table 4a

Diploid genotypes for seven loci of goslings and male and female nest attendants for three families of the snow goose.

		RFLP locus						
		A	B	C	D	E	F	G
Family 1	Male attendant	2,2	2,2	2,3	1,2	1,1	1,1	1,4
	Female attendant	2,2	2,2	2,2	1,1	1,1	1,1	1,3
	Gosling 1	2,2	2,2	2,2	1,2	1,1	1,1	1,1
	Gosling 2	2,2	2,2	2,2	1,1	1,1	1,1	3,4
	Gosling 3	2,2	2,2	2,2	1,2	1,1	1,1	1,3
	Gosling 4	2,3*	2,2	1,1§	1,1	1,1	1,1	1,2*
Family 3	Male attendant	1,2	2,2	2,4	1,1	1,1	1,1	1,1
	Female attendant	2,2	2,2	1,2	1,1	2,2	1,1	1,2
	Gosling 9	2,2	2,2	1,2	1,1	1,2	1,1	1,1
	Gosling 10	2,2	2,2	2,4	1,1	1,2	1,1	1,2
	Gosling 11	1,2	2,2	2,4	1,1	1,1†	1,1	1,1
	Gosling 12	2,3*	2,2	2,2	1,1	1,1†	1,1	2,2‡
	Gosling 13	1,2	2,2	2,2	1,1	1,2	1,1	1,2
Family 4	Male attendant	2,2	2,2	3,3	1,2	1,1	1,1	1,1
	Female attendant	2,2	1,2	1,1	1,1	1,1	1,1	1,1
	Gosling 14	2,2	1,2	1,3	1,2	1,1	1,1	1,1
	Gosling 15	2,2	1,2	1,3	1,2	1,1	1,1	1,1
	Gosling 16	2,2	2,2	1,3	1,2	1,1	1,1	1,1
	Gosling 17	2,2	2,2	1,3	1,2	1,1	1,1	1,1

* Excludes one unspecified parent; † excludes putative mother; ‡ excludes putative father; § excludes both putative parents. Data from Quinn et al. 1987 and Avise 1994.

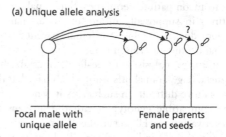

(a) Unique allele analysis

Focal male with unique allele — Female parents and seeds

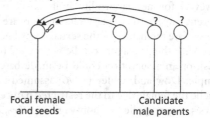

(b) Paternity analysis

Focal female and seeds — Candidate male parents

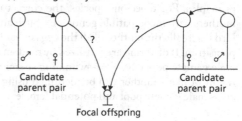

(c) Parentage analysis

Candidate parent pair — Candidate parent pair — Focal offspring

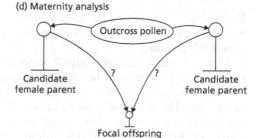

(d) Maternity analysis

Outcross pollen

Candidate female parent — Candidate female parent — Focal offspring

Fig. 4.2

Schematic representations of different types of direct analysis of dispersal in plant populations (from Ennos 2001). (a) Using unique alleles to measure pollen dispersal. (b) Using paternity analysis to measure pollen dispersal. (c) Using parentage analysis to measure pollen and seed dispersal. (d) Using maternity analysis to measure seed dispersal.

Paternity, maternity, and parentage analysis

It is possible to work out three types of parent/offspring relationship using genetic exclusion methods (Fig. 4.2).

• Paternity analysis: For progeny where the mother (and her genotype) are known (e.g., seeds on trees, newly born mammals), the identity of the father can be established by comparing the genotype of the progeny with potential males in the population, once the maternal genotype has been subtracted. This process can involve screening progeny for a unique allele carried by only one male individual in the population, and provides a good estimate of the mating success of this individual. An alternative method is to use information from all available loci to exclude males, due to incompatible genotypes. Such analyses may be used to provide evidence for extra-pair matings in supposedly monogamous animals. The possibility of autogametic or apomictic progeny does not exclude the female parent from also being the male parent in some species (e.g., hermaphrodites and apomicts).

• Parentage analysis: For free-living or established progeny where the identity of both parents is required (e.g., animal offspring which have left the maternal home or plant seedlings), the problem is more difficult. In such cases it is necessary to show that two potential parent individuals have compatible genotypes with the progeny (i.e., have not been excluded by the available marker variation). However, it is also necessary to show that the contribution of both individuals (or only a single individual for selfed progeny) is required to establish the full genotype of the progeny (Schnabel 1998). For dioecious species, this pair must clearly comprise different sexes. However, for hermaphrodite individuals, molecular markers exclusively inherited through different gametes (e.g., organelle markers are often only maternally inherited) could be used to differentiate the nature of the sexual contribution of parents or specific maternal tissue still associated with the propagule can be genotyped (e.g., pericarp or megagametophyte tissue of plant seeds), or an assumption could be made based on the proximity of progeny and parents. For example, Dow and Ashley (1996) assumed that the closest identified parent to oak seedlings must be the female due to the restricted nature of acorn dispersal. Another principle of parentage analysis, when applied to non-vagile species, is that the distance between the two parents corresponds to the dispersal distance of the male gamete.

• Maternity analysis: For progeny where the mother is unknown, she can be deduced using two principles. For dioecious species, the correct mother can be identified by excluding potential mothers with incompatible genotypes. Such analysis could be used to identify whether a female bird is actually the mother of all the eggs in her brood or whether there is evidence for intrabrood parasitism. If offspring are fertilized by random mating events involving a homogeneous outcross male gamete pool (e.g., free-swimming sperm of marine species or wind-dispersed pollen) then the most likely mother can be identified using genotype information of the offspring, the outcross male gamete pool and potential female parents.

Problems with parental identification

Parental analysis by genetic exclusion across loci results in one of three outcomes:
• A single parent (or parent pair for parentage cases) may be found, in which case parental contribution is documented.
• All parents may be excluded leaving no potential parent(s). There are several explanations for this. First, both parents may be outside the sampled plot. Second, there may have been a mutation between the parent and progeny generation, consequently the true parent has been excluded by incompatibility (depending on the marker this may be a significant factor, although it is rarely assessed; Dow and Ashley 1996). Third, one or both parents may have died, the probability of which increases if an individual's life span is short or if the progeny analyzed are no longer young juveniles.
• Several potential parents within a plot may be identified. Although highly polymorphic molecular markers (e.g., microsatellites) exhibit high exclusion probabilities (Gerber et al. 2000), unique assignment is rarely possible for all offspring (Chase et al. 1996, Schnabel 1998). Therefore to select one individual as the correct parent from several possibilities, a model for exclusion has to be adopted. Several assumptions are made using exclusion models, however, simulation approaches allow an estimation of the confidence of assignment. The use of exclusion models usually increases the proportion of offspring for which parents can be allocated. This advantage is important. Parental assignment is usually a means to an end, such as investigating pollinator ecology or gene flow changes. Discarding data in the form of offspring, because no parent can be assigned, wastes effort and time in the data sampling, and leads to a reduction in the explanatory power of results (Snow and Lewis 1993), which in extreme cases may lead to incorrect conclusions.

Box
4.9

Estimating cryptic gene flow

Sometimes foreign gametes are indistinguishable from local gametes (i.e., have identical genotypes for marker used). In these cases, external gene flow events are not detected, and this is known as cryptic gene flow. Exclusion analysis allows an estimation of the paternal (or maternal) contribution that could not have been produced by the population from which the offspring is taken, and is referred to as apparent gene flow, which is a minimal estimate of gene flow and underestimate of total gene flow.

Cryptic gene flow can be calculated according to Devlin and Ellstrand (1990). The assumptions of the model are: that surrounding populations are considered as one single large population; the set of successful foreign gametes received by a target population is a random sample of individuals in the outlying population, the probability of obtaining this set is a function of allele frequencies; and that screened loci are at linkage equilibrium.

Method 1

Φ = probability that a gamete produced in the outlying population could not be produced by any individual in the target population.

$$\Phi = \sum IPg$$

where $g \in \Omega$ (set of all possible gametes in the outlying population) and Pg, the probability that member g of Ω is formed.

$I = 1$ if the gamete could not be produced, and $I = 0$ otherwise.

$$\Phi/(1 - \Phi) = x/y$$

where x = apparent pollen flow, and y = cryptic pollen flow. $x + y$ is an estimate of total gene flow by pollen.

Method 2

If τ is the number of seeds in the sample fathered by foreign gametes and x is the number of apparent gene flow events, then:

$$L(\tau/x) = P(X = x/\tau)$$

where X is a random variable.

Using a Monte Carlo simulation to obtain $P(X = x/\tau)$, it is then possible to obtain an approximate maximum likelihood estimate of τ (Devlin and Ellstrand 1990).

Method 3

Estimation of cryptic gene flow according to Dow and Ashley (1996). Let P_m = probability that an adult matched an unrelated offspring over the entire population of size n:

$$P_m = 1 - (1 - x)^n$$

where $x = \sum(2p_i q_i + q_i^2)$, probability of having a given allele i of frequency q_i by chance for each of the loci considered.

P_m is the probability that a foreign offspring is attributed a parent in the target population and represents an estimation of the probability of cryptic gene flow for each parent–offspring match.

Another problem with exclusive assignment of parents is the possibility of assigning false positives. This problem increases significantly as the exclusion probability decreases below 1, and can significantly skew gene dispersal estimates. For example, if an individual within a plot is identified as a progeny's parent but the real parent is actually located outside the plot, then the amount of gene flow inferred by this misidentification is underestimated (i.e., the real level of gene flow is much higher than the analysis indicates). This phenomenon is referred to as "cryptic gene flow" because the real gene dispersal event has been missed. Fortunately the magnitude and variance of cryptic gene flow have a close relationship with the exclusion probability and can be estimated statistically (see Box 4.9).

4.5.3 Models of exclusion

Unique alleles and simple exclusion analysis

Slatkin's private allele principle (1985) can be applied to directly assess gene dispersal both between populations and within populations. If after sufficient individual and population sampling unique alleles can be identified then analysis of progeny arrays from other individuals/ populations can screen for the presence of such markers. This method can be very effective for calculating dispersal parameters, although it does require the existence of unique alleles, and an exhaustive sampling strategy to ensure that alleles are truly unique. The effects of cryptic gene flow can also be estimated using exclusion probabilities.

Simple exclusion analysis was first introduced as a method of paternity analysis by Ellstrand (1984), and has since been used in many other studies (e.g., Devlin and Ellstrand 1990, Ellstrand, Devlin, and Marshall 1989, Ellstrand and Marshall 1985), including extension to the parentage analysis of dispersed progenies (Dow and Ashley 1996, 1998, Schnabel and Hamrick 1995).

The method is straightforward. For given loci, the genotype of an offspring and its mother are known, potential male parents are then excluded as fathers if they cannot have produced gametes bearing a genotype compatible with that of the progeny, given the genotype of the mother (Ellstrand 1984). A worked example of simple exclusion is presented in Box 4.8.

However, whilst straightforward, this type of analysis does have limitations (Schnabel 1998). For example, there is no way to discriminate between two males that could have produced gametes bearing the same genotype. Also the problem of cryptic gene flow is not dealt with directly by this model and has to be calculated separately (Box 4.9).

Problems with the simple exclusion method are enhanced when applied to dispersed progeny for which neither of the parents is known in advance. In this case, exclusion of a possible parent requires that no allele be shared between the two potential parents at one or more loci (Dow and Ashley 1996, Schnabel 1998). Moreover, the subset of immigration events inferred through the exclusion of parent pairs is always a composite of gene flow by gametes and progeny (Schnabel 1998). Hence estimation of gene flow by gametes requires further assumptions, such as prior assignment of mother to dispersed, established progeny (Dow and Ashley 1998, Schnabel and Hamrick 1995).

Maximum-likelihood method and fractional paternity assignment

This model was first introduced by Meagher (1986) and is derived from Bayes theorem (chapter 3). For a set of genetic markers, it considers the statistical likelihood that a male is the true paternal parent, given the genotypes of the offspring and the known maternal parent. The distribution of "log of the odds (LOD) ratio" scores (Meagher, 1986) of the genotype (i.e., sampled) candidates is compared to the distribution of LOD scores that would be generated by a random sample sharing the same allele frequencies to dtermine a threshold for rejection. The test of paternity will reject all candidates with a LOD score below the chosen threshold and will not categorically assign paternity to one of the likely fathers in case of unresolved paternity. It, thus, provides a statistical estimate of effective pollen flow. Determining confidence levels for rejecting a male with a compatible genotype as potential father usually relies on a simulation procedure (e.g., Gerber et al. 2000). The maximum likelihood method does not require any prior assumption of fertility (Box 4.10).

The method solely identifies mother–father–offspring triplets and assesses the probability of males being the father (Schnabel 1998). A measure of fertility variation within the population can be obtained, but only under the assumption of population isolation (i.e., no gene flow; Meagher 1991). Nevertheless, three major criticisms of the method have been expressed. First, even if the assumptions of the model are met, there is a bias in respect to individuals which are homozygous

Box 4.10

Expansions of maximum likelihood analysis and fractional paternity assignment

Maximum likelihood

Let R be the relationship between three individuals B, C, and D, and g_i the genotype of individual i.

The likelihood of relationship R_1 that D is the father given offspring B and known mother C is:

$$L(R_1) = P(g_B, g_C, g_D/R)$$
$$= P(g_B/\text{mother } g_C, \text{father } g_D) P(g_C) P(g_D)$$
$$= T(g_B/g_C, g_D) P(g_C) P(g_D)$$

where $P(g_i)$ is the population probability of genotype g_i and $T(g_B/g_C, g_D)$ is the

Mendelian segregation probability from parent to offspring or transition probability.

The likelihood of relationship R_2 in which C is mother of B but D is an unrelated male is:

$$L(R_2) = T(g_B/g_C) P(g_C) P(g_D)$$

Hence, LOD $(R_1{:}R_2) = \log[T(g_B/g_C, g_D)/T(g_B/g_C)]$

Attributing log of likelihood ratio (LOD) scores to each male allows a comparison of the likelihood of their paternity of a given offspring (Meagher 1986). The male with the highest LOD score is accepted as the most likely male parent of the offspring (Meagher 1986, Meagher and Thompson 1987).

---

at many loci, and whose fertility is overestimated. Second, the individual with the highest LOD score is arbitrarily chosen as the likely parent with little regard to the relative likelihood of other possible parents. Third, the model does not give any answer to tied LOD scores (Devlin, Roeder, and Ellstrand 1988, Schnabel 1998).

The fractional paternity assignment method (Devlin, Roeder, and Ellstrand 1988) is very similar to the maximum-likelihood method of Meagher (1986) but, instead of attributing paternity to the most likely male, a fraction of the progenies' paternity is assigned amongst all non-excluded males (Box 4.10). The fraction of paternity assigned is proportional to a male's likelihood of paternity. However, unlike the maximum-likelihood method, all progeny are assigned to a parent but some may not be assigned a single father (Devlin, Roeder, and Ellstrand 1988). Thus the fractional paternity assignment method focuses on population patterns of paternity and is a good model for estimating gene flow patterns (Schnabel 1998).

The maximum-likelihood and fractional paternity methods described above are most accurate when the numbers of loci and alleles per locus progeny scored are high, and when the number of potential parents is low. A limitation of likelihood-based methods is that they assume that the entire set of potential parent has been sampled (Marshall et al. 1998) and hence require census surveying to provide unambiguous and unbiased estimates of parentage. Therefore, while parentage analysis provides very good information on reproductive success and patterns of pollen and seed dispersal, it is only really applicable to small populations. An example of the use of maximum-likelihood methods to calculate paternity is shown in Box 4.11.

Using likelihood methods to estimate fertility

Assigning prior probability of paternity to individuals is a problem (Devlin, Roeder, and Ellstrand 1988). The fractional paternity assignment method assumes equal fertility, which is highly unlikely; moreover it is often one of the parameters that one aims to test or measure in a parentage analysis. Roeder et al. (1989) addressed this question and took a likelihood-model approach and modeled the probabilistic structure of an entire sample of progeny as a function

Box
4.11

Pollen dispersal distance in continuous and fragmented Amazonian forest populations of the canopy emergent, *Dinizia excelsa*

Habitat fragmentation is expected to have a detrimental impact on tree populations experiencing such landscape level changes. Problems include change in microenvironment and lack of reproductive assurance due to gene dispersal changes. For insect-pollinated tree species, fragmentation is expected to disrupt mutualistic relationships with native pollinators and lower viability.

(a)

(c)

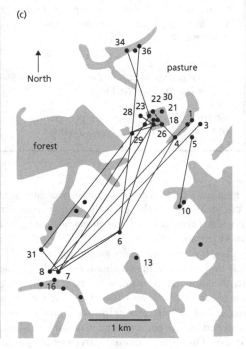

(b)

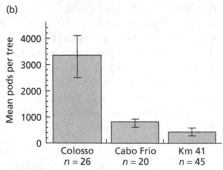

Fig. 4f

(a) Collecting seed arrays for genetic analysis. (b) Mean pod set of remnant trees in pasture (Colosso) compared to neighboring forest (Cabo Frio and Km 41). (c) Network of inferred pollen dispersal events using a paternity maximum likelihood exclusion analysis. White areas are pasture and secondary vegetation and shaded areas are primary forest.

To test this hypothesis and study the impact that fragmentation has on breeding system and gene dispersal parameters, Dick (2001) studied populations of *Dinizia excelsa* (Fabaceae), a canopy-emergent tree, in Amazonian pastures, forest fragments, and intact, continuous forest blocks. Microsatellite analysis of open-pollinated seed collected from pasture and forest trees (Fig. 4f(a)), indicated that genetic diversity within seed arrays was maintained across habitat. However, in addition trees within habitat fragments produced, on average, over three times as many seeds as trees in continuous forest (Fig. 4f(b)), and paternity analysis revealed extremely long gene dispersal distances of up to 3.2 km for isolated trees in pasture (Fig. 4f(c)), the most distant pollination event so far recorded for any plant species. In addition, the differentiation of progeny arrays was compared to that of adult tree populations (the TWOGENER approach, Austerlitz and Smouse 2001), and average pollen dispersal distances were estimated as 1509 m in open pasture compared to 212 m in undisturbed forest (Dick, Etchelecu, and Austerlitz 2003).

Canopy observations (Dick 2002) indicate that African honeybees (*Apis mellifera scutellata*) are the predominant floral visitors to trees in fragmented habitats and have completely replaced native insects in isolated pasture trees. Usually considered as dangerous exotics, African honeybees appear to have become important pollinators in degraded tropical forests, and may significantly expand genetic neighborhood areas, allowing genetic connectivity between fragmented and continuous forest populations, and genetic rescue of isolated individuals previously considered as the "living dead".

of the fertility of the parent pairs (Roeder, Devlin, and Lindsay 1989). More recently, Schnabel et al. (1998) tackled the problem of assigning maternity to dispersed progenies. A maternity analysis model based on maximum-likelihood estimates was used to infer maternal fertilities from genotypic data on dispersed seeds and seedlings. The model is applicable only if random mating can be assumed and was developed for a dioecious species (Schnabel et al. 1998).

4.6 The importance of biological and environmental factors on gene flow

Table 4.4 summarizes the findings of a number of studies that have tried to correlate the magnitude of indirect measures of gene flow calculated from population differentiation parameters with various life-history and environmental variables. This has been done most successfully with the meta-analyses of Hamrick and co-workers (e.g., Loveless and Hamrick 1984) who have considered a range of plant case studies. Some similar work has also been conducted on animal taxa but results are only available by taxonomic group (Ward et al. 1992).

The predominant influence on gene flow appears to be the vagility of the organism, in the case of animals (insects and birds have very high between-population gene flow estimations), or the vagility of propagules for plants (high estimates of gene flow are obtained for plants with dispersive pollen and seed). A number of intuitive correlations are also highlighted in Table 4.4, for example, plants which are widespread, outcrossing, tree form, or with synchronous flowering have higher gene flow potential than plants which are endemic, selfed, herbaceous, or with asynchronous flowering.

A number of studies measuring direct gene dispersal in plants are now available. However, it has been difficult to summarize the results of these as the scale of the different studies is very different and often several different parameters were being investigated, making derivation of comparable statistics difficult. Overall the trend is that gene flow by pollen is idiosyncratic. Indeed significant differences between species (Stacy et al. 1996), populations (Goodell et al.

Table 4.4

Magnitude of gene flow correlated by biological characteristics of plant and animal groups.

Factor	State	Gene flow within populations	Gene flow between populations
Animals			
Taxonomic group:	Mammals	–	Medium
Vertebrates	Birds	–	Very high
	Reptiles	–	Medium
	Amphibians	–	Low
	Fishes	–	High
Taxonomic group:	Insects	–	Very high
Invertebrates	Crustaceans	–	High
	Molluscs	–	Medium
	Others	–	Very high
Plants			
Breeding system	Self-fertilization	Low	Low
	Mixed mating	Medium	Medium
	Outcrossing	High	High
Floral morphology	Monecious	Depends on % selfing	Depends on % selfing
	Dioecious	High	High
Reproductive mode	Apomictic	Depends	Potentially low
	Sexual	Depends	Depends
Pollination	Sedentary animal	Potentially low	Low
mechanism	Dispersive animal	High	High
	Wind	High	High
Seed dispersal	Limited	Potentially low	Low
	Long range	High	High
Seed dormancy	Absent	Depends	Depends
	Present	Increases potential	Increases potential
Phenology	Asynchronous	Reduces potential	Reduces potential
	Synchronous	Increases potential	Increases potential
Life form	Annual	Reduces potential	Reduces potential
	Long-lived	Increases potential	Increases potential
Timing of	Monocarpic	Reduces potential	Reduces potential
reproduction	Polycarpic	Increases potential	Increases potential
Successional stage	Early	Depends	Reduces potential
	Late	Depends	Increases potential
Geographic range	Narrow endemic	High	Low
	Widespread	Depends	Depends
Population size,	High	Depends	Depends
density	Low	Depends	Depends

Animal data are taken from Ward et al. (1992) and Avise (1994), and based on Nei's coefficient of gene differentiation $(H_T - H_S)/H_T$; plant data are taken from Loveless and Hamrick (1984) and Avise (1994), and based on the population genetic structure parameter F_{ST}.

1997), and even individuals (Dow and Ashley 1996, Ellstrand and Marshall 1985, Schnabel and Hamrick 1995, Schnabel et al. 1998, Streiff et al. 1999) are reported. In most cases, gene flow appears sufficient to counteract the effects of drift (Schnabel and Hamrick 1995), but few studies report any correlation between the magnitude of gene flow or dispersal and a change in distance (Ellstrand et al. 1989, Isagi et al. 2000) or phenology (Devlin and Ellstrand 1990, Goodell et al. 1997, Meagher 1986).

Such generalizations are, however, difficult. For future work a concentration on the spatial context of gene flow is expected to be an expanding area of investigation, and the interaction of dispersal and selection at different life stages is likely to be a fruitful and important area for gene flow study.

REFERENCES

Adams, W.T. 1992. Gene dispersal within forest tree populations. *New Forests*, 6: 217–40.

Adams, W.T. and Birks, D.S. 1991. Estimating mating patterns in forest tree populations. In S. Fineschi, M.E. Malvolti, F. Cannaata, and H.H. Hattemer, eds. *Biochemical Markers in the Population Genetics of Forest Trees*. The Hague: SPB Academic Publishing, pp. 157–72.

Allendorf, F.W. 1983. Isolation, gene flow and genetic differentiation among populations. In C.M. Schonewalk-Cox, S.M. Chambers, B. MacBryde, and L. Thomas, eds. *Genetics and Conservation*. London: Benjamin/Cummings, pp. 51–65.

Austerlitz, F. and Smouse, P.E. 2001. Two-generation analysis of pollen flow across a landscape. II. Relation between Phi_{ft}, pollen dispersal and interfemale distance. *Genetics*, 157: 851–7.

Avise, J.C. 1994. *Molecular Markers, Natural Selection, and Evolution*. New York: Chapman and Hall.

Avise, J.C. and Nelson, W.S. 1989. Molecular genetic relationships of the extinct Dusky Seaside Sparrow. *Science*, 243: 646–8.

Bacilieri, R., Labbe, T., and Kremer, A. 1994. Intraspecific genetic structure in a mixed population of *Quercus petraea* (Matt.) Liebl and *Q. robur* L. *Heredity*, 73: 130–41.

Baker, C.S., Palumbi, S.R., Lambertsen, R.H., Weinrich, M.T., Calambokidis, J., and O'Brien, S.J. 1990. Influence of seasonal migration on geographic distribution of mitochondrial DNA haplotypes in humpback whales. *Nature*, 34: 238–40.

Baker, H.G. 1955. Self-compatibility and establishment after "long distance" dispersal. *Evolution*, 9: 347–8.

Baker, H.G. 1967. Support for Baker's law – as a rule. *Evolution*, 21: 853–6.

Barton, N.H. and Slatkin, M. 1986. A quasi-equilibrium theory of the distribution of rare alleles in a subdivided population. *Heredity*, 56: 409–15.

Berg, E. and Hamrick, J. 1995. Fine-scale genetic structure of a Turkey oak forest. *Evolution*, 49: 110–20.

Bermingham, E., Forbes, S.H., Friedland, K., and Pla, C. 1991. Discrimination between Atlantic salmon (*Salmo salar*) of North American and European origin using restriction analyses of mitochondrial DNA. *Canadian Journal of Fish and Aquatics*, 48: 884–93.

Bossema, I. 1979. Jays and oaks: An eco-ethological study of a symbiosis. *Behaviour*, 70: 1–117.

Bowen, B.W., Meylan, A.B., Ross, J.P., Limpus, C.J., Balazs, G.H., and Avise, J.C. 1992. Global population structure and natural history of the green turtle (*Chelonia mydas*) in terms of matriarchal phylogeny. *Evolution*, 46: 865–81.

Burczyk, J., Adams, W.T., Moran, G.F., and Griffin, A.R. 2002. Complex patterns of mating revealed in a *Eucalyptus regnans* seed orchard using allozyme markers and the neighborhood model. *Molecular Ecology*, 11: 2379–91.

Burland, T.M., Bennett, N.C., Jarvis, J.U.M., and Faulkes, C.G. 2002. Eusociality in African mole-rats: new insights from patterns of genetic relatedness in the Damaraland mole-rat (*Cryptomys damarensis*). *Proceedings of the Royal Society of London, Series B*, 269: 1025–30.

Cain, M.L., Milligan, B.G., and Strand, A.E. 2000. Long-distance seed dispersal in plant populations. *American Journal of Botany*, 87: 1217–27.

Campbell, D.R. 1985. Pollen and gene dispersal: the influences of competition for pollination. *Evolution*, 39: 418–31.

Campbell, D.R. 1991. Comparing pollen dispersal and gene flow in a natural population. *Evolution*, 45: 1965–8.

Campbell, D.R. and Waser, N.M. 1989. Variation in pollen flow within and among populations of *Ipomopsis aggregata*. *Evolution*, 43: 1444–55.

Carvalho, G. 1998. Molecular ecology: origins and approach. In G. Carvalho, ed. *Advances in Molecular Ecology*. Amsterdam: IOS Press, pp. 1–28.

Cavers, S., Navarro, C., and Lowe, A.J. 2003a. A combination of molecular markers (cpDNA, PCR-RFLP, AFLP) identifies evolutionarily significant units in *Cedrela odorata* L. (Meliaceae) in Costa Rica. *Conservation Genetics*.

Cavers, S., Navarro, C., and Lowe, A.J. 2003b. Chloroplast DNA phylogeography reveals colonization history of a Neotropical tree, *Cedrela odorata* L., in Mesoamerica. *Molecular Ecology*, 12: 1451–60.

Chakraborty, R., Meagher, T., and Smouse, P.E. 1998. Parentage analysis with genetic markers in natural populations. I. The expected proportion of offspring with unambiguous paternity. *Genetics*, 118: 527–36.

Chase, M., Moller, C., Kesseli, R., and Bawa, K. 1996. Distant gene flow in tropical trees. *Nature*, 383: 398–9.

Cottrell, J.E., Munro, R.C., Tabbener, H.E., Milner, A.D., Forrest, G.I., and Lowe, A.J. 2003. Comparison of fine-scale genetic structure within two British oakwoods using microsatellites: consequences of recolonisation dynamics and past management. *Forest Ecology and Management*, **176**: 287–303.

Devlin, B., and Ellstrand, N.C. 1990. The development and application of a refined method for estimating gene flow from angiosperm paternity analysis. *Evolution*, **44**: 248–59.

Devlin, B., Roeder, K., and Ellstrand, N. 1988. Fractional paternity assignment: Theoretical development and comparison to other methods. *Theoretical and Applied Genetics*, **76**: 369–80.

Dick, C.W. 2001. Genetic rescue of remnant tropical trees by an alien pollinator. *Proceedings of the Royal Society of London, Series B*, **268**: 2391–6.

Dick, C.W. 2002. Effects of pollinator composition on the breeding structure of tropical timber trees. In B. Degen, M.D. Loveless, and A. Kremer, eds. *Modelling and Experimental Research on Genetic Processes in Tropical and Temperate Forests, Conference Proceedings, Kourou, French Guiana, 2000*. Belem, PA, Brazil: Embrapa Amazônia Oriental, pp. 146–58.

Dick, C.W., Etchelecu, G., and Austerlitz, F. 2003. Pollen dispersal of tropical trees (*Dinizia excelsa*: Fabaceae) by native insects and African honeybees in pristine and fragmented Amazonian rainforest. *Molecular Ecology*, **12**: 753–64.

Dow, B.D. and Ashley, M.V. 1996. Microsatellite analysis of seed dispersal and parentage of saplings in bur oak, *Quercus macrocarpa*. *Molecular Ecology*, **5**: 615–27.

Ellstrand, N.C. 1984. Multiple paternity within the fruits of the wild radish, *Raphanus sativus*. *The American Naturalist*, **123**: 819–28.

Ellstrand, N.C., Devlin, B., and Marshall, D.L. 1989. Gene flow by pollen into small populations: data from experimental and natural stands of wild radish. *Population Biology*, **86**: 9044–7.

Ellstrand, N.C. and Elam, D.R. 1993. Population genetic consequences of small population size: implications for plant conservation. *Annual Review of Ecology and Systematics*, **24**: 217–42.

Ellstrand, N.C. and Marshall, D.L. 1985. Interpopulation gene flow by pollen in wild radish, *Raphanus sativus*. *American Naturalist*, **126**: 606–16.

Endler, J.A. 1977. *Geographic Variation, Speciation, and clines*. New Jersey: Princeton University Press.

Ennos, R.A. 1994. Estimating the relative rates of pollen and seed migration among plant populations. *Heredity*, **72**: 250–9.

Ennos, R.A. 2001. Inferences about spatial processes in plant populations from the analysis of molecular markers. In J. Silvertown and J. Antonovics, eds. *Integrating Ecology and Evolution in a Spatial Context*. Oxford: Blackwell Science, pp. 45–71.

Epperson, B.K. and Allard, R.W. 1989. Spatial autocorrelation analysis of the distribution of genotypes within populations of lodgepole pine. *Genetics*, **121**: 369–77.

Estoup, A. and Angers, B. 1998. Microsatellites and minisatellites for molecular ecology. Theoretical and empirical considerations. In G. Carvalho, ed. *Advances in Molecular Ecology*. Amsterdam: IOS Press, pp. 55–86.

Excoffier, L., Smouse, P.E., and Quattro, J.M. 1992. Analysis of molecular variance inferred from metric distances among DNA haplotypes: application to human mitochondrial DNA restriction data. *Genetics*, **131**: 479–91.

Faulkes, C.G., Abbott, D.H., O'Brien, H.P., Lau, L., Roy, M.R., Wayne, R.K., and Bruford, M.W. 1997. Micro- and macrogeographical genetic structure of colonies of naked mole-rats *Heterocephalus glaber*. *Molecular Ecology*, **6**: 615–28.

Felsenstein, J. 1975. A pain in the torus: some difficulties with models of isolation-by-distance. *American Naturalist*, **109**: 359–68.

Felsenstein, J. 1976. The theoretical population genetics of variable selection and migration. *Annual Review of Genetics*, **10**: 253–80.

Felsensein, J. 1992. Estimating effective population size from samples of sequences: inefficiency of pairwise and segregating sites as compared to phylogenetic estimates. *Genetic Research, Cambridge*, **59**: 139–47.

Franceschinelli, E.V. and Bawa, K. 2000. The effect of ecological factors on the mating system of a South American shrub species (*Helicteres brevispira*). *Heredity*, **84**: 116–23.

Gerber, S., Mariette, S., Streiff, R., Bodénès, C., and Kremer, A. 2000. Comparison of microsatellites and amplified fragment length polymorphism markers for parentage analysis. *Molecular Ecology*, **9**: 1037–48.

Gillies, A.C.M., Cornelius, J.P., Newton, A.C., Navarro, C., Hernandez, M., and Wilson, J. 1997. Genetic variation in Costa Rican populations of the tropical timber species *Cedrela odorata* L., assessed using RAPDs. *Molecular Ecology*, **6**: 1133–45.

Goodell, K., Elam, D.R., Nason, J.D., and Ellstrand, N.C. 1997. Gene flow among small populations of a self-incompatible plant: an interaction between demography and genetics. *American Journal of Botany*, **84**: 1362–71.

Gram, W. and Sork, V. 2001. Associations between environmental and genetic heredity in forest tree populations. *Ecology*, **82**: 2012–21.

Hamilton, W.D. 1964. The genetical evolution of social behaviour. *Journal of Theoretical Biology*, **7**: 1–52.

Hardy, O.J. and Vekemans, X. 1999. Isolation-by-distance in a continuous population: reconciliation between spatial autocorrelation analysis and population genetic models. *Heredity*, **83**: 145–54.
Hollingsworth, M.L. and Bailey, J.P. 2000. Evidence for massive clonal growth in the invasive weed *Fallopia japonica* (Japanese knotweed). *Botanical Journal of the Linnean Society*, **133**: 463–72.
Hollingsworth, M.L., Hollingsworth, P.M., Jenkins, G.I., Bailey, J.P., and Ferris, C. 1998. The use of molecular markers to study patterns of genotypic diversity in some invasive alien *Fallopia* spp. (Polygonaceae). *Molecular Ecology*, **7**: 1681–91.
Hu, X.-S. and Ennos, R. 1997. On estimation of the ratio of pollen to seed flow among plant populations. *Heredity*, **79**: 541–52.
Hudson, R.R., Slatkin, M., and Madison, W.P. 1992. Estimation of levels of gene flow from DNA sequence data. *Genetics*, **132**: 283–589.
Isagi, Y., Kanazashi, T., Suzuki, W., Tanaka, H., and Abe, T. 2000. Microsatellite analysis of the regeneration process of *Magnolia obovata* Thunb. *Heredity*, **84**: 143–51.
Jarvis, J.U.M. 1981. Eusociality in mammals: cooperative breeding in naked mole-rat colonies. *Science*, **212**: 571–3.
Jamieson, A. and Taylor, C. 1997. Parentage exclusion. *Animal Genetics*, **28**: 397–400.
Jensen, J.S., Olrik, D.C., Siegismund, H.R., and Lowe, A.J. 2003. Within-population genetics and spatial autocorrelation within an intensive studied plot of *Quercus petraea* (Matt.) Liebl. in Denmark. *Scandinavian Journal of Forestry*.
Karl, S.A., Bowen, B.W., and Avise, J.C. 1992. Global population genetic structure and male-mediated gene flow in the green turtle (*Chelonia mydas*): RFLP analyses of anonymous nuclear loci. *Genetics*, **131**: 163–73.
Kimura, M. 1953. "Stepping-stone" model of population. *Annual Report of the National Institute of Genetics, Japan*, **3**: 62–3.
Lansman, R.A., Avise, J.C., Aquadro, C.F., Shapira, J.F., and Daniel, S.W. 1983. Extensive genetic variation in mitochondrial DNAs among geographic populations of the deer mouse, *Peromyscus maniculatus*. *Evolution*, **37**: 1–16.
Latter, B.D.H. 1973. The island model of population differentiation: a general solution. *Genetics*, **73**: 147–57.
Loveless, M.D. and Hamrick, J.L. 1984. Ecological determinants of genetic structure in plant populations. *Annual Review of Ecology and Systematics*, **15**: 65–95.
Lowe, A.J., Cavers, S., and Wilson, J. 2001. Final Scientific Report 1997–2001. In Wilson, J. ed. *Assessment of Levels and Dynamics of Intra-Specific Genetic Diversity of Tropical Trees*. European Commission.
Lynch, M. and Crease, T.J. 1990. The analysis of population survey data on DNA sequence variation. *Molecular Biology and Evolution*, **7**: 377–94.
Lynch, M. and Milligan, B.G. 1994. Analysis of population genetic structure with RAPD markers. *Molecular Ecology*, **3**: 91–9.
Marshall, T.C., Slate, J., Kruuk, L.E.B., and Pemberton, J. 1998. Statistical confidence for likelihood-based paternity inference in natural populations. *Molecular Ecology*, **7**: 639–55.
Meagher, T.R. 1986. Analysis of paternity within a natural population of *Chamaelirium luteum*. I. Identification of most-likely parents. *American Naturalist*, **128**: 199–215.
Meagher, T.R. 1991. Analysis of paternity within a natural population of *Chamaelirium luteum*. II. Patterns of male reproductive success. *American Naturalist*, **137**: 738–52.
Meagher, T.R. and Thompson, E. 1987. Analysis of parentage for naturally established seedlings of *Chamaelirium luteum* (Lilliaceae). *Ecology*, **68**: 803–12.
Michod, R.E. and Anderson, W.W. 1979. Measures of genetic relationship and the concept of inclusive fitness. *American Naturalist*, **114**: 637–47.
Moran, P.A.P. 1950. Notes on continuous stochastic phenomena. *Biometrika* **37**: 17–23.
Nathan, R. and Muller-Landau, H.C. 2000. Spatial patterns of seed dispersal, their determinants and consequences for recruitment. *Trends in Ecology and Evolution*, **15**: 278–85.
Nei, M. 1973. Analysis of gene diversity in subdivided populations. *Proceedings of the National Academy of Sciences USA*, **70**: 3321–3.
Neigel, J.E. 1997. A comparison of alternative strategies for estimating gene flow from genetic markers. *Annual Review of Ecology and Systematics*, **28**: 105–28.
Neigel, J.E., Ball, R.M., and Avise, J.C. 1991. Estimation of single generation migration distances from geographic variation in animal mitochondrial DNA. *Evolution*, **45**: 423–32.
Pamilo, P. and Crozier, R.H. 1982. Measuring genetic relatedness in natural populations: methodology. *Theoretical Population Biology*, **21**: 171–93.
Pannel, J.R. and Barrett, C.H. 1998. Baker's law revisited: reproductive assurance in a metapopulation. *Evolution*, **52**: 657–68.
Parker, P., Snow, A., Schug, M., Booton, G., and Fuerst, P. 1998. What molecules can tell us about populations: choosing and using a molecular marker. *Ecology*, **79**: 361–82.

Pons, O. and Petit, R.J. 1996. Measuring and testing genetic differentiation with ordered versus unordered alleles. *Genetics*, **144**: 1237.

Provan, J., Soranzo, N., Wilson, N., McNicol, J., Morgante, M., and Powell, W. 1999. The use of uniparentally inherited simple sequence repeat markers in plant population studies and systematics. In P. Hollingsworth, R. Bateman, and R. Gornall, eds. *Molecular Systematics and Plant Evolution*. London: Taylor & Francis, pp. 35–56.

Queller, D.C. and Goodnight, K.F. 1989. Estimating relatedness using genetic markers. *Evolution*, **43**: 258–75.

Quinn, T.W., Quinn, J.S., Cooke, F., and White, B.N. 1987. DNA marker analysis detects multiple maternity and paternity in single broods of the lesser snow goose. *Nature*, **326**: 392–4.

Rassmann, K., Tautz, D., Trillmich, F., and Gliddon, C. 1997. The microevolution of the Galapagos marine iguana *Amblyrhynchus cristatus* assessed by nuclear and mitochondrial genetic analyses. *Molecular Ecology*, **6**: 437–52.

Richards, A.J. 1996. *Plant Breeding Systems*, 2nd edn. London: George Allen and Unwin.

Ritland, K. 1986. Joint maximum likelihood estimation of genetic and mating structure using open-pollinated progenies. *Biometrics*, **42**: 25–43.

Ritland, K. 2002. Extensions of models for the estimation of mating systems using n independent loci. *Heredity*, **88**: 221–8.

Ritland, K. and Jain, S. 1981. A model for the estimation of outcrossing rate and gene frequencies using n independent loci. *Heredity*, **47**: 35–52.

Roeder, K., Devlin, B., and Lindsay, B. 1989. Application of maximum likelihood methods to population genetic data for the estimation of individual fertilities. *Biometrics*, **45**: 365–79.

Roubik, D.W. 2002. Tropical bee colonies, pollen dispersal and reproductive gene flow in forest trees. In B. Degen, M.D. Loveless, and A. Kremer, eds. *Modelling and Experimental Research on Genetic Processes in Tropical and Temperate Forests, Conference Proceedings, Kourou, French Guiana, 2000*. Belem, PA, Brazil: Embrapa Amazônia Oriental, pp. 31–41.

Sawyer, S. 1967. Branching diffusion processes in population genetics. *Advances in Applied Probability*, **8**: 659–89.

Schlötterer, C. 1998. Genome evolution: are microsatellites really simple sequences? *Current Biology*, **8**: R132–4.

Schnabel, A. 1998. Parentage analysis in plants: mating systems, gene flow and relative fertilities. In G. Carvalho, ed. *Advances in Molecular Ecology*. Amsterdam: IOS Press, pp. 173–89.

Schnabel, A. and Hamrick, J.L. 1995. Understanding the population genetic structure of *Gleditsia triacanthos* L.: The scale and pattern of pollen gene flow. *Evolution*, **49**: 921–31.

Skellam, J.G. 1951. Random dispersal in theoretical populations. *Biometrika*, **38**: 196–218.

Slatkin, M. 1985. Rare alleles as indicators of gene flow. *Evolution*, **39**: 53–65.

Slatkin, M. 1991. Inbreeding coefficients and coalescence times. *Genetic Research, Cambridge*, **58**: 167–75.

Slatkin, M. and Barton, N.H. 1989. A comparison of three indirect methods for estimating average levels of gene flow. *Evolution*, **43**: 1349–68.

Slatkin, M. and Maddison, W.P. 1989. A cladistic measure of gene flow inferred from the phylogenies of alleles. *Genetics*, **123**: 603–13.

Slatkin, M. and Maddison, W.P. 1990. Detecting isolation by distance using phylogenies of genes. *Genetics*, **126**: 249–60.

Smouse, P.E., Dyer, R.J., Westfall, R.D., and Sork, V.L. 2001. Two-generation analysis of pollen flow across a landscape. I. Male gamete heterogeneity among females. *Evolution*, **55**: 260–71.

Smouse, P.E. and Peakall, R. 1999. Spatial autocorrelation analysis of individual multi-allelic and multilocus genetic structure. *Heredity*, **82**: 561–73.

Snow, A. and Lewis, P. 1993. Reproductive traits and male fertility in plants: empirical approaches. *Annual Review of Ecology and Systematics*, **24**: 331–51.

Sork, V. 1984. Examination of seed dispersal and survival in Red Oak, *Quercus rubra* (Fagaceae), using metal-tagged acorns. *Ecology*, **65**: 1020–2.

Sork, V., Nason, J., Campbell, D., and Fernandez, J. 1999. Landscape approaches to historical and contemporary gene flow in plants. *Trends in Ecology and Evolution*, **14**: 219–24.

Squirrel, J., Hollingsworth P.M., Bateman, R.M., Dickson, J.H., Light, M.H.S., MacConaill, M. and Tebbitt, M.C. 2001. Partitioning and diversity of nuclear and organelle markers in native and introduced populations of *Epipactis Helleborine Corchidaceae*. *American Journal of Botany*, **88**: 1409–18.

Stacy, E.A., Hamrick, J.L., Nason, J.D., Hubbell, S.P., Foster, R.B., and Condit, R. 1996. Pollen dispersal in low-density populations of three neotropical tree species. *American Naturalist*, **148**: 275–98.

Streiff, R., Ducousso, A., Lexer, C., Steinkellner, H., Glossl, J., and Kremer, A. 1999. Pollen dispersal inferred from paternity analysis in a mixed oak stand of *Quercus robur* L. and *Q. petraea* (Matt.) Liebl. *Molecular Ecology*, **8**: 831–41.

Streiff, R., Labbe, T., Bacilieri, R., Steinkellner, H., Glossl, J., and Kremer, A. 1998. Within-population genetic structure in *Q. robur* L. and *Q. petraea* (Matt.) Liebl. assessed with isozymes and microsatellites. *Molecular Ecology*, **7**: 317–28.

Takahata, N. and Palumbi, S.R. 1985. Extranuclear differentiation and gene flow in the finite island model. *Genetics*, **109**: 441–57.

Wainhouse, D. and Jukes, M.R. 1997. Geographic variation within populations of *Panolis flammea* (Lepidoptera: Noctuidae) in Britain. *Bulletin of Entomological Research*, **87**: 98–9.

Ward, R.D., Skibinski, D.O.F., and Woodwark, M. 1992. Protein heterozygosity, protein structure and taxonomic differentiation. *Evolutionary Biology*, **26**: 73–159.

Weiss, S., Persat, H., Eppe, R., Schlötterer, C., and Uiblein, F. 2002. Complex patterns of colonization and refugia revealed for European grayling *Thymallus thymallus*, based on complete sequencing of the mitochondrial DNA control region. *Molecular Ecology*, **11**: 1393–407.

Whitlock, M.C. and McCauley, D.E. 1999. Indirect measures of gene flow and migration: $F_{ST} \neq 1/(4Nm + 1)$. *Heredity*, **82**: 117–25.

Wolfe, A.D. and Liston, A. 1998. Contributions of PCR-based methods to plant systematics and evolutionary biology. In D.E. Soltis, P.S. Soltis, and J.J. Doyle, eds. *Molecular Systematics of Plants II. DNA Sequencing*. London: Kluwer, pp. 43–86.

Wright, S. 1931. Evolution in Mendelian populations. *Genetics*, **16**: 97–159.

Wright, S. 1940. Breeding structure of population in relation to speciation. *American Naturalist*, **74**: 232–48.

Wright, S. 1943. Isolation-by-distance. *Genetics*, **28**: 114–38.

Wright, S. 1946. Isolation-by-distance under diverse systems of mating. *Genetics*, **28**: 139–56.

Wright, S. 1951. The genetic structure of populations. *Annals of Eugenics*, **15**: 354.

Wright, S. 1965. The interpretation of population structure by F-statistics with special regard to systems of mating. *Evolution*, **19**: 395–420.

Ziegenhagen, B., Scholz, F., Madaghiele, A., and Vendramin, G.G. 1998. Chloroplast microsatellites as markers for paternity analysis in *Abies alba*. *Canadian Journal of Forest Research*, **28**: 317–21.

5

Intraspecific phylogenies and phylogeography

"What seest thou else in the dark backward and abysm of time?"
W. Shakespeare (1611) *The Tempest*

Summary

1 Intraspecific phylogenies can only be constructed from intraspecifically variable markers with a phylogenetic signal, that is, the markers must be ordered rather than unordered, for example, DNA sequences or RFLP data.

2 The application of the coalescent makes it possible to study population level processes within a temporal framework: microevolution can be studied as a dynamic, historical process, changing over time within a species.

3 Intraspecific phylogenies are influenced by historical and contemporary population size, gene flow, and population fragmentation.

5.1 Introduction

Evolution is a central principle of biology, and phylogeny is a central principle of evolution. With the advent of molecular methods that allow the phylogenetic analysis of mutations that differ among genetic variants, it has become possible to trace evolutionary relationships not only among species but also among combinations of genetic markers within and among populations. Furthermore, with appropriate data sampling and analysis it is possible to investigate the impacts of selection, changes in population size, and population substructuring on genealogical relationships among these alleles. Such investigations have been described by Avise (2000) as phylogeography, a "field of study concerned with the principles and processes governing the geographic distributions of genealogical lineages, especially those within and among closely related species." That is, phylogeography is a syncretic discipline contended to be at the interface of population and systematic biology.

Avise (2000), who provides a history of its development, argues that phylogeography emphasizes the role of contemporary ecology in shaping the spatial distribution of an organism's traits. However, selection is not the only force capable of generating geographic patterns in genetic attributes. For small, scattered populations of a species that have experienced little or no gene flow over long periods, divergence also occurs in selectively neutral markers (Barrett and Kohn 1990). Phylogeography interprets the historical processes that may have left their evolutionary signatures on the present geographic distributions of genetic traits.

There are two essential steps in phylogeographic analysis: (i) the generation of a gene genealogy; and (ii) the interpretation of this genealogy. Thus, phylogeographic markers must be intraspecifically variable and have a phylogenetic signal associated with them, that is, they must be ordered rather than unordered. Ordered data, for example, DNA sequences or restriction sites, contain information about genealogical relationships between samples. In contrast, unordered data, for example, allozymes and RAPDs, provide information about absolute differences between samples but no information about genealogical relationships (Harris 1999). The need for a phylogenetic signal limits the techniques that can be used to retrieve phylogeographic data (see chapter 2). Thus, allozymes have almost no value in phylogeographic studies since the historical relationships between alleles revealed on a gel are unknown and alleles are defined by state rather than descent. Similar criticisms can be produced against RAPDs and AFLPs (see chapter 2), although these markers have been used to infer the existence of glacial refugia and postglacial spread (e.g., RAPDs in *Saxifraga cespitosa*, Tollefsrud et al. 1998; allozymes in *Fagus crenata*, Tomaru et al. 1997). Traditionally, variation in organelle (mitochondria and chloroplasts) genomes has been detected as RFLPs, defined by restriction site mutations, although this technique is usually now used in a modified form as PCR-RFLP analysis. The maximum amount of phylogenetically useful information is provided by DNA sequence analysis. Thus, restriction site and DNA sequence data provide the best sources of phylogeographic data. In addition to the technical advances made in the generation of relationships between sequences based on networks (Posada and Crandall 2001; Box 5.5) and the identification of ordered genetic data, conceptual advances have been made in the application of the coalescent process, a framework for the study of the effects of population level processes (e.g., population size changes, selection, and gene flow) on the expected time to common ancestry within a gene tree (Emerson, Paradis, and Thébaud 2001; Box 5.6).

Animal mtDNA has proven to be an ideal molecule for phylogeographic analysis due to its high rate of sequence evolution, uniparental inheritance, and lack of recombination (see chapter 2). In fact, 70% of all studies that claim to have a phylogeographic component are from animals (e.g., Box 5.1; Avise 2000). In contrast, plant mtDNA, which has a slow rate of sequence evolution (see chapter 2), has rarely been used in phylogeographic studies (e.g., Sinclair, Morman, and Ennos 1998), unlike cpDNA (e.g., Box 5.2; Newton et al. 1999). The low levels of plant mtDNA sequence variation and the slow rate of cpDNA sequence evolution conspire to limit the resolution of phylogeographic patterns revealed in studies of plants (Schaal and Olsen 2000). Despite limitations, phylogeographic structure has been found in plants (e.g., Soltis et al. 1997). However, many cpDNA mutations are length changes, which can have profound impacts on the structure of gene genealogies reconstructed from RFLP data (Box 5.4), particularly in the determination of fragment homology. It is also assumed that organelle genomes are uniparentally inherited, although in higher plants a broad range of inheritance patterns are displayed by those species for which investigations have been undertaken (Harris and Ingram 1991; Table 5.1), and the mode of cpDNA inheritance should ideally be determined before an investigation is started (chapter 2). More recently, nDNA sequences have been suggested as suitable for the analysis of

Table 5.1

Examples of the cpDNA and mtDNA inheritance patterns displayed by higher plants.

Species	Family	cpDNA	mtDNA	References
Actinidia deliciosa	Actinidiaceae	P	M	Chat et al. (1999)
Pinus taeda	Pinaceae	P	M	Neale and Sederoff (1989)
Musa acuminata	Musaceae	M	P	Fauré et al. (1994)
Cucumis melo	Cucurbitaceae	M	P	Havey et al. (1998)
Medicago sativa	Fabaceae	PM	M	Forsthoefel et al. (1992)
Turnera ulmifolia	Turneraceae	PM	?	Shore and Triassi (1998)
Quercus robur	Fagaceae	M	M	Dumolin-Lapègue et al. (1998)

M, paternal transmission; P, paternal transmission; PM, mixture of maternal and paternal transmission.

Box
5.1

Structure and history of African elephant (*Loxodonta africana*) populations (Georgiadis et al. 1994)

African elephants (*Loxodonta africana*) once populated the entire African continent. However, throughout this range, populations have become fragmented and reduced in size as human populations have expanded. Elephant reintroduction programs over hundreds of kilometers have been attempted but most existing populations have not been reduced in size for long enough for there to have been severe depletion of genetic variation.

Aims of study. The aim of the study was to reconstruct a mtDNA genealogy so that historical patterns of gene flow between eastern and southern Africa could be inferred.

Methodology. 270 African elephants were sampled from 10 protected areas in five countries: Kenya (43 individuals); Tanzania (35 individuals); Zimbabwe (74 individuals); Botswana (84 individuals); and South Africa (34 individuals), representing two regions (southern Africa and East Africa). In addition, 15 Asian elephants (*Elephas maximus*) from North American zoos were included for comparative purposes. Samples comprised skin from biopsy darts, scraps from tusk bases, muscle from culled individuals, and blood samples. Total DNA extracts were PCR-amplified using primers for the mtDNA ND5–6 region (c. 2450 bp) before cleavage with nine 4bp-cutting and one 6bp-cutting restriction enzymes. End labeling was used to map restriction sites.

Analytical methods. The degree of genetic differentiation among sample sites was quantified using Cockerham and Weir's (1987) identity-by-descent definition of F_{st}, treating each haplotype as an allele at a single locus. Phylogenetic trees were constructed using maximum parsimony and the minimum number of migration events was estimated by overlaying geographic distance onto the haplotype phylogeny. *Nm* was estimated from the minimum number

of migration events at each sampling locality using Slatkin and Madison's (1989) cladistic measure of gene flow.

Results and importance. The equivalent of 194 bp of the mitochondrial genome was screened and 47 restriction sites were mapped, of which 23 were polymorphic within African elephants and six were fixed between the two species. There were 10 mtDNA haplotypes among the African elephants, of which two haplotypes were widespread and the remainder were localized. Across the continental range, African elephant populations were differentiated [F_{ST} (95% CI) = 0.39 (0.19 – 0.58)] but there was no significant subdivision at the regional levels. The 10 African elephant mtDNA haplotypes were monophyletic, and within this clade, there were two monophyletic haplotype clades, one of haplotypes from Tanzania, Botswana, and Zimbabwe and the other from across the continental distribution (Fig. 5a). The co-occurrence of haplotypes from divergent mtDNA clades is unexpected given that there is no evidence that a geographic barrier has isolated populations for sufficient time (millions of years) to produce these divergent lineages. Georgiadis et al. (1994) suggested that a hypothesis of mtDNA lineage exchange with extinct elephants (e.g., *L. atlantica*) could be invoked, although this would not explain the divergence within each mtDNA lineage. Predictions from coalescence theory (Box 5.6) indicate that it is possible that the two mtDNA clades could have persisted by chance alone for more than four million years in the absence of a geographic barrier. Furthermore, under the assumption of a coalescence time of four million years and a generation time of 20 years it is estimated that *Ne* is about 50,000 females. These data support the view that populations of savannah elephants have a complex population history, characterized by intermittent gene flow between subdivided populations, and illustrate the value of a coalescent approach to the analysis of mtDNA data. Fleischer et al. (2001) have undertaken similar phylogeographic investigations on the Asian elephant.

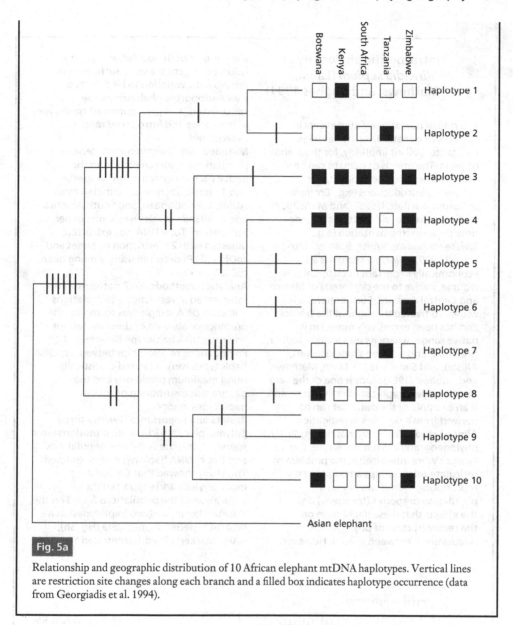

Fig. 5a

Relationship and geographic distribution of 10 African elephant mtDNA haplotypes. Vertical lines are restriction site changes along each branch and a filled box indicates haplotype occurrence (data from Georgiadis et al. 1994).

phylogeographic structure (e.g., Box 5.3; Olsen, 2002), since it is possible to analyze numerous loci and hence complex evolutionary hypotheses can be tested. Furthermore, biparental inheritance of nDNA means that genealogical patterns are likely to be more representative of the entire population history than uniparentally inherited organelle genomes. However, a significant drawback with nDNA loci is the problem of paralogy (Martin and Burg 2002).

The identification of intraspecific variation in the organelle genomes of plants and animals has meant that it is possible to infer relationships between populations as well as between species. That is, the genealogies of genomes can be reconstructed, and inferences made about phenomena

Intraspecific phylogeny in *Gliricidia sepium* (Lavin, Mathews, and Hughes 1991)

The role of human-mediated events in patterns of genetic diversity has been recognized, albeit implicitly, for thousands of years. The effects of humans can be profound and include accidental or deliberate introductions (e.g., *Bromus tectorum*, Bartlett, Novak, and Mack 2002) and interference with patterns of gene flow through fragmentation (e.g., *Swietenia humilis*, White, Boshier, and Powell 1999). *Gliricidia sepium* is an economically important neotropical tree legume, native to the dry forests of Mexico and Central America. Furthermore, it is widely cultivated throughout the tropics and has been extensively moved in its native range, where its uses include fodder, firewood, and a coffee shade (Stewart, Allison, and Simons 1996). Lavin, Mathews, and Hughes' (1991) study is one of the very first phylogeographic studies in plants, and is an example of the data that can be derived from a rigorous investigation.
Aims of study. The study was not explicitly phylogeographic in nature. Rather the authors were interested in the problem of reticulate relationships among sexually reproducing individuals within a population or species (tokogeny) and the effects that these might have on the reconstruction of phylogenetic relationships between species. However,

sampling of both populations and the chloroplast genome were sufficient for intraspecific variation to be detected, a well-supported phylogeny to be constructed, and the chloroplast haplotypes in the native and introduced ranges determined.
Methodology. Twenty-nine *G. sepium* populations were sampled from the native Mesoamerican range of species and 13 populations were sampled from Africa, the Caribbean, and South America; one to 10 individuals were sampled per population. Total DNA was extracted, digested with 21 restriction enzymes and cpDNA RFLPs detected using a mung bean cpDNA library.
Analytical methods. RFLP patterns were interpreted as restriction site mutations and each DNA sample was scored for the presence or absence of these mutations and its cpDNA haplotype determined. Phylogenetic relationships between cpDNA haplotypes were analyzed cladistically using maximum parsimony and the geographic distribution of cpDNA haplotypes mapped.
Results and importance. Twenty-three polymorphic restriction site mutations were found among the *G. sepium* populations and five cpDNA haplotypes were defined. This study showed that *Gliricidia* is monophyletic and supported the separation of the populations found on the Yucatan Peninsula (two haplotypes) as a separate species, *G. maculata* (Fig. 5b); RAPD markers also differentiated Yucatan

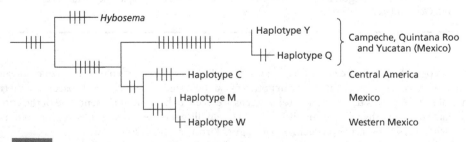

Fig. 5b

Parsimony cladogram of *Gliricidia sepium* showing the geographic distribution of chloroplast DNA haplotypes in Central America and Mexico. Vertical lines indicate restriction site mutations (after Lavin et al. 1991).

populations from the rest of the distribution (Chalmers et al. 1992). Three haplotypes were found over the rest of the Mesoamerican distribution. One haplotype was found in Central America, one was found in western Mexico, and a third, which differed from the western Mexican haplotype by a single restriction site mutation, occurred in populations in southern Mexico and northern Central America. Such north–south differentiation of cpDNA haplotypes has been seen in other Mesoamerican species, for example, *Cedrela odorata*. Chloroplast DNA

haplotype variation was also useful in hypothesizing the origins of *G. sepium* introductions outside of Mesoamerica. For example, Central American populations were the most likely source of seed for introductions to Cuba, Dominican Republic, Puerto Rico, and north South America. In contrast, Thai populations are polymorphic for cpDNA haplotype, which suggests that there were either multiple introductions or a single introduction of seed from a population polymorphic for the Central American and west Mexican haplotypes.

Box 5.3

Intraspecific nDNA phylogeny of black tiger prawn (*Penaeus monodon*; Duda and Palumbi 1999)

The Indo-West Pacific is considered one of the most diverse assemblages of marine species. Whereas some groups, for example, corals and nemerteans, have related species that occur in either the Indian or Pacific Oceans, others, for example, cone snails and prawns, are distributed throughout the region. Genetic studies using allozyme and mtDNA analyses suggest that *Penaeus monodon* populations on the east and west coasts of Australia are genetically differentiated, despite appearing to have a continuous distribution from southeast Africa to eastern Australia and Japan. There is the potential for extensive gene flow among populations, since the prawn has an offshore planktonic larval phase, estuarine post-larval and juvenile phases, and an inshore adult and spawning phase.
Aims of study. The authors were interested in the issue of whether genetic differences reflect a regional discontinuity among populations separated by the Indo-Australian Archipelago. Specifically, they tested the hypothesis that there were no significant nuclear DNA differences among populations from the Indian and Pacific Oceans.
Methodology. Six populations, three from each of the western Pacific (Malaysia

(12 individuals), Indonesia (13 individuals), Philippines (11 individuals)) and western Indian (Madagascar (12 individuals), Mauritius (20 individuals), Tanzania (6 individuals)) Oceans, were sampled. The intron (c. 200 bp) from nDNA-encoded elongation factor 1-α (*ef1-α*) was amplified from purified DNA samples, the PCR products cloned and sequenced.
Analytical methods. Relationships between aligned allele sequences (indels were treated as single characters) were estimated using maximum parsimony and F_{st} calculations were made based on the sequence data. Nucleotide diversity was calculated between individuals.
Results and importance. The 112 cloned sequences (60 different alleles) sequenced from 74 individuals showed 34 phylogenetically informative characters, including six indels. The western Pacific populations formed a clade distinct from the western Indian populations (Fig. 5c), which was confirmed by a statistically significant difference in pairwise F_{st} (= 0.505) between the Pacific and Indian populations. These data show that western Pacific and western Indian prawn populations have separate evolutionary histories. Combined with data from other species, for example, *Linckia laevigata* (Williams and Benzie 1998), there is evidence that dissimilar marine taxa share similar vicariant histories. Average sequence diversity was significantly greater within the Pacific than within the Indian

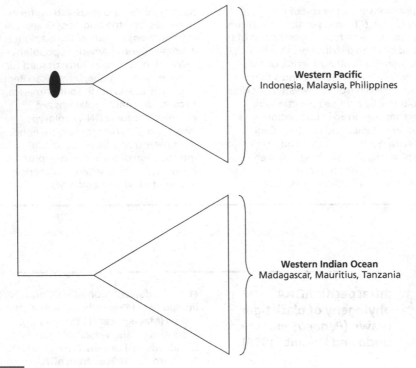

Western Pacific
Indonesia, Malaysia, Philippines

Western Indian Ocean
Madagascar, Mauritius, Tanzania

Fig. 5c

Major relationships between populations of *Penaeus monodon* using maximum parsimony analysis of the nDNA-encoded elongation factor $1 - \alpha$ intron sequence. The filled ellipse indicates a statistically significant clade (summarized from Duda and Palumbi 1999).

Ocean populations. Low diversity in the Indian Ocean populations may be due to small effective population sizes, recent or historical population bottlenecks in the Indian Ocean, or a recent colonization of the Indian from the Pacific Ocean. More detailed, coalescent-based analyses of the genealogical data may provide the opportunity to differentiate between these different hypotheses. Furthermore, when two alleles were sequenced from the same individual they were more likely to be different and this was interpreted as indicating that more than one locus was present within the data set. This study is one of only a few phylogeographic investigations that have used nDNA, and illustrates the value of the approach but emphasizes some of the potential difficulties that might be encountered, for example, cloning of PCR products and identification of additional loci.

such as hybridization, lineage sorting, and the origins of autopolyploids and allopolyploids (see chapter 6); in addition to the consequences of historical changes in population size and gene flow (Avise 2000, Emerson, Paradis, and Thébaud 2001). Furthermore, organelle genealogies are increasingly important in making conservation decisions, especially for species where resources limit the conservation of all populations (e.g., Moritz and Faith 1998, Newton et al. 1999).

This chapter begins with an introduction to the structure of phylogenetic trees and continues with a summary of the means by which trees are generated from molecular data. The final part of this chapter concerns the interpretation of phylogenies within geographic and demographic contexts.

5.2 Homology, gene trees, and species trees

Homology, similarity due to descent from a common ancestor, is a principal biological concept (Hall 1994). Thus, two DNA sequences are homologous if they are derived from the same DNA sequence in a common ancestor. Despite frequent references to percentages of homology in the molecular biological literature, homology is an all-or-nothing concept: two DNA sequences are either homologous or non-homologous (Doyle and Gaut 2000). All sequences in a multigene family (e.g., histones) are homologous, although divergence between any two sequences is termed paralogy or orthology (Doyle and Gaut 2000). If the two sequences were generated by duplication, then they are paralogous, for example, primate globin genes (Doolittle 1987), in contrast, orthologous sequences are the result of speciation events (Fig. 5.1).

Definitive evidence of paralogy is provided by the occurrence of sequences from the same gene family in a single genome. In contrast, orthology must be inferred from a variety of sources (Doyle and Gaut 2000), including: (i) common function between sequences from the same gene family in different species; (ii) phylogenetic demonstration that sequences from different species belong to the same gene family; and (iii) demonstration of shared chromosomal position and linkage relationships between species. If paralogous sequences diverge after duplication events, there will be a clear distinction between orthologous and paralogous sequences.

Naively, one might believe that once the evolutionary history of a gene (gene tree) is known, then the evolutionary history of the organism (species tree) is known. However, gene trees may only estimate species tree relationships since paralogs occur within individuals that can confound phylogeny reconstruction. That is, species trees can be recovered only if complete sets of genes are available, else it is necessary to restrict analysis to those genes that have not undergone duplication, that is, orthologous comparisons must be made.

Undetected paralogy, lineage sorting, and complications due to lineage reticulation (hybridization) are important factors that can influence gene tree and species tree relationships

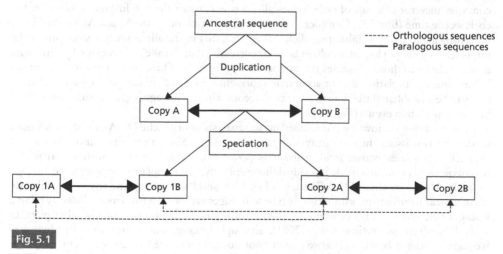

Fig. 5.1

Paralogy, similarity due to gene duplication (A and B), and orthology, similarity due to speciation (1 and 2).

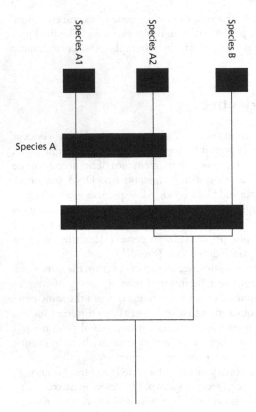

Species A1

Species A2

Species B

Species A

Fig. 5.2

Differences between species (filled boxes) and genes (lines). Gene divergence usually predates species divergence, hence a gene phylogeny of the relationships between species A1, A2, and B would show that species A2 was more closely related to species B rather than species A1.

(e.g., Doyle 1992, Page and Charleston 1997, Sanderson and Doyle 1992; chapter 6). The occurrence of ancestral polymorphism and differential allele extinction, in orthologous comparisons, can result in the allele phylogeny not matching the species phylogeny (Fig. 5.2). The most recent common ancestor of a pair of orthologous alleles occurs when the two lineages coalesce at the coalescence time (Box 5.6). Consider a case where speciation of A into A_1 and A_2 resulted in A_1 inheriting allele 1 and A_2 inheriting allele 2 (Fig. 5.2), that is, the allele lineages were sorted. An intraspecific gene phylogeny would indicate (incorrectly) that the alleles in species A_2 were more closely related to those of species B than species A_1 (Fig. 5.2). Thus, alleles present in a lineage before lineage speciation do not accurately represent species relationships. Lineage sorting is likely to be a problem if the time to allele coalescence within a lineage is greater than the interval between speciation events (Fig. 5.3).

Gene genealogical investigations based on the analysis of organelle DNA can be confounded further by two issues: heteroplasmy and paternal leakage. Since there are multiple copies of organellar genomes within individuals, it is possible that two or more distinct organellar haplotypes will occur in a single individual (heteroplasmy). Heteroplasmy appears to be rare, for example, *Mysotis mysotis* mtDNA (Petri, von Haeseler, and Paabo 1996), and any variants that do occur appear to arise from within closely related lineages rather than from more distantly related lineages (see chapter 2). Furthermore, sorting of organelle genotypes usually takes place in several hundred generations (Birky 2001), although homoplasmy is not absolute since low frequency variants continually arise and are not normally detected unless stringent assays are used (e.g., Comas, Paabo, and Bertranpetit 1995). Similarly, paternal leakage of animal mtDNA is not usually detected unless highly sensitive PCR procedures are used. Avise (2000) has argued

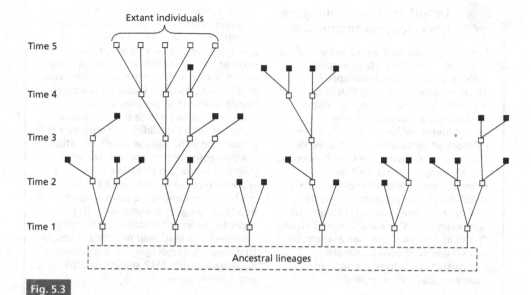

Fig. 5.3

Lineage sorting in a uniparentally inherited DNA sequence over time. Filled boxes indicate extinction.
All of the extant lineages coalesce to one individual at time 1.

that paternal leakage and other rare sources of heteroplasmy (e.g., *de novo* mutation) do not compromise the utility of mtDNA in phylogeographic studies. The majority of investigations assume maternal organelle DNA transmission, and whilst this is true in the majority of cases, paternal and biparental organelle transmission can also occur (see chapter 2). Undetected paternal or biparental inheritance of an organelle genome may complicate the phylogeographic interpretation of data.

5.3 Tree form and building

5.3.1 Tree terminology

A tree, which consists of nodes connected by branches, is a mathematical structure that can be used to model the evolutionary history of a group of samples, that is, a hypothesis of relationships. Depending on the investigation, these relationships can be between organisms (taxa), haplotypes, or sequences. In contrast, a phylogeny (or evolutionary tree) is the actual pattern of historical relationships. For trees based on interspecific comparisons, the terminal taxa (or operational taxonomic units (OTUs)) are the data (e.g., DNA sequences), whilst the internal nodes are hypothetical ancestors. For intraspecific trees, ancestral haplotypes may exist in the sample (see section 5.5). At the base of this tree is the root, which is the ancestor of all terminal taxa that comprise the tree (Fig. 5.4).

Different types of tree have been used to summarize evolutionary data, and different researchers often have rigid views on the validity, or otherwise, of different techniques and the types of information that can be obtained from them (Stuessy 1990, Swofford et al. 1996). The three most common types of tree construction, used for interspecific studies, are: (i) cladograms; (ii) additive trees; and (iii) ultrametric trees (Fig. 5.5). However, networks may also be appropriate for the construction of intraspecific trees (Box 5.5).

Length mutations and gene genealogy reconstruction

Length mutations appear to be particularly common types of mutations in cpDNAs (Palmer 1985), whether intraspecific variation is detected using PCR-RFLP, SSR, or sequencing approaches. Length mutations are becoming increasingly important for the identification of variation in the chloroplast genome through the use of chloroplast microsatellites since they evolve quickly compared to restriction sites (Provan, Powell, and Hollingsworth 2001; chapter 2). Length mutations have been widely advocated as useful phylogenetic characters (e.g., Lloyd and Calder 1991, van Dijk et al. 1999) and are used as characters in intraspecific cpDNA phylogeny reconstruction, despite the homoplasy problems that can be created.

Consider the example shown in Fig. 5d. A region of cpDNA from four taxa (A–D) is digested with two restriction enzymes, X and Y. In case I, the 0.5 kb length mutations occur at the same positions in taxa A and B, whilst, in case II, the length mutations occur at different positions. Enzyme Y produces identical restriction profiles for both cases, leading to the tree shown for case I. However, with the addition of enzyme X it is clear that although the length mutations are the same size they are in different positions in the two cases. In case I, a fully resolved tree is produced, whilst for case II, the length mutations are autapomorphic and the phylogeny is unresolved. Thus, in order to use length mutations effectively it is necessary to map their positions to ensure that they are homologous. This is necessary whether one uses RFLP analysis or DNA sequence analysis.

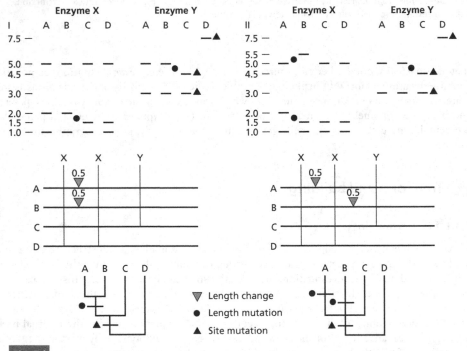

Fig. 5d

Length mutations and phylogeny reconstruction. DNA was extracted from four taxa (A–D), a cpDNA sequence amplified and the product digested with two restriction enzymes (X and Y). The top of the diagram shows the restriction patterns that result when DNA fragments represented in the middle panel are digested with enzymes X and Y. The tree at the bottom is constructed based on fragment data. The closed triangle represents a site mutation, whilst the closed circle represents a length mutation. (I) Length mutation occurs in the same enzyme X restriction fragments of taxa A and B. (II) Length mutation occurs in different enzyme X restriction fragments of taxa A and B.

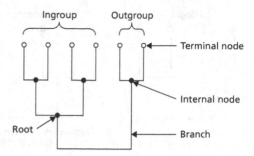

Fig. 5.4

Morphology of an evolutionary tree.

Box 5.5

Networks

The possibilities of reticulation, multifurcation, recombination, and persistence of ancestral alleles in gene genealogies mean that conventional approaches to phylogeny reconstruction based on interspecific models may be inappropriate. Networks (Fig. 5e), which incorporate cycles into trees rather than being strictly bifurcating, allow for such events and can account for both species-level processes and incorporate predictions from population genetic theory (Posada and Crandall 2001). Network methods are usually based on distance approaches, where the distance among haplotypes is minimized (i.e., the likelihood function is maximized). Network approaches are recent developments in phylogenetic

analysis, although numerous methods have been developed based on distance, maximum parsimony, and maximum-likelihood approaches (Posada and Crandall 2001). A method that is frequently used for the analysis of phylogeographic data is based on the identification of a minimum spanning network (e.g., Smouse 1998), that minimizes the length of the branches that connect all of the haplotypes in a network.

One of the difficulties with networks is that they can be difficult to root, since outgroups are often separated from the ingroup by many mutations. However, it is expected that the oldest ancestral haplotype is the root of the tree, and based on the coalescent it is the ancestral haplotype that is expected to be the most frequent haplotype (Castelloe and Templeton 1994, Hudson 1990).

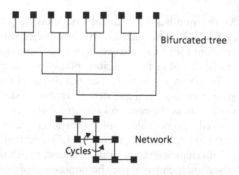

Fig. 5e

Comparison of a bifurcated tree and a network. Filled boxes indicate the terminal taxa.

Cladograms show the relative recency of common ancestry, thus A and B share a more recent common ancestor than either does with C (Fig. 5.5). In contrast, additive trees have information associated with the branch lengths: the longer the branch the more evolutionary change there has been along it. An ultrametric tree is a special type of additive tree, where all

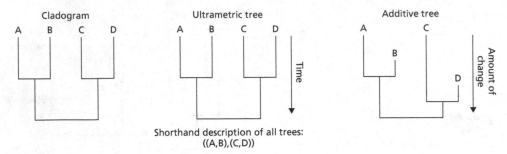

Comparison of the information content of a cladogram, ultrametric, and additive trees. No meaning attaches to the horizontal axis in each of these trees.

the terminal taxa are equidistant from the tree root, and can be used to depict evolutionary time, either directly as years or indirectly as the amount of sequence divergence using a molecular clock (see section 5.4.3).

Cladograms and additive trees can be rooted or unrooted, whilst ultrametric trees are rooted by definition. In a rooted tree, a node is identified from which all other nodes descend; hence, the tree has direction. In the absence of a root, ancestor–descendant relationships cannot be identified and therefore evolutionary relationships are not specified. If one considers four terminal taxa, then each of the three unrooted trees corresponds to a set of five rooted trees (since roots can be placed on any of the five branches that connect the nodes (Fig. 5.6)). The number of possible unrooted trees (U_n) for n terminal taxa is:

$$U_n = (2n - 5)(2n - 7) \ldots (3)(1) = \frac{(2n - 5)!}{2^{n-3}(n - 3)!}$$

for $n \geq 2$, and the number of rooted trees (R_n) is:

$$R_n = (2n - 3)(2n - 5) \ldots (3)(1) = \frac{(2n - 3)!}{2^{n-2}(n - 2)!} = (2n - 3)U_n$$

As the number of terminal taxa increases, the number of possible trees increases rapidly (Table 5.2). This increase in tree number with the number of terminal taxa is a fundamental problem for tree construction, but has been overcome, to some extent, in recent years with the availability of easily accessible computing power.

In a phylogenetic tree, ancestral character states are represented by internal nodes and for any given tree, ancestral and derived character states can be determined; if a terminal node has the same state as the common ancestor then it is primitive (plesiomorphic), whilst the alternative is derived (apomorphic). Unique characters states are called autapomorphies, whilst shared derived states are synapomorphies (Fig. 5.7). For two identical character states, similarity may be due to inheritance directly from an ancestor with the same state (homology) or similarity may have arisen independently (i.e., homoplasy). Homoplasy may be distinguished as: (i) convergent (similarity from different ancestral conditions) and parallel (similarity from the same ancestral conditions) evolution which results in the independent evolution of the same feature in two unrelated sequences; and (ii) secondary loss of a derived trait which results in apparent reversion to the ancestral condition (Fig. 5.8).

Relationships between alleles sampled from different species are hierarchical, since they result from reproductive isolation and population differentiation over long time scales. However,

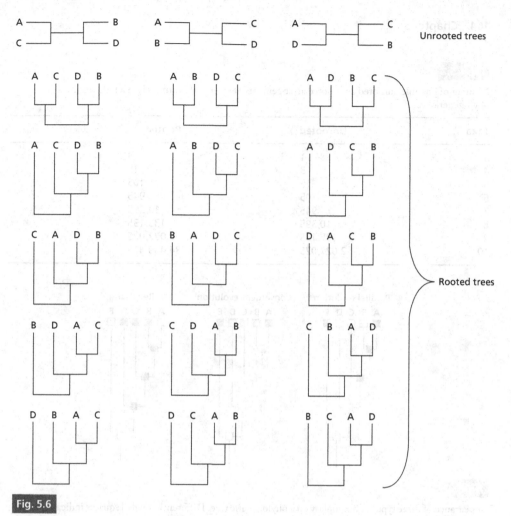

Fig. 5.6

All possible rooted and unrooted trees between four terminal taxa (A, B, C, and D).

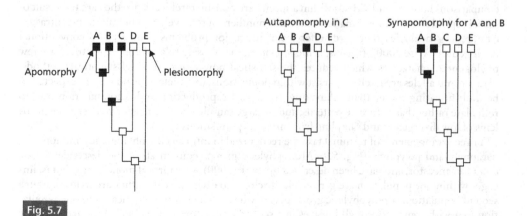

Fig. 5.7

Definition of synapomorphy and autapomorphy on a hypothetical tree of five taxa. Differently shaded squares indicate different character states for one character.

Table 5.2

Number of possible unrooted and rooted arrangements of three to 10 terminals in a fully resolved phylogenetic tree.

Taxa	Unrooted	Rooted
3	1	3
4	3	15
5	15	105
6	105	945
7	945	10,395
8	10,395	135,135
9	135,135	2,027,025
10	2,027,025	34,459,425

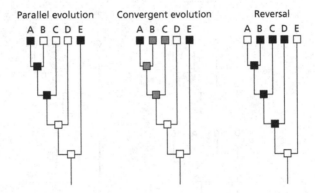

Fig. 5.8

Appearance of three types of homoplasy on a phylogenetic tree. Differently shaded squares indicate different character states for one character.

comparisons between individuals within a species are non-hierarchical since they are the result of sexual reproduction, recombination, and small numbers of relatively recent mutations. Intraspecific gene genealogies therefore suffer from five major problems compared to conventional phylogenies (Avise 2000): (i) conspecific individuals are closely related and hence there are few phylogenetic characters which reduces the statistical power of traditional statistical methods; (ii) ancestral alleles are likely to persist within populations; (iii) allele genealogies are expected to be multifurcating rather than bifurcating; (iv) sexual reproduction and recombination lead to reticulate rather than bifurcate patterns; and (v) large sample sizes are needed to overcome problems of low divergence and sampling from multiple populations.

Three arrangements of terminal taxa are recognized in intraspecific phylogenies: monophyly, paraphyly and polyphyly (Fig. 5.9). Monophyletic groups contain all of the descendants of a common ancestor, and have been described by Avise (2000) as reciprocal monophyly; that is, lineages within one population are genealogically closer to each other than they are to lineages in a second population. A polyphyletic group occurs where some but not all lineages in one population join with some but not all lineages in a second population to form a clade. A paraphyletic group is where all lineages in one population form a monophyletic group that is nested within a second population. Interspecific phylogeny would consider that polyphyletic groups are typically

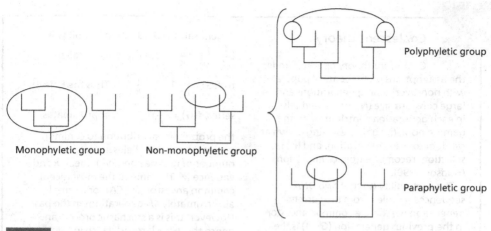

Fig. 5.9

Monophyletic and the two types of non-monophyletic group: paraphyletic and polyphyletic.

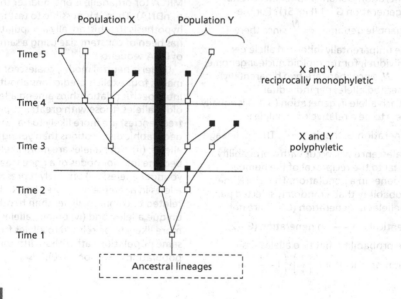

Fig. 5.10

Matrilinear sorting in two populations (X and Y) separated by a barrier to gene flow. Isolated daughter populations initially appear to be polyphyletic and then are eventually reciprocally monophyletic.

based on convergent characters, whilst paraphyletic groups are based on shared primitive characters. However, in intraspecific comparisons populations may progress through polyphyly, paraphyly to monophyly (Fig. 5.10), which may be an effect of incomplete gene lineage sorting. However, polyphyly and paraphyly of lineages may be the result of gene flow following separation of the populations, although it is difficult to distinguish between competing hypotheses of incomplete lineage sorting and secondary gene flow (see Section 5.4.2), since similar patterns can be produced depending upon whether separation is historical or contemporary.

Box
5.6

Coalescent theory

Coalescent theory operates under the assumptions of: (i) identical populations with non-overlapping generations and large constant size (N); (ii) individuals in each generation contributing to the gamete pool; (iii) gametes being drawn at random for next generation; and (iv) no selection, recombination, or migration (Hudson 1990).

The probability (v) that two DNA sequences sampled from the current generation (G) share a common ancestor in the previous generation (G – 1) is the same for any two sampled sequences. Let a population sample comprise κ alleles: what is the probability that two alleles in generation G share a common ancestor in generation G – 1 (Fig. 5f)? For the organelle genome, $\kappa = \dfrac{N}{2}$, since there is one uniparentally inherited allele per individual, for the diploid nuclear genome, $\kappa = 2N$, since there are two biparentally inherited alleles per individual.

Each allele in generation (G – 1) is equally likely to be a relative of an allele in generation G, that is, $v = \dfrac{1}{\kappa}$. That is, lineage coalescence will occur with a probability equal to the reciprocal of the number of genes in a population. Therefore, the probability that a randomly selected pair of alleles in generation (G – 1) are not identical is $1 - \dfrac{1}{\kappa}$. In generation (G – 2), the probability that two alleles share a common ancestor is $\left(\dfrac{1}{\kappa}\right) \cdot \left(1 - \dfrac{1}{\kappa}\right)$, and

in generation (G – 3) this probability is $\left(\dfrac{1}{\kappa}\right) \cdot \left(1 - \dfrac{1}{\kappa}\right)^2$. This can be generalized to $f(G) = \left(\dfrac{1}{\kappa}\right) \cdot \left(1 - \dfrac{1}{\kappa}\right)^{G-1}$. Thus the Poisson series, $f(G) = \left(\dfrac{1}{\kappa}\right) e^{-\left(\frac{(G-1)}{\kappa}\right)}$, approximates the probability distribution to common ancestry between alleles in terms of the number of generations, with mean κ and variance κ^2. The time to the most recent common ancestor (MRCA) for a sample is approximately $4N$ generations in the past. However, this is a stochastic process and hence there is a large variance in the total coalescence time: if only a very small sample is taken then the time to MRCA will be less than $4N$ generations (the mean time to MRCA for organelle is one-quarter that of nDNA). Thus, it is possible to test the hypothesis that historically a population has been of constant size using a sample of DNA sequences.

Under a neutral model, coalescent theory makes four explicit predictions about the intraspecific relationships among alleles: (i) older alleles (those with greater population frequencies) are more likely to have wider geographic distributions than younger alleles; (ii) older alleles are more likely to become interior nodes of a gene tree than younger alleles; (iii) uniquely represented alleles in a sample are more likely to be related to common alleles than to other unique alleles; and (iv) unique alleles are more likely to be related to alleles from the same population rather than different populations (Hudson 1990).

166 Fig. 5f

Allele coalescence in a population of five individuals (a–e). The probability (p) that alleles from a and b (sampled in generation G) did not coalesce in generation G–1 is $1 - (1/\kappa)$, where κ is the number of alleles sampled and the probability that they did coalesce in generation G–2 is $1/\kappa$. Therefore the total probability is $(1/\kappa)(1 - (1/\kappa))$.

5.3.2 Tree construction

Before data are analyzed using a particular method, one should be aware of the method's limitations and the type of information that it is capable of producing. Broad groups of methods for tree construction are introduced in 'Essential methods information,' pp. 181–4.

An ideal tree-building method will be: (i) efficient, that is, fast; (ii) powerful, that is, little data produces a reasonable result; (iii) consistent, that is, it converges on one result if there is enough data; (iv) robust, that is, minor violations of the assumptions do not produce poor phylogeny estimates; and (v) falsifiable, that is, the method indicates which assumptions are violated (Nei and Kumar 2000). However, all of the current methods emphasize one or more criteria at the expense of the remainder. It is also important to appreciate the difference between accuracy and precision: accuracy is how close the tree is to the true tree and precision is how many trees are excluded. A method that finds only one tree is very precise but if this tree is different from the true tree then it is not very accurate.

One of the most widely used means of dividing the methods of tree construction is to consider whether distance or discrete information is used (Fig. 5.11). Distance-based methods convert aligned sequences or comparisons between gel-banding patterns (e.g., RAPDs and AFLPs) into a pairwise distance matrix and then this matrix is used in a tree-building method (e.g., neighbor-joining). In contrast, discrete methods consider each character as an independent source of information (e.g., maximum parsimony and maximum likelihood). Discrete data can be converted to distance data but distance data cannot be converted to discrete data.

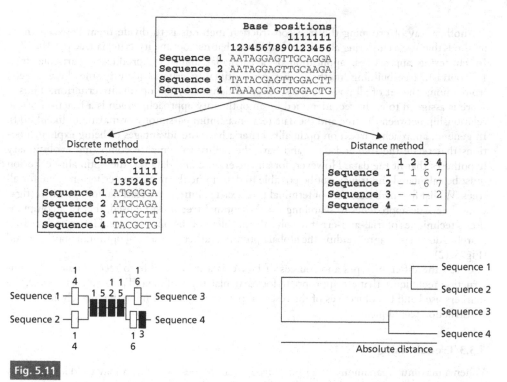

Fig. 5.11

Comparison of distance versus discrete characters as data for tree construction. DNA sequence from four taxa are compared. For the distance method, the absolute number of base differences between taxa are calculated. For the discrete method, variable base positions are used as characters to generate a parsimony tree. Shaded and unshaded boxes on the discrete tree indicate character states.

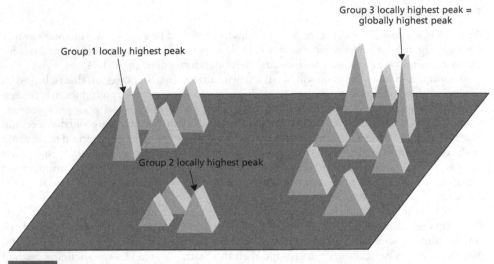

Group 3 locally highest peak =
globally highest peak

Group 1 locally highest peak

Group 2 locally highest peak

Fig. 5.12

Landscape of different height peaks to illustrate the concept of local and global optima. Three isolated groups of peaks are shown, each of which has a locally highest peak (local optimum). Only in one group is the locally highest peak also the globally highest peak (global optimum).

Another way of grouping the tree construction methods is to divide them between those methods that use a clustering algorithm and those that use optimality criteria (see pp. 181–84). In clustering approaches, application of a particular algorithm produces a particular tree. In contrast, tree-building methods that use optimality criteria identify the best tree(s) from among the set of all possible trees; "best" is defined by the optimality criterion. Thus, a score is assigned to each tree, identified by an optimality approach, which is a function of the relationship between the tree and the data (e.g., maximum parsimony or maximum likelihood). In general, approaches based on optimality criteria have the advantages of being explicit functions that relate the tree and data, and have the ability to compare competing evolutionary hypotheses that fit the data. However, for any given tree and data set, the optimality criterion must be calculated and it may not be possible to determine the maximum criterion value for all trees. Whilst for small numbers of terminal taxa, exact solutions are feasible, for most investigations, heuristic approaches to searching for the optimal tree must be used, that is, a computer uses a techniques of trial-and-error to solve the problem of the optimal tree. However, heuristic searches do not guarantee finding the global optimum; rather only a local optimum may be found (Fig. 5.12).

Given the different types and sources of DNA data it is possible to recommend tree construction techniques that are appropriate for particular types of data (Table 5.3). However, the markers used and the objectives of the research program will determine the final choice of tree construction technique.

5.3.3 Tree choosing

When a maximum parsimony tree is produced, even the best set of data may yield a set of very similar trees (e.g., with the same tree length) and it may be useful to represent the common features of this set of trees, that is, produce a consensus tree (Rohlf 1982). There are two popular types of consensus tree: (i) strict consensus; and (ii) majority rule (usually 50%) consensus. In strict consensus trees, only those groups that occur in all trees are considered (Fig. 5.13). For

Table 5.3

Examples of methods of tree construction that may be appropriate for the analysis of different types and sources of molecular data. The final decision regarding the appropriateness of a particular approach will be determined by the investigation.

Data type	Data source	Tree construction			
		Distance method examples		Discrete method examples	
		UPGMA	NJ	MP	ML
Discrete	DNA sequence	Yes	Yes	Yes	Yes
	RFLPs (mapped sites)	Yes	Yes	Yes	No
Distance	RAPDs	Yes	Yes	No	No
	AFLPs	Yes	Yes	No	No
	RFLPs (fragments)	Yes	Yes	No	No
	SSRs	Yes	Yes	No	No

Tree construction types: ML, maximum likelihood; MP, maximum parsimony; NJ, neighbor-joining; UPGMA, unweighted pair-group analysis using arithmetic means.
Data source: AFLPs, amplified fragment length polymorphisms; RAPDs, randomly amplified polymorphic DNAs; RFLPs, restriction fragment length polymorphisms based on either scoring of fragment positions or the relative positions of restriction sites within a DNA fragment; SSRs, simple sequence repeats or microsatellites.

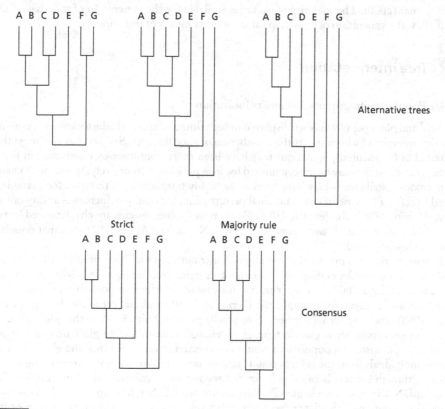

Fig. 5.13

The effect of strict and majority rule consensus approaches on three equally probable hypotheses of relationships between seven terminal taxa.

50% majority rule consensus trees, only groups that occur in 50% or more of the trees are included, that is, any group in the strict consensus will appear in the majority rule tree (Fig. 5.13). It is important to note that if characters are going to be mapped onto a phylogenetic tree then one of the equal length trees must be chosen for this purpose; characters cannot be mapped onto consensus trees.

As with any statistical procedure it is necessary to have some way of estimating the precision of an estimate and this should be undertaken for phylogenetic analysis, that is, the phylogenetic tree should be accompanied by some estimate of its confidence interval (Felsenstein 1985). One of the most widely used methods for doing this is to take multiple samples from the data set with replacement and to calculate a bootstrap (Felsenstein 1985). The bootstrap generates a pseudoreplicate sample of the same size as the original data, since the sampling occurs with replacement it is possible that some of the characters occur more than once in the pseudoreplicate and others are not represented at all. The new data matrix is then used to generate a tree and the whole process is repeated many times (usually 100–1000 times). The majority rule consensus of the population of trees is then calculated and the data presented as the percentage of support for each clade. The statistical properties of the bootstrap are complex (e.g., Li and Zharkikh 1995), although there is an emerging consensus that as a "rule of thumb", clades with ≥75% bootstrap support are strongly supported, those with 50–74% support are weakly supported, and those with <50% support are unsupported (e.g., Cameron et al. 1999). Other methods, including jack-knifing (where resampling occurs without replacement), to test the statistical support of trees are available and these are reviewed by Li and Zharkikh (1995) and summarized by Nei and Kumar (2000). The majority of packages available for the generation of trees have features that allow the generation of bootstrap values or other means of tree support.

5.4 Tree interpretation

5.4.1 Population fragmentation and colonization

Animal and plant populations are dispersed in both time and space, the historical and contemporary demography of which affects the distribution of gene lineages. Species that are currently distributed as fragmented populations may have been more continuously distributed in the past, thus genetic contact may have been limited for long periods. Alternatively, species with currently continuous distributions may have been more highly fragmented in the past, for example, isolated in refugia because of environmental perturbation, for example, increased aridity or glaciation (Hewitt 2000, Taberlet et al. 1998). Furthermore, other species are characterized by recent range expansions (e.g., *Bromus tectorum*; Bartlett, Novak, and Mack 2002) such that populations may be closely related.

Fragmentation of a previously continuously distributed species has a multiplicity of ecological and genetic effects depending on the amount of gene flow between the isolated populations (Young and Clarke 2000). Consider an extreme case of a continuous population fragmented into small, isolated refugia, for example, the Iberian and Balkan glacial refugia in Europe (Taberlet et al. 1998), and a maternally inherited organelle genome (Fig. 5.14). If the initial population were monomorphic for organelle type, given enough time following glaciation, the organelle genomes of the isolated populations would be expected to diverge. Once the glaciers retreated north, individuals from the refugia could colonize new habitats, meet each other, and exchange genes, although borders between colonization waves may be maintained for centuries/millennia (e.g., cpDNA in *Quercus*; chapter 7). Populations would therefore appear to be a mixture of organelle genotypes. Vicariance has been extensively invoked to explain patterns of organelle haplotype variation across species' ranges, particularly in the analysis of the effects of Quaternary climatic change (e.g., Comes and Kadereit 1998, Schneider, Cunningham, and Moritz 1998). Neutral gene diversity (H_e) in a population of size N is governed by the balance between mutation

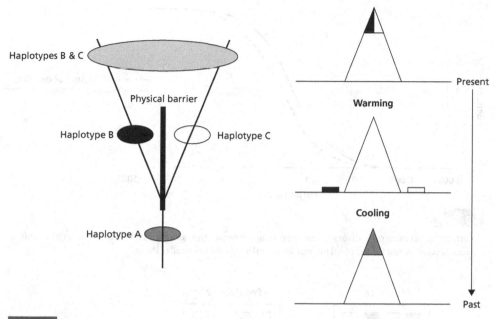

Haplotypes B & C

Physical barrier

Haplotype B   Haplotype C

Haplotype A

Present

Warming

Cooling

Past

Fig. 5.14

Illustration of the effects of fragmentation and post-fragmentation on the patterns of haplotype variation expected for an organelle genome, using glaciation as an example.

(μ), which increases diversity, and genetic drift, which reduces diversity. The equilibrium gene diversity for a diploid nuclear locus is:

$$H_e = \frac{4N\mu}{(4N\mu + 1)}$$

and for a haploid organelle locus is:

$$H_e = \frac{2N\mu}{(2N\mu + 1)}$$

It is clear that for equivalent values of $N\mu$ gene diversity is always lower for haploid markers than diploid markers (Fig. 5.15; Birky, Fuest, and Maruyama 1989). In addition, the generally lower mutation rate of organelle genomes compared to the nuclear genome has provided the rationale for the use of organelle genomes in the identification of glacial refugia. High allele diversity at both organelle and nuclear loci are used to define refugia, together with fossil or geological data. However, refugia may not harbor high levels of nuclear genetic diversity because of the different ways in which genetic diversity measures treat those populations with intermediate numbers of intermediate frequency alleles and those with high numbers of low frequency alleles (see chapter 3).

Differentiation between populations is typically estimated from the frequencies of unordered alleles (e.g., allozymes; chapter 2). However, with the widespread availability of ordered genetic markers, measures have been proposed that enable ordered data to be used. Pons and Petit (1996) proposed one such measure, where two values for genetic differentiation are calculated from the

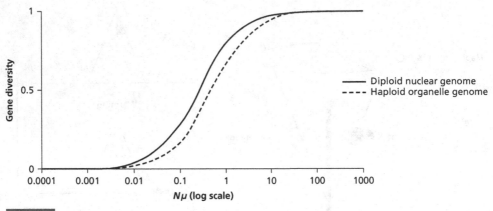

Fig. 5.15

Relationship between gene diversity and the product of population size (N) and mutation rate (μ) at drift mutation equilibrium for a diploid nuclear locus and a haploid organellar locus.

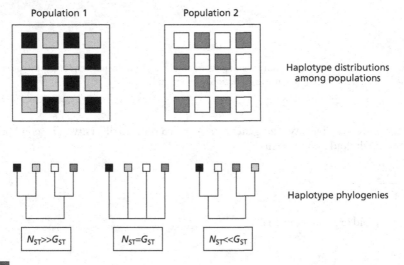

Fig. 5.16

Haplotype phylogeny and geographic distribution. The distribution of organellar haplotypes (shaded squares) between two populations is shown at the top of the diagram. Three types of relationship between these haplotypes are shown in the bottom of the diagram. If the haplotype phylogeny and geographic distribution coincide, then differentiation based on the relationships between haplotypes (N_{ST}) is greater than the differentiation based on haplotype frequency (G_{ST}). When haplotypes have no specific relationships, $N_{ST} = G_{ST}$. When the most closely related haplotypes are never found in the same population, $N_{ST} \ll G_{ST}$.

same data set; G_{ST} considers only haplotype frequencies, whilst N_{ST} takes into account similarities between haplotypes. The occurrence of significant genetic structure in the data may be tested by comparing N_{ST} to zero, whilst statistically significant differences between G_{ST} and N_{ST} provide information about haplotype phylogenies and their geographic distribution (Fig. 5.16). For example, when $N_{ST} \gg G_{ST}$ there is evidence that closely related haplotypes are more likely

Interpretation of differences between haplotypes and nucleotide diversities.

		Haplotype diversity Low (<0.5)	**High (≥0.5)**
Nucleotide diversity	Low (<0.5%)	Recent population bottleneck. Founder effect with single or few organelle lineages.	Bottleneck followed by rapid population growth and mutation accumulation.
	High (≥0.5%)	Divergence between geographically subdivided populations.	Large stable population with long evolutionary history. Secondary contact between differentiated lineages.

After Grant and Bowen (1998).

to be found in the same rather than different populations, although such differences may be due to sampling artifacts or unequal mutation rates between lineages, rather than the existence of phylogeographic structure.

Grant and Bowen (1998) compared animal mtDNA haplotype and nucleotide diversities as a means of assessing the demographic history of populations (Table 5.4). Haplotype diversity summarizes information on the number and frequency of different variants at a locus regardless of their sequence relationships. Nucleotide diversity is the weighted sequence divergence between individuals in a population, regardless of the number of different haplotypes. Populations with both low haplotype and nucleotide diversities may have recently experienced prolonged or severe bottlenecks. In contrast, high haplotype and nucleotide diversities are expected for stable populations with large long-term effective population sizes or mixed samples from historically split populations. High haplotype and low nucleotide diversities suggest that there has been rapid population growth from a small population, assuming that there has been sufficient time for the recovery of haplotype variation via mutation but too short for the accumulation of large sequence differences. Low haplotype and high nucleotide diversities may be the result of transient bottlenecks in large ancestral populations since short crashes eliminate many haplotypes but do not severely affect nucleotide diversity.

Intraspecific phylogeny and phylogeography have generated an impressive body of descriptive data for individual species from similar geographic areas and have shown that these species often have their variation structured in similar manners. These results imply that lineage distributions within different species are influenced by similar factors, for example, the glacial histories of northwest North America and Europe (Soltis et al. 1997, Taberlet et al. 1998), and potentially allow information to be obtained about the evolutionary factors that have influenced the distribution of species across landscapes. Such data highlight the potential value of comparative phylogeography, where multiple species comparisons are made from disparate evolutionary groups (Bermingham and Moritz 1998), although the statistical independence of the data used in such analyses must be determined through undertaking the analysis within a phylogenetic framework.

5.4.2 Gene flow estimates

Traditionally, gene flow estimation (chapter 4) has been made indirectly based on genetic measures of population structure using unordered markers, for example, allozyme or SSR, applied

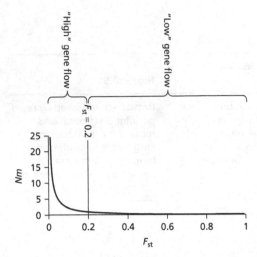

Relationship between F_{st} and Nm according to Wright's island model. An arbitrary cut-off is made at $F_{ST} = 0.2$ to indicate high and low gene flow taxa.

within a neutral model framework, where gene flow has a homogenizing effect on population structure and drift has a diversifying effect (Bossart and Prowell 1998). Geographical variation in population allele frequencies is used to calculate a combined parameter, Nm, which is usually interpreted as the mean per generation estimate of the absolute number of migrants exchanged among populations. Nm values that are greater than one indicate that the effect of gene flow is greater than the effect of drift (i.e., "high gene flow"), whilst the converse is true when $Nm < 1$ (i.e., "low gene flow"; Bossart and Prowell 1998).

The two most common methods of Nm estimation are Wright's island model (Wright 1951) and Slatkin's private allele method (Slatkin 1985). In Wright's method, the equilibrium value of Nm is related to F_{ST} according to

$$Nm = \frac{(1 - F_{ST})}{4 F_{ST}}$$

(see Fig. 5.17). Thus, if F_{ST} can be determined from empirical allele frequency data, then Nm can be calculated. The inflexion point of the curve indicates the distinction between "high gene flow" and "low gene flow". Slatkin's private allele method focuses on alleles restricted to single populations, and is based on the rationale that private alleles will only reach high frequencies when gene flow (i.e., Nm) is low. Computer simulations have shown that Nm is related to the average frequency of private alleles $[p(1)]$ by (Slatkin, 1985):

$$\ln (Nm) = \frac{\ln [p(1) + 2.44]}{-0.505}$$

However, the use of Nm as an indirect measure of contemporary gene flow has been severely criticized (e.g., Bossart and Prowell 1998, Whitlock and McCauley 1999) based on: (i) model expectations and presumptions of equilibrium; (ii) the curvilinear relationship between Nm and F_{ST} and $p(1)$; and (iii) the failure to distinguish historical from contemporary gene flow.

Nm estimation models are limited by the assumption that populations are at equilibrium between gene flow and random genetic drift. However, this can be biologically misleading, particularly for species in which recent history is a major determinant of population structure, where non-equilibrium outcomes are more likely. In rapidly evolving molecules, mutations that delineate particular lineages may not be dispersed at rates sufficient to attain equilibrium between drift and gene flow (Avise 2000).

The curvilinear relationships between Nm and F_{ST} and $p(1)$ can conceal important differences in the magnitude of interpopulation gene exchange. For example, the relationship between Nm and F_{ST} is almost flat in two parts of the curve, that is, small differences in one measure produce large differences in others. Values of $F_{ST} < 0.2$ indicate high gene flow but the population may comprise few to several thousand individuals, values of $F_{ST} > 0.2$ give low but indistinguishable values of Nm (Fig. 5.17). Furthermore, calculation of F_{ST} is scale-dependent and has a wide variance among loci (Whitlock and McCauley 1999). That is, F_{ST} values have large standard errors, although gene flow theory assumes that F_{ST} is parametric.

Nm values fail to distinguish between contemporary and historical gene flow. Thus, a large Nm may mean a high contemporary gene flow at equilibrium, a recent history that involved high gene flow, but where there is zero gene flow at present, or a mixture of these. Furthermore, estimates of Nm have different demographic impacts on large and small populations: the effect of gene flow on a small population is likely to be greater than on a large population.

The inability of standard estimates of Nm to differentiate contemporary from historical gene flow events has been instrumental in the use of gene genealogies in gene flow estimation. Thus, historical gene flow and population fragmentation can be considered in an explicit genealogical framework. The statistical methods that incorporate demographic–phylogenetic approaches into analyses of spatial variation in intraspecific lineages are in a preliminary stage of development – the two most widely known are the Neigel method (Neigel and Avise 1993) and the Templeton method (Templeton, Routman, and Phillips 1995).

The Neigel method of gene flow estimation considers that lineages are distributed under an isolation-by-distance scenario, that is, in a continuously distributed species, restricted gene flow means that older lineages are more widely distributed than younger ones. A continuum is envisaged in which DNA lineages are dispersed by a random walk, and for any generation the standard dispersal distance can be estimated between parent and progeny, according to any age or location. In this method, absolute dispersal distances are estimated rather than the rates estimated from other approaches. However, Barton and Wilson (1995) have highlighted four reservations over the Neigel method. Firstly, it is assumed that independent lineages are a random geographical sample, although the majority of genetic surveys are geographically biased. Secondly, information on lineage coalescence times is needed and this can only be obtained from application of the molecular clock. Thirdly, the composite estimate of lineage dispersal distance fails to identify particular genealogical units in a species. Fourthly, the assumption of an unconstrained random walk is unrealistic since species have range limitations, which will weaken the correlation between lineage age and geographic distance.

In Templeton's method of gene flow estimation, otherwise known as nested clade analysis, geographic information is overlain on an established gene tree within a rigorous statistical framework designed to measure the strength of any association and interpret the evolutionary processes responsible (Templeton 1998). Firstly, an unrooted cladogram is estimated using statistical parsimony procedures, so that a series of nested clades within the gene tree are produced; younger clades are nested within older clades. Geographical information is overlaid on the cladogram to generate clade distances and nested-clade distances. Clade distance is the mean spatial distance of clade members from the geographical clade center, whilst nested-clade distance is the mean spatial distance of nested-clade members from the geographical nested-clade center. Permutation tests are then used to determine which clade and nested clade are statistically large or small. A clade and nested-clade analysis of terminal versus interior clades contrasts geographic dispersion pattern for younger and older clades, and a permutation test can be used to determine support. The great advantage of the method is that it allows the rigorous testing of the null hypothesis of no association between geography and inferred structure from the tree. The failure to reject the null hypothesis may be due to biological factors (e.g., high contemporary

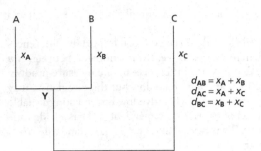

Fig. 5.18

Illustration of the relative rate test to estimate the difference in the number of substitutions between two closely related taxa (A and B) in comparison to a third (C) more distantly related taxon. If taxon A and taxon B evolve according to a molecular clock then $d_{AC} = d_{BC}$, that is, $d_{AC} - d_{BC} = 0$. Y is the common ancestor of taxa A and B.

gene flow or recent historical association) or insufficient power of the tests due to small sample size or inadequate geographic sampling (Bossart and Prowell 1998).

5.4.3 Dating divergence

Zuckerkandl and Pauling (1965) observed that the rate of amino acid substitution in primate hemoglobin was approximately constant, which led to the proposition of the molecular clock hypothesis. The molecular clock hypothesis asserts that the rate of nucleotide (or amino acid) substitution is approximately constant over evolutionary time. The hypothesis is controversial (Easteal, Collet, and Betty 1995, Li et al. 1996), and objections to a universal clock have been raised through the observation that gene function may change over time and that mechanisms of DNA damage and repair vary among different groups of organisms. However, the molecular clock hypothesis has remained attractive since it provides a means of estimating divergence time, especially when calibrated against paleontological or geological events.

Numerous methods have been proposed to test the molecular clock hypothesis (review by Nei and Kumar 2000), although one of the simplest is Fitch's (1976) relative rate test. This approach tests the accuracy of the molecular clock by estimating the number of substitutions between two closely related taxa and a third, more distantly related taxon (Fig. 5.18). The method does not require knowledge of divergence times, and can make use of either distance data or nucleotide substitutions. Let d_{AB}, d_{AC}, and d_{BC} be the distances between taxa A, B, and C, respectively. Therefore, the number of substitutions along each branch of the tree (x_A, x_B and x_C in Fig. 5.18), can be estimated according to:

$$x_A = \frac{(d_{AB} + d_{AC} - d_{BC})}{2}, x_B = \frac{(d_{AB} - d_{AC} + d_{BC})}{2}, \text{ and } x_C = \frac{(d_{BC} + d_{AC} - d_{AB})}{2}$$

If the substitution rates are constant, then x_A and x_B will be the same, therefore it is necessary to test the null hypothesis that $E(x_A) = E(x_B)$. There are three major problems associated with the relative rate test: (i) a known phylogeny is necessary, since taxa A and B must be closely related; (ii) if the relative rate test is applied to the same sequence from multiple sets of taxa there are problems with independence and hence statistical analysis; (iii) taxon C must not be too distantly related else problems associated with multiple substitution are increased and any rate differences are likely to have smaller impacts on the results. Multiple substitutions potentially occur between sequences that are widely separated: as sequences diverge it becomes increasingly likely that mutations will occur at the same site as previous mutations and therefore go unrecorded. For example, if G does not change in one lineage but changes to an A then back to a C in another lineage, these will not be apparent for the sequence. Consequently, the actual number of evolutionary changes along a lineage will be underestimated.

Attempts to calibrate molecular clocks have met with only limited success (e.g., Gibbons 1998), whilst the complexity of DNA substitution has led Muse (2000) to warn that a time-calibrated molecular clock is unlikely and that extreme caution should be exercised in cases of clock-based dating of phylogenetic events, especially at higher levels of the taxonomic hierarchy. Furthermore, even among closely related lineages pronounced rate heterogeneity has been reported, suggesting caution in establishing separation times from observed sequence differences (e.g., Zhang and Ryder 1995). Rate calibrations are often provisional because of uncertain geological or fossil evidence concerning separation times of related taxa.

5.4.4 The coalescent process

The majority of ecological genetic studies have relied on taking samples from a population, determining allele frequencies, and calculating "summary statistics", for example, mean gene diversity. These data can then be used to answer biological questions that are often associated with the history of the population. However, "summary statistics" ignore the majority of information in the data, especially the relationships between alleles, since alleles are usually identified by state rather than by descent. Very different evolutionary processes can generate similar patterns of allele frequencies that are very difficult to differentiate with summary statistics; for example, contemporary or historical gene flow may be responsible for F_{ST} differences between pairs of populations (Bossart and Prowell 1998). However, efficient generation of ordered, intraspecific genetic data (e.g., DNA sequences or mapped restriction sites) means detailed information is available about evolutionary relationships among individuals in a sample. Careful analysis of branching structure can provide information about the processes responsible for present-day structure. Thus, it is possible to study population level process within a temporal, non-equilibrium framework, that is, microevolution studied as a dynamic, historical process, changing over time within a species, where Hardy–Weinberg equilibria need not be assumed.

If DNA sequence data (or similar ordered data) are available then coalescent theory can be applied to the sample. Coalescent theory describes what happens when a small sample of alleles is taken from a population and is the formal mathematical and statistical treatment of gene genealogies that provides a framework for the study of the effects of population level processes (e.g., population size fluctuations, selection, and gene flow) on the expected time to common ancestry within a gene tree (Hudson 1990).

Over successive generations, new alleles arise through mutation and go extinct through genetic drift, all extant alleles being derived from common ancestral alleles that existed in the past. If alleles are sampled from two individuals in a population of constant size that is subject only to drift, it is possible to trace the lineages to the point at which the alleles shared a common ancestor, that is, the point at which they coalesce (Fig. 5.19). If a large sample of genes is studied, with random mating but no recombination, then the coalescence of any two lineages in a generation is equally likely (Box 5.6). As one proceeds back, the number of lineages is reduced at each coalescence (i.e., tree node), until a single ancestral allele (most recent common ancestor, MRCA) is reached. The coalescent is effectively genetic drift traced backwards through time, whilst in traditional population genetic approaches, drift runs forward through time to allele fixation. Coalescent events become less likely as one moves back: half of the total coalescence time occurs in the last two lineages (Box 5.6).

For a maternally inherited genome, if each individual produced only one progeny there would be no lineage sorting, no hierarchical branching, and no coalescent. However, in reality, there is variance in individual contributions to the progeny pool. Thus, in the interpretation of coalescent events for a maternally inherited DNA sequence, coalescence to a single ancestor does not imply that only one female was alive in the coalescent generation, rather other females did not have matrilinear descendants. Furthermore, organelle coalescence does not imply that any other females in the ancestral population did not contribute to later generations, for example, nDNA from other females may be represented by alleles from the non-maternal pedigree.

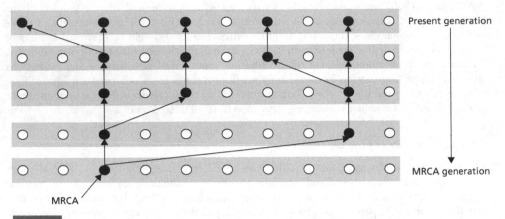

Coalescence. A gene genealogy that links a sample of five individuals (black circles) from a population of 10 individuals investigated at different points in time. The coalescence approach allows lineages to be traced back to the point when all of the individuals shared a common ancestor (MRCA).

5.4.5 Population size

The coalescent assumes that lineage survival and extinction are independent events, and small changes in fitness alter population demography and coalescent probabilities. Population growth inhibits lineage extinction and dramatically raises the probability of lineage survival over time. Conversely, lineage survival is normally dramatically reduced in declining populations.

For populations affected only by genetic drift, the coalescent indicates how gene genealogies are expected to appear if populations have different demographic histories (i.e., how populations are affected by changes in population size and structure; Fig. 5.20). For neutral mutations, since there is no alteration in the individual fitness and there will be no alteration in the numbers of offspring produced by an individual, changes in population size are easier to detect, than using adaptive markers (Ennos, Worrell, and Malcolm 1998). However, coalescent analyses are complicated by selection (fit alleles tend to produce more copies and therefore more branches) and recombination (one sequence can represent multiple evolutionary histories; Emerson, Paradis, and Thébaud 2001).

At its simplest, coalescent theory assumes that the populations are of constant size. However, if populations are expanding, coalescent events are more likely to occur in the past, when the populations were small, rather than in the present, when the populations are large (i.e., more coalescent events occur towards the root of the genealogy). The use of coalescent theory can show whether populations have been historically large or whether they have recently expanded, in contrast to summary statistics. Evolutionary processes, other than drift, leave different signatures on a gene genealogy since allelic coalescence times are affected. For example, positive selection produces genealogies with very short terminal branches (i.e., all lineages coalesce rapidly to produce a "star phylogeny") and with very little branching at the base compared to that expected under genetic drift. This arises since any variation must have arisen since the advantageous mutations had swept to fixation (i.e., "selective sweep") and purged all previous polymorphism. Similar effects are produced if bottlenecks occur, followed by population expansion. In order to distinguish selective sweeps from bottlenecks it is necessary to have information from multiple unlinked loci, since selective sweeps affect variation at single loci, whilst bottlenecks will affect numerous loci. The comparison of single and multiple unlinked loci can also be used to

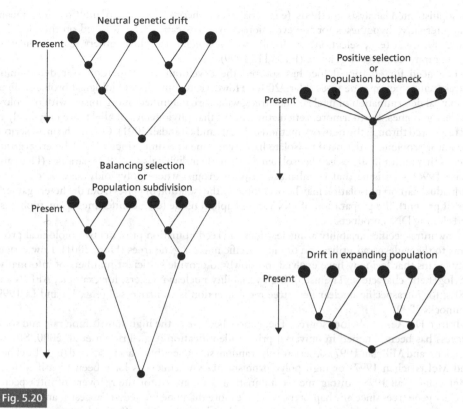

Fig. 5.20

Expectation of coalescent branching patterns for different evolution processes within and among populations.

distinguish between the consequences of stabilizing selection and population subdivision. Stabilizing selection (overdominance or frequency-dependent selection) will tend to maintain alleles for longer than expected under drift, that is, some of the coalescent events are likely to be older than $4N$ generations. Similarly, the same pattern is expected for subdivided populations, since coalescence between alleles can only occur when migration occurs. Migration involves all the alleles in a genome and selection is likely to involve only one or a few alleles, thus patterns influenced by stabilizing selection will be found using single loci and population subdivision will be detected using multiple loci.

5.5 Organelle versus nuclear intraspecific phylogenies

Organelle genomes have provided the primary data for investigations of the historical relationships among alleles and the differentiation of contemporary and historical processes, for example, gene flow (Avise 2000). However, uniparental inheritance and the absence of recombination mean that organelle genomes effectively behave as single loci, as is the case for paternally inherited nuclear sequences in heterogametic species, for example, the mammalian Y chromosome. Thus, an intraspecific organelle phylogeny may be unrepresentative of a species population history (Harpending et al. 1998) since contemporary genetic structure may be the result of a mosaic of population processes, that is, more loci are needed, despite the availability of sophisticated analysis methods (e.g., coalescent theory). Access to multiple loci means that alternative hypotheses for the evolutionary patterns can be tested based on the idea that some processes (e.g., selection) act locally within a genome, whilst others (e.g., population demography) act across a genome (Fu and Li 1999).

The need for additional loci has spurred the development of methods for determining intraspecific nuclear gene trees (Hare 2001). However, technical and biological problems limit empirical determination of nuclear gene trees, which are magnified in organisms with polyploid nuclear genomes. Furthermore, reticulation means that phylogenies of alleles are only likely to be recovered through the network methods (Posada and Crandall 2001). One of the most serious technical problems is the need to isolate haplotypes one at a time, since nDNA heterozygosity means that multiple alleles can be isolated from single individuals as well as paralogs (Hare and Avise 1996) – an issue that is enhanced in plant groups where polyploidy may be common. Individual haplotype isolation may be overcome by the isolation of genes from the heterogametic sex (if present), the preparation of DNA from haploid tissue (e.g., conifer megagametophytes), and cloning DNA products.

Low intraspecific variability at nuclear loci and recombination presents two biological problems to the widespread utilization of intraspecific nuclear gene trees (Hare 2001). Low variability means that loci may have evolved too slowly to provide sufficient numbers of informative phylogenetic characters. The use of high variability nuclear markers, for example, AFLPs and SSRs, for intraspecific nuclear gene tree reconstruction is controversial (e.g., Fu and Li 1999, Sunnocks 2000).

Intron DNA is an obvious source of sequences because of the high substitution rate and some success has been obtained in universal primer identification (e.g., Friesen et al. 2000, Strand, Leebens, and Milligan 1997). Alternatively, random sequences have been isolated (e.g., Lockhart and McLenachan 1997) or high polymorphism nDNA sequences have been identified (e.g., Bagley and Gall 1998). Intragenic recombination is a concern for the recovery of intraspecific nuclear gene trees since any haplotype will have more than one immediate ancestor and different segments within a haplotype will have independent histories. The effects of intragenic recombination include molecular clock disruption, homoplasy introduction, and the potential to bias the gene-tree shape to that expected for an expanding population (Schierup and Hein 2000). However, methods have been developed for the estimation and identification of recombinants in aligned sequences (Crandall and Templeton 1999). Hare (2001) argues that intragenic recombination may not be a significant problem since reduced interpopulational gene flow can restrict the effects of recombination, allowing phylogenetic structure to be built up between populations, whilst heterogeneous recombination rates across chromosomes mean that it may be possible to choose recombination rates appropriate to the problem of interest. The reconstruction of phylogenetic relationships among taxa using data from nuclear sequences requires a clear understanding of gene-tree relationships before inferring species relationships (e.g., Avise 1989, Doyle 1992, Pamilo and Nei 1988, Sanderson and Doyle 1992).

5.6 Further reading

Nei and Kumar (2000) is an excellent introduction to the methods of phylogeny reconstruction in molecular evolutionary analysis. Hall (2001) is a practical "cookbook" to phylogeny reconstruction from DNA sequence data, particularly parsimony and maximum-likelihood approaches. Posada and Crandall (2001) review network approaches to phylogeny reconstruction and Hudson (1990) summarizes the coalescent and its application in evolutionary biology. Templeton (1998) provides a very good background to the use of intraspecific phylogenies in gene flow and population size problems. Avise (2000) summarizes the use of intraspecific phylogenies in phylogeography. Li (1997) and Page and Holmes (1998) are good introductions to molecular evolution.

Distance methods

Basis of the method. Evolutionary distances are calculated for all pairs of terminal taxa, and a tree is constructed based on the relationships between these distance values. It is important to distinguish between the methods used to obtain the distance matrix and the methods used to construct the trees. If the estimate of the distance matrix is poor then the tree building may be ill affected. For a distance measure to be effective in phylogeny reconstruction, it must be metric and additive (Fig. 5g).

Distances are expressed as the fraction of characters that differ between pairs of terminal taxa. Consider the case of four terminal taxa (a, b, c, d) and let the distances between pairs of terminal taxa be represented in the form $D(taxon\ 1, taxon\ 2)$. For the distance to be metric it must be: (i) positive, that is, $D(a,b) \geq 0$; (ii) symmetrical, that is, $D(a,b) = D(b,a)$; (iii) distinctive, that is, $D(a,b) = 0$, if and only if $a = b$; and (iv) with a triangular inequality, that is, $D(a,c) \leq D(a,b) + D(b,c)$. A metric is ultrametric if it satisfies the additional criterion known as the three-point condition, that is, $D(a,b) \leq \max[D(a,c), D(b,c)]$. This criterion implies that the two largest distances are equal, thus defining an isosceles triangle and a constant rate of evolution. However, to be valid as a measure of evolutionary change, a metric or ultrametric distance must be additive and satisfy the four-point condition, that is, $D(a,b) + D(c,d) \leq \max[D(a,c) + D(b,d), D(a,d) + D(b,c)]$.

The two most widely used methods for constructing distance trees are both algorithmic: unweighted pair-group method with arithmetic means (UPGMA; Sneath and Sokal 1973) and neighbor-joining (NJ; Saitou and Nei 1987). In the UPGMA method, the pair of taxa with the smallest distance is located, a node placed halfway between them, a cluster produced, which is then used to rewrite the distance matrix, and the clustering process continued. UPGMA produces an ultrametric tree, where all the terminal taxa are equidistant from the root. In contrast, with the NJ method, distances to internal nodes are calculated directly and it does not assume that all of the taxa are equidistant from the root.

Advantages. Distance methods are easy to implement and fast computer programs are available to generate, what is usually, a single unambiguous tree. The tree generated may provide a suitable starting point for trees that involve model-based analyses, for example, maximum-likelihood approaches (Swofford et al. 1996). The short genetic distances involved in intraspecific phylogenies mean that sequence data are unlikely to be saturated with superimposed nucleotide substitutions, hence multiple substitutions are unlikely to be a significant problem.

Disadvantages. There are limitations in the computer software for distance analysis, for example, the recovery of only a single tree, the difficulty of identifying alternative tree solutions based on a particular distance matrix and idiosyncrasies of particular programs, which are a reflection of the software rather than the approach (Backlejau et al. 1996). In particular, the trees produced can be determined by the order of terminal taxon input and the occurrence of ties within the distance matrix. The major objections to distance methods are: (i) information is lost when a matrix of DNA sequences is summarized as

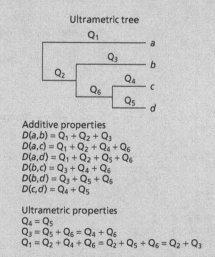

Additive tree

Additive properties
$D(a,b) = Q_1 + Q_2 + Q_3$
$D(a,c) = Q_1 + Q_2 + Q_4$
$D(a,d) = Q_1 + Q_5$
$D(b,c) = Q_3 + Q_4$
$D(b,d) = Q_3 + Q_2 + Q_5$
$D(c,d) = Q_4 + Q_2 + Q_5$

Ultrametric tree

Additive properties
$D(a,b) = Q_1 + Q_2 + Q_3$
$D(a,c) = Q_1 + Q_2 + Q_4 + Q_6$
$D(a,d) = Q_1 + Q_2 + Q_5 + Q_6$
$D(b,c) = Q_3 + Q_4 + Q_6$
$D(b,d) = Q_3 + Q_5 + Q_6$
$D(c,d) = Q_4 + Q_5$

Ultrametric properties
$Q_4 = Q_5$
$Q_3 = Q_5 + Q_6 = Q_4 + Q_6$
$Q_1 = Q_2 + Q_4 + Q_6 = Q_2 + Q_5 + Q_6 = Q_2 + Q_3$

Fig. 5g

Features of additive and ultrametric trees. Branch lengths between the four taxa (a, b, c, d) are represented by Q.

a pairwise distance matrix; and (ii) branch lengths may not be evolutionarily interpretable. Whilst parsimony analysis will enable particular character changes to be located on a tree, a distance tree only gives information about how much change occurred along a particular branch. The assumption of a molecular clock in the UPGMA approach may be unrealistic. Since the NJ method is a minimum-change approach there is no guarantee that the tree generated is the one with the smallest overall distance.

Types of data. All sources of molecular data can be converted to distance data, although the loss of information during the conversion means that if discrete data are available they should be analyzed using discrete methods. Where possible distance data types should be analyzed using a method that does not assume a molecular clock.

Further information. Excellent explanations of the methods of building trees using distance measures can be found in Swofford et al. (1996) and Nei and Kumar (2000).

Parsimony methods

Basis of the method. Parsimony methods are based on the assumption that the most likely tree is one that requires the fewest changes to explain the aligned data, and when conflicts with this assumption occur, they are explained as homoplasy. Consider four terminal taxa with the following aligned DNA sequences shown in Fig. 5h. There are three possible unrooted trees for these taxa, of which one is shown in Tree 1, along with two possible reconstructions of the character states for site 5 (Trees 2 and 3). It is assumed that nucleotide substitutions occur along the branch, although one does not know where along the branch this happens. In one case, one substitution is proposed, whilst in the second five substitutions are proposed. Thus, under the principle of parsimony the reconstruction that requires a single nucleotide substitution is preferred. To calculate the total number of evolutionary changes for a tree (tree length), the total number of changes at all sites is calculated. Invariant characters (e.g., sites 3 and 6) and characters that occur in only one terminal taxon (e.g., sites 1 and 7) are ignored.

Advantages. Parsimony is easily understood, appears to make relatively few assumptions, has much mathematical theory associated with it, and there is powerful computer software available to implement the approach. Parsimony may be very useful for intraspecific studies since relatively short branch lengths overcome the problem of extreme rate heterogeneity and long-branch

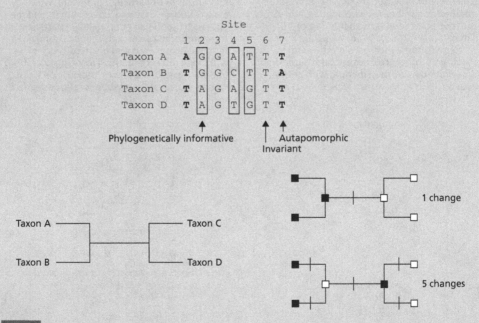

Fig. 5h

Parsimony analysis of four taxa using a seven-base sequence. Phylogenetically informative sites are shown in gray. One of the three possible unrooted trees is shown on the left of the diagram, whilst on the right are the two possible arrangements of characters at site 5 (filled box is T and unfilled box is G: vertical bars indicate character changes). The fewest number of changes is indicated by the upper tree, hence this is the one that would be accepted using maximum parsimony.

attraction that can compromise parsimony approaches at higher levels.

Disadvantages. Under some evolutionary models, parsimony is inconsistent, that is, independent of the addition of new data the wrong tree may still be recovered. One reason for this is long-branch attraction (Swofford et al. 1996), where long branches in a tree are likely to accumulate similar mutations at a base position and hence appear together in a parsimony analysis. Long-branch attraction is more likely to occur when rates of evolution show considerable variation among sequences, or when the sequences being analyzed are very divergent. One strategy to reduce the problem is to add sequences that break the long branches. However, long-branch attraction is unlikely to be a major problem in intraspecific phylogenies. More problematic may be low resolution data and the recovery of many hundreds of possible minimum-length trees.

Types of data. Maximum parsimony methods are appropriate for all types of discrete data.

Further information. Excellent explanations of the parsimony methods can be found in Swofford et al. (1996) and Nei and Kumar (2000; chapter 7).

Maximum-likelihood method

Basis of the method. Maximum-likelihood methods try to infer evolutionary trees by finding the tree that maximizes the probability of observing the data. In order to construct the tree, assumptions must be made about the rates of character change (e.g., nucleotide substitution), which are stated in the form of a model. Therefore the method is most practical for sequence data, although restriction site data can be used if a suitable model is available (e.g., DeBry and Slade 1985). There are many different models of nucleotide substitution, including the Jukes–Cantor one-parameter model, which proposes that all nucleotide substitution rates are equal and Kimura's two-parameter model, which assumes that transversions and transitions occur at different rates (Li 1997).

Consider an evolutionary model of rates of nucleotide substitution at one base position in a DNA sequence and a hypothetical tree of four terminal taxa. There are three possible unrooted trees of four taxa (Fig. 5i, Tree 1), which can be rooted at any node (Tree 2). The nucleotides at the two internal nodes are unknown, although since one of four different nucleotides can occur at each node, there are 16 possible arrangements. The probability of the arrangement shown in Tree 2, assuming that P and Q are A and A, respectively, is the probability of observing A at the root (P_A) and probability of each change along the branches leading to the terminal taxa; the probability of changing from A to T (P_{AT}) is calculated from the nucleotide substitution rate matrix in the chosen model and the length of the branch from A to T. Therefore, the probability of

```
                    Site
              1 2 3 4 5 6 7
    Taxon A   A A G G A T T T
    Taxon B   T G G C T T A
    Taxon C   T A G A G T T
    Taxon D   T A G T G T T
```

Model of nucleotide substitution

	A	C	G	T
A	P_{AA}	P_{AC}	P_{AG}	P_{AT}
C	P_{CA}	P_{CC}	P_{CG}	P_{CT}
G	P_{GA}	P_{GC}	P_{GG}	P_{GT}
T	P_{TA}	P_{TC}	P_{TG}	P_{TT}

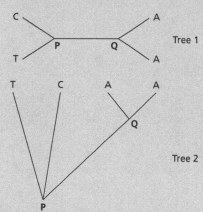

$$\text{Prob}(P = A; Q = A) = P_A + P_{AT} + P_{AC} + P_{AA} + P_{AA} + P_{AA}$$
$$\text{Prob}(P = A; Q = C) = P_A + P_{AT} + P_{AC} + P_{AC} + P_{CA} + P_{CA}$$
$$\text{Prob}(P = A; Q = G) = P_A + P_{AT} + P_{AC} + P_{AG} + P_{GA} + P_{GA}$$
$$\text{Prob}(P = A; Q = T) = P_A + P_{AT} + P_{AC} + P_{AT} + P_{TA} + P_{TA}$$

$$\text{Prob}(P = T; Q = T) = P_T + P_{TT} + P_{TC} + P_{TT} + P_{TA} + P_{TA}$$
$$\text{Prob}(\text{Tree 2}) = \text{Prob}(P = A; Q = A) + \text{Prob}(P = A; Q = C) + \ldots + \text{Prob}(P = T; Q = T)$$

Fig. 5i

Maximum-likelihood analysis of four taxa using site 4 and the model of nucleotide substitution indicated. Tree 1 is one of three possible unrooted trees between the four taxa, whilst tree 2 is one of the possible rooted trees. To determine the likelihood of tree 2 it is necessary to sum the probabilities of the 16 possible trees based on the nucleotides at P and Q.

this arrangement is $P_A \times P_{AT} \times P_{AC} \times P_{AA} \times P_{AA} \times P_{AA}$. Since there are 16 possible nucleotide arrangements at the internal nodes, then the probability of each arrangement must be determined and added together to get the probability of Tree 2, that is, $P_{arrangement\ 1}$ + $P_{arrangement\ 2}$ + . . . $P_{arrangement\ 16}$. The probability of observing all the data at all the base positions (i) is the product of the probabilities for each position, i.e., $P_{tree} = \prod_{i=1}^{N} P_i$. However, this probability is usually expressed in terms of the log likelihood of the tree ($\ln L_{tree}$), that is the sum of the log likelihoods for each base position ($\ln L_i$), that is,

$$\ln L_{tree} = \sum_{i=1}^{N} \ln L_i.$$ This expresses the log likelihood

of observing the data and the chosen evolutionary model, given a particular tree, and maximum-likelihood program search for trees that maximize log likelihood values. In the example of four taxa there are 15 rooted trees (and their associated maximum-likelihood values) which must be evaluated. As the number of terminal taxa becomes larger the numbers of possible trees increase and hence it is usual to have either a parsimony or neighbor-joining tree as a starting point.

Advantages. The requirement for an explicit evolutionary model means that the assumptions made are explicit, rather than implicit. Generally, a single maximum-likelihood tree is produced, although there are maximum-likelihood procedures, based around Bayes theorem, that produce a group of trees with similar maximum-likelihood values (Huelsenbeck and Ronquist 2001).

Disadvantages. Dependence on an evolutionary model means that it is necessary to find an appropriate model and suitable model parameters, although to determine the most appropriate model and parameters a tree is necessary. One approach is to choose the model and parameter combination that maximizes the likelihood value, although this is computationally intensive. Furthermore, the possibility of more than one maximum-likelihood value for any given tree means that it is difficult to guarantee that this value is maximal.

Types of data. All molecular markers that can be described in terms of an appropriate evolutionary model can be analyzed using maximum-likelihood techniques. This usually means that it is necessary to have DNA sequence data.

Further information. Excellent explanations of the maximum-likelihood procedures can be found in Swofford et al. (1996) and Nei and Kumar (2000; chapter 8).

REFERENCES

Avise, J.C. 1989. Gene trees and organismal histories: a phylogenetic approach to population biology. *Evolution*, **43**: 1192–208.

Avise, J.C. 2000. *Phylogeography. The History and Formation of Species*. Cambridge, MA: Harvard University Press.

Backeljau, T., de Bruyn, L., de Wolf, H., Jordaens, K., van Dongen, S., and Winnepenninckx, B. 1996. Multiple UPGMA and neighbor-joining trees and the performance of computer packages. *Molecular Biology and Evolution*, **13**: 309–13.

Bagley, M.J. and Gall, G.A.E. 1998. Mitochondrial and nuclear DNA sequence variability among populations of rainbow trout (*Oncorhynchus mykiss*). *Molecular Ecology*, **7**: 945–61.

Barrett, S.C.H. and Kohn, J.R. 1990. Genetic and evolutionary consequences of small population size in plants: implications for conservation. In D.A. Falk and K.E. Holsinger, eds. *Genetics and Conservation of Rare Plants*. Oxford: Oxford University Press, pp. 3–30.

Bartlett, E., Novak, S.J., and Mack, R.N. 2002. Genetic variation in *Bromus tectorum* (Poaceae): differentiation in the eastern United States. *American Journal of Botany*, **89**: 602–12.

Barton, N.H. and Wilson, I. 1995. Genealogies and geography. *Philosophical Transactions of the Royal Society of London Series B*, **349**: 49–59.

Bermingham, E. and Moritz, C. 1998. Comparative phylogeography: concepts and applications. *Molecular Ecology*, **7**: 367–9.

Birky, C.W. 2001. The inheritance of genes in mitochondria and chloroplasts: laws, mechanisms, and models. *Annual Review of Genetics*, **35**: 125–48.

Birky, C.W., Fuest, P., and Maruyama, T. 1989. Organelle gene diversity under migration, mutation and drift: equilibrium expectations, approaches to equilibrium, effects of heteroplasmic cells and comparison of nuclear genes. *Genetics*, **121**: 613–27.

Bossart, J.L. and Prowell, D.P. 1998. Genetic estimates of population structure and gene flow: limitations, lessons and new directions. *Trends in Ecology and Evolution*, **13**: 202–6.

Cameron, K.M., Chase, M.W., Whitten, W.M., Kores, P.J., Jarrell, D.C., Albert, V.A., Yukawa, T., Hills, H.G., and Goldman, D.H. 1999. A phylogenetic analysis of the Orchidaceae – evidence from *rbc*L nucleotide sequences. *American Journal of Botany*, **86**: 208–24.

Castelloe, J. and Templeton, A.R. 1994. Root probabilities for intraspecific gene trees under neutral coalescent theory. *Molecular Phylogenetics and Evolution*, **3**: 102–13.

Chalmers, K.J., Waugh, R., Sprent, J.I., Simons, A.J., and Powell, W. 1992. Detection of genetic variation between and within populations of *Gliricidia sepium* and *G. maculata* using RAPD markers. *Heredity*, **69**: 465–72.

Chat, J., Chalak, L., and Petit, R.J. 1999. Strict paternal inheritance of chloroplast DNA and maternal inheritance of mitochondrial DNA in intraspecific crosses of kiwifruit. *Theoretical and Applied Genetics*, **99**: 314–22.

Cockerham, C.C. and Weir, B.S. 1987. Correlations of descent measures: drift with migration and mutation. *Proceedings of the National Academy of Sciences USA*, **84**: 8512–14.

Comas, D., Paabo, S., and Bertranpetit, J. 1995. Heteroplasmy in the control region of human mitochondrial DNA. *Genome Research*, **5**: 89–90.

Comes, H.P. and Kadereit, J.W. 1998. The effect of Quaternary climatic changes on plant distribution and evolution. *Trends in Plant Science*, **3**: 432–8.

Crandall, K.A. and Templeton, A.R. 1999. Statistical methods for detecting recombination. In K.A. Crandall, ed. *The Evolution of HIV*. Baltimore: Johns Hopkins University Press, pp. 153–76.

DeBry, R.W. and Slade, N.A. 1985. Cladistic analysis of restriction endonuclease cleavage maps within a maximum-likelihood framework. *Systematic Zoology*, **34**: 21–34.

Doolittle, R.F. 1987. The evolution of vertebrate plasma proteins. *Biological Bulletin*, **172**: 269–83.

Doyle, J.J. 1992. Gene trees and species trees: molecular systematics as one-character taxonomy. *Systematic Botany*, **17**: 144–63.

Doyle, J.J. and Gaut, B.S. 2000. Evolution of genes and taxa: a primer. *Plant Molecular Biology*, **42**: 1–23.

Duda, T.F. and Palumbi, S.R. 1999. Population structure of the black tiger prawn, *Penaeus monodon*, among western Indian Ocean and western Pacific populations. *Marine Biology*, **134**: 705–10.

Dumolin-Lapègue, S., Pemonge, M.H., and Petit, R.J. 1998. Association between chloroplast and mitochondrial lineages in oaks. *Molecular Biology and Evolution*, **15**: 1321–31.

Easteal, S., Collet, C., and Betty, D. 1995. *The Mammalian Molecular Clock*. Austin, TX: R.G. Landes.

Emerson, B.C., Paradis, E., and Thébaud, C. 2001. Revealing the demographic histories of species using DNA sequences. *Trends in Ecology and Evolution*, **16**: 707–16.

Ennos, R.A., Worrell, R., and Malcolm, D.C. 1998. The genetic management of native species in Scotland. *Forestry*, **71**: 1–23.

Fauré, S., Noyer, J.-L., Carreel, F., Horry, J.-P., Bakry, F., and Lanaud, C. 1994. Maternal inheritance of chloroplast genome and paternal inheritance of mitochondrial genome in bananas (*Musa acuminata*). *Current Genetics*, **25**: 265–9.

Felsenstein, J. 1985. Confidence limits on phylogenies: an approach using the bootstrap. *Evolution*, **39**: 783–91.

Fitch, W.M. 1976. The molecular evolution of cytochrome c in eukaryotes. *Journal of Molecular Evolution*, 8: 18–40.

Fleischer, R.C., Perry, E.A., Muralidharan, K., Stevens, E.E., and Wemmer, C.W. 2001. Phylogeography of the Asian elephant (*Elephas maximus*) based on mitochondrial DNA. *Evolution*, **55**: 1882–92.

Forsthoefel, N.R., Bohnert, H.J., and Smith, S.E. 1992. Discordant inheritance of mitochondrial and plastid DNA in diverse alfalfa genotypes. *Journal of Heredity*, **83**: 342–5.

Friesen, V.L., Congdon, B.C., Kidd, M.G., and Birt, T.P. 2000. Polymerase chain reaction (PCR) primers for the amplification of five nuclear introns in vertebrates. *Molecular Ecology*, **8**: 2147–9.

Fu, Y.-X. and Li, W.-H. 1999. Coalescing into the 21st century: an overview and prospects of coalescent theory. *Theoretical Population Biology*, **56**: 1–10.

Georgiadis, N., Bischof, L., Templeton, A., Patton, J., Karesh, W., and Western, D. 1994. Structure and history of African elephant populations: I. Eastern and southern Africa. *Journal of Heredity*, **85**: 100–4.

Gibbons, A. 1998. Calibrating the mitochondrial clock. *Science*, **279**: 28–9.

Grant, W.S. and Bowen, B.W. 1998. Shallow population histories in deep evolutionary lineages of marine fishes: insights from sardines and anchovies and lessons for conservation. *Journal of Heredity*, **89**: 415–26.

Hall, B.G. 2001. *Phylogenetic Trees Made Easy. A How-to Manual for Molecular Biologists*. Sunderland, MA: Sinauer.

Hall, B.K. 1994. *Homology. The Hierarchical Basis of Comparative Biology*. London: Academic Press.

Hare, M.P. 2001. Prospects for nuclear phylogeography. *Trends in Ecology and Evolution*, **16**: 700–6.

Hare, M.P. and Avise, J.C. 1996. Molecular genetic analysis of a stepped multilocus cline in the American oyster (*Crassostrea virginica*). *Evolution*, **50**: 2305–15.

Harpending, H.C., Batzer, M.A., Gurven, M., Jorde, L.B., Rogers, A.R., and Sherry, S.T. 1998. Genetic traces of ancient demography. *Proceedings of the National Academy of Sciences USA*, **95**: 1961–7.

Harris, S.A. 1999. RAPDs in systematics – A useful methodology? In P.M. Hollingsworth, R.M. Bateman, and R.J. Gornall, eds. *Advances in Molecular Systematics*. London: Taylor & Francis, pp. 211–28.

Harris, S.A. and Ingram, R. 1991. Chloroplast DNA and biosystematics: the effects of intraspecific diversity and plastid transmission. *Taxon*, **40**: 393–412.

Havey, M.J., McCreight, J.D., Rhodes, B., and Taurick, G. 1998. Differential transmission of the *Cucumis* organellar genomes. *Theoretical and Applied Genetics*, **97**: 122–8.

Hewitt, G. 2000. The genetic legacy of Quaternary ice ages. *Nature*, **405**: 907–13.

Hudson, R.R. 1990. Gene genealogies and the coalescent process. *Oxford Surveys in Evolutionary Biology*, **7**: 1–44.

Huelsenbeck, J.P. and Ronquist, F. 2001. MRBAYES: Bayesian inference of phylogenetic trees. *Bioinformatics-Oxford*, **17**: 754–5.

Lavin, M., Mathews, S., and Hughes, C. 1991. Chloroplast DNA variation in *Gliricidia sepium* (Leguminosae): intraspecific phylogeny and tokogeny. *American Journal of Botany*, **78**: 1576–85.

Li, W.-H. 1997. *Molecular Evolution*. Sunderland, MA: Sinauer.

Li, W.-H. and Zharkikh, A. 1995. Statistical tests for DNA phylogenies. *Systematic Biology*, **44**: 49–63.

Li, W.-H., Ellsworth, D.L., Krushkal, J., Chang, B.H.-J., and Hewett-Emmett, D. 1996. Rates of nucleotide substitution in primates and rodents and the generation-time effect hypothesis. *Molecular Phylogenetics and Evolution*, **5**: 182–7.

Lloyd, D.G. and Calder, V.L. 1991. Multiresidue gaps, a class of molecular characters with exceptional reliability for phylogenetic analysis. *Journal of Evolutionary Biology*, **4**: 9–21.

Lockhart, P.J. and McLenachan, P.A. 1997. Isolating polymorphic plant DNA fragments identified using AFLP technology without acrylamide gels: markers for evolutionary studies. *Focus*, **19**: 70–1.

Martin, A.P. and Burg, T.M. 2002. Perils of paralogy: using HSP70 genes for inferring organismal phylogenies. *Systematic Biology*, **51**: 570–87.

Moritz, C. and Faith, D.P. 1998. Comparative phylogeography and the identification of genetically divergent areas for conservation. *Molecular Ecology*, **7**: 419–29.

Muse, S.V. 2000. Examining rates and patterns of nucleotide substitution in plants. *Plant Molecular Biology*, **42**: 25–43.

Neale, B.B. and Sederoff, R.R. 1989. Paternal inheritance of chloroplast DNA and maternal inheritance of mitochondrial DNA in loblolly pine. *Theoretical and Applied Genetics*, **77**: 212–16.

Nei, M. and Kumar, S. 2000. *Molecular Evolution and Phylogenetics*. Oxford: Oxford University Press.

Neigel, J.E. and Avise, J.C. 1993. Application of a random walk model to geographic distributions of animal mitochondrial DNA variation. *Genetics*, **135**: 1209–20.

Newton, A.C., Allnutt, T.R., Gillies, A.C.M., Lowe, A.J., and Ennos, R.A. 1999. Molecular phylogeography, intraspecific variation and the conservation of tree species. *Trends in Ecology and Evolution*, **14**: 140–5.

Olsen, K.M. 2002. Population history of *Manihot esculenta* (Euphorbiaceae) inferred from nuclear DNA sequences. *Molecular Ecology*, **11**: 901–11.

Page, R.D.M. and Charleston, M.A. 1997. From gene to organismal phylogeny: reconciled trees and the gene tree/species tree problem. *Molecular Phylogenetics and Evolution*, **7**: 231–40.

Page, R.D.M. and Holmes, E.C. 1998. *Molecular Evolution. A Phylogenetic Approach*. Oxford: Blackwell Science.

Palmer, J.D. 1985. Comparative organisation of chloroplast genomes. *Annual Review of Genetics*, **19**: 325–54.

Pamilo, P. and Nei, M. 1988. Relationship between gene trees and species trees. *Molecular Biology and Evolution*, **5**: 568–83.

Petri, B., von Haeseler, A., and Paabo, S. 1996. Extreme sequence heteroplasmy in bat mitochondrial DNA. *Biological Chemistry*, **377**: 661–7.

Pons, O. and Petit, R. 1996. Measuring and testing genetic differentiation with ordered versus unordered alleles. *Genetics*, **144**: 1237–45.

Posada, D. and Crandall, K.A. 2001. Intraspecific gene genealogies: trees grafting into networks. *Trends in Ecology and Evolution*, **16**: 37–45.

Provan, J., Powell, W., and Hollingsworth, P.M. 2001. Chloroplast microsatellites: new tools for studies in plant ecology and evolution. *Trends in Ecology and Evolution*, **16**: 142–7.

Rohlf, F.J. 1982. Consensus indices for comparing classifications. *Mathematical Bioscience*, **59**: 131–44.

Saitou, N. and Nei, M. 1987. The neighbor-joining method: a new method for reconstructing phylogenetic trees. *Molecular Biology and Evolution*, **4**: 406–25.

Sanderson, M.J. and Doyle, J.J. 1992. Reconstruction of organismal and gene phylogenies from data on multigene families: concerted evolution, homoplasy, and confidence. *Systematic Biology*, **41**: 4–17.

Schaal, B.A. and Olsen, K.M. 2000. Gene genealogies and population variation in plants. *Proceedings of the National Academy of Sciences USA*, **97**: 7024–9.

Schierup, M.H. and Hein, J. 2000. Consequences of recombination on traditional phylogenetic analysis. *Genetics*, **156**: 879–91.

Schneider, C.J., Cunningham, M., and Moritz, C. 1998. Comparative phylogeography and the history of endemic vertebrates in the Wet Tropics rainforests of Australia. *Molecular Ecology*, **7**: 487–98.

Shore, J.S. and Triassi, M. 1998. Paternally biased cpDNA inheritance in *Turnera ulmifolia* (Turneraceae). *American Journal of Botany*, **85**: 328–32.

Sinclair, W.T., Morman, J.D., and Ennos, R.A. 1998. Multiple origins of Scots pine (*Pinus sylvestris* L.) in Scotland: evidence from mitochondrial DNA variation. *Heredity*, **80**: 233–40.

Slatkin, M. 1985. Rare alleles as indicators of gene flow. *Evolution*, **39**: 53–65.

Slatkin, M. and Madison, W.P. 1989. A cladistic measure of gene flow inferred from the phylogenies of alleles. *Genetics*, **123**: 603–13.

Smouse, P.E. 1998. To tree or not to tree. *Molecular Ecology*, **7**: 399–412.

Sneath, P.H.A. and Sokal, R.R. 1973. *Numerical Taxonomy*. San Francisco: Freeman.

Soltis, D.E., Gitzendanner, M.A., Strenge, D.A., and Soltis, P.S. 1997. Chloroplast DNA intraspecific phylogeography of plants from the Pacific Northwest of North America. *Plant Systematics and Evolution*, **206**: 353–73.

Stewart, J.S., Allison, G.E., and Simons, A.J. 1996. Gliricidia sepium: *Genetic Resources for Farmers*. Oxford: Oxford Forestry Institute.

Strand, A.E., Leebens, M.J., and Milligan, B.G. 1997. Nuclear DNA-based markers for plant evolutionary biology. *Molecular Ecology*, **6**: 113–18.

Stuessy, T.F. 1990. *Plant Taxonomy*. New York: Columbia University Press.

Sunnocks, P. 2000. Efficient genetic markers for population biology. *Trends in Ecology and Evolution*, **15**: 199–203.

Swofford, D.L., Olsen, G.J., Waddell, P.J., and Hillis, D.M. 1996. Phylogenetic inference. In D.M. Hillis, C. Moritz, and B.K. Mable, eds. *Molecular Systematics*. Sunderland, MA: Sinauer, pp. 407–514.

Taberlet, P., Fumagalli, L., Wust-Saucy, A.-C., and Cosson, J.-F. 1998. Comparative phylogeography and post-glacial colonization routes in Europe. *Molecular Ecology*, **7**: 453–64.

Templeton, A.R. 1998. Nested-clade analyses of phylogeographic data: testing hypotheses about gene flow and population history. *Molecular Ecology*, **7**: 381–97.

Templeton, A.R., Routman, E., and Phillips, C.A. 1995. Separating population structure from population history: a cladistic analysis of the geographical distribution of mitochondrial DNA haplotypes in the tiger salamander, *Ambystoma tigrinum*. *Genetics*, **140**: 767–82.

Tollefsrud, M.M., Bachmann, K., Jakobsen, K.S., and Brochmann, C. 1998. Glacial survival does not matter. II: RAPD phylogeography of Nordic *Saxifraga cespitosa*. *Molecular Ecology*, **7**: 1217–32.

Tomaru, N., Mitsutsuji, T., Takahashi, M., Tsumura, Y., Uchida, K., and Ohba, K. 1997. Genetic diversity in *Fagus crenata* (Japanese beech): influence of the distributional shift during the late-Quaternary. *Heredity*, **78**: 241–51.

van Dijk, M.A.M., Paradis, E., Catzeflis, F., and De Jong, W.W. 1999. The virtues of gaps: xenarthran (edentate) monophyly supported by a unique deletion in a-crystallin. *Systematic Biology*, **48**: 94–106.

White, G.M., Boshier, D.H., and Powell, W. 1999. Genetic variation within a fragmented population of *Swietenia humilis* Zucc. *Molecular Ecology*, **8**: 1899–909.

Whitlock, M.C. and McCauley, D.E. 1999. Indirect measures of gene flow and migration: $F_{st} \neq 1/(4Nm + 1)$. *Heredity*, **82**: 117–25.

Williams, S.T. and Benzie, J.A.H. 1998. Evidence of a biogeographic break between populations of a high dispersal starfish: congruent regions within the Indo-West Pacific defined by color morphs, mtDNA, and allozyme data. *Evolution*, **52**: 87–99.

Wright, S. 1951. The genetical structure of populations. *Annals of Eugenics*, **15**: 323–54.

Young, A.G. and Clarke, G.M. 2000. *Genetics, Demography and Viability of Fragmented Populations*. Cambridge: Cambridge University Press.

Zhang, Y. and Ryder, O.A. 1995. Different rates of mitochondrial DNA sequence evolution in Kirk's Dik-dik (*Madoqua kirkii*) populations. *Molecular Phylogenetics and Evolution*, **4**: 291–7.

Zuckerkandl, E. and Pauling, L. 1965. Evolutionary divergence and convergence in proteins. In V. Bryson and H.J. Vogel, eds. *Evolving Genes and Proteins*. New York: Academic, pp. 97–166.

6

Speciation and hybridization

Summary

1 Species concepts differentiate taxonomic groups either by pattern (e.g., morphological or genetic variation) or process (e.g., reproductive isolation or ecological differentiation). Most of the available concepts make a useful contribution to the discussion of what a species constitutes, and several such concepts may have to be satisfied to prove a newly evolved species (neospecies).

2 Modes of speciation can be classified either as gradual (e.g., allopatric, sympatric, and divergent) or sudden (e.g., saltational, mating system change, hybridization, and polyploidy). All are effective speciation mechanisms although the frequency of modes varies between taxonomic groups (e.g., angiosperms vs. gymnosperms and vertebrates vs. invertebrates).

3 Several methods can be employed to demonstrate speciation events, particularly for neospecies where it may be possible to infer a process by analyzing variation in extant populations, including demonstrating reproductive isolation, ecological differentiation, normal chromosome segregation, and morphological and/or molecular genetic differentiation.

4 The outcomes of interspecific hybridization are diverse and include inviable zygotes, infertile progeny, hybrid zones, introgression, formation of stabilized introgressants, and the evolution of new homoploid or polyploid hybrid taxa.

5 The products of hybridization are generally highly variable both at the taxonomic and individual levels. Thus studies examining such processes will need to demonstrate the range of variation present. A range of individual-based methods (and particularly multivariate statistics) is recommended for presenting the morphological and molecular genetic variation of hybrids.

6.1 Introduction

Speciation can be viewed as the ultimate consequence of ecological and evolutionary genetic processes. Indeed in the absence of gene flow between populations speciation may be inevitable (Wolpoff 1989). The study of mechanisms that generate new species or cause their breakdown (e.g., via hybridization) is therefore of considerable interest to evolutionary biologists and ecologists alike. Species are also the base unit of evolutionary biology and for the most part are discontinuous entities. However, realizing the limits and definition of species entities is important as the processes of population biology and ecological genetics (i.e., gene flow and drift) are hypothesized to act within species units rather than between them. This emphasizes the importance of accurately defining the species level of classification.

Whilst species are the base unit of evolution, their definition has proven to be a particularly difficult task. Many different concepts of species have been offered. Some reflect the specific research background of their originators but several unifying themes run through many different concepts (e.g., reproductive mechanisms). The first part of this chapter concentrates on the theoretical basis of some of the available species concepts and what they add to our understanding of the species debate. The most common mechanisms of reproductive isolation, which form a common theme in several species concepts, are presented together with a summary of the main ecological and genetic processes that are associated with speciation. The most common modes of speciation are described and classified into two groups, gradual and sudden. Examples of each of the modes of speciation are presented, but it is notable that the frequency of the mode varies considerably between taxonomic groups.

In addition to covering speciation, the process of hybridization, being both a speciation mechanism and a process by which species barriers breakdown is described. Hybridization can lead to a range of outcomes from ephemeral, sterile hybrid products, through long-lived hybrid zones that facilitate the introgression of adaptive genes across species barriers, to the production of new geologically long-lived lineages. The incidence of hybridization also varies considerably between taxonomic groups, and despite being viewed as maladaptive by some workers it is a common occurrence with considerable evolutionary potential.

Whilst the first part of this chapter covers the theory of speciation and hybridization, the second part of the chapter seeks to deliver guidance on methods of analysis that may be used to investigate these processes, and is presented with examples of how such techniques have been applied in a research context. In general, a range of techniques is suggested to properly investigate the phenomena of speciation or hybridization, including examination of breeding mechanisms, chromosome complement, ecological differentiation, morphological variation, molecular genetic markers, and genomics. In particular, the use of multivariate statistics are emphasized as useful tools for investigating suspected cases of speciation and hybridization as these analytical tools make no or few evolutionary assumptions and tend to present variation at the individual, rather than population level. Individual variation is an important component of many speciation and hybridization models, where the rare variant may have tremendous evolutionary potential.

6.2 Species

6.2.1 Concepts of species

To understand the processes that mediate the generation and fusion of species, it is first necessary to classify the great diversity of organisms. In an attempt to produce order from a world of myths and uncertainty, the ancient Greeks devised systems for the classification of nature. For example, Pliny developed a threefold system for animals living on land, in water, and in the air, whilst Aristotle distinguished the main classes of creature based on multiple morphological characteristics (Russell 1961). Linnaeus who devised a system of grouping plants and animals based mainly on floral and body structures made one of the most important contributions towards classification. Furthermore, his binomial nomenclature and view that every organism must belong to the lowest taxonomic entity, the species, remain largely valid today. Linneaus assigned labels to nearly all the species of flora and fauna known in the western world at that time (i.e., genus and species names, although plant family names came largely from later work by de Jussieu), and variants within the species system were also recognized (e.g., subspecies and variety). Initially, Linnaeus held rigid ideas on species, viewing them as invariant since he believed them all to be well-defined unchanging entities created by a single act of God. However, later in his life these views changed as he became increasingly aware of instances of instances of hybridization (Dobzansky et al. 1977). Darwin and Lamark also believed that species could change by splitting or fusing over time (Darwin 1859); a view that was strongly reinforced by the difficulties encountered in classifying some animal and plant groups (Roberts 1929).

A workable species concept is the primary requisite for taxonomic classification. However, the definition of a species is a contentious issue and several concepts have been proposed. These concepts can be broadly classified into pattern based and process based. Pattern-based concepts,

Box 6.1

Overview of the different species concepts

Pattern-based concepts

Taxonomic species concept

Although not formally defined, it has been applied by taxonomists to differentiate species based on the shared morphological characters that distinguish a group of individuals from other such coherent groups. A useful and practical definition, probably most widely applied. This concept concerns morphological features (usually reproductive organs) as a basis for identifying members of the same species (similar characters) and for differentiating them from members of other species (divergent characters). It falls into problems with sister and cryptic species where there is little or no morphological differentiation. The basis of the concept is not directly related to a shared breeding system or reproductive isolation.

Numerical taxonomic species concept

Emphasizes the phenetic differences (i.e., morphological and or genetic) between species (Sokal and Crovello 1970). Similar in principle to the taxonomic species concept, and straightforward to apply, but tries to constrain genetic differences too rigidly.

Phylogenetic species concept (PSC)

Defined as "the smallest discernible cluster of individual organisms within which there is a parental pattern of ancestry and descent" (Cracraft 1983). Emphasizes the distinctive taxonomic differences between species (mainly morphological, but also biochemical, physiological, and behavioral), as it is these characters that prevent interbreeding with other species. Despite its strengths, the PSC fails to establish at what level of character differentiation a new species should be recognized. For example, DNA sequencing or fingerprinting of certain genomic regions could differentiate every individual of a population. The power to distinguish individuals using such molecular genotyping tools obviously does not warrant their classification as separate species.

Evolutionary species concept (ESC)

Simpson (1961) based this concept on the idea that the evolutionary lineage of a species includes a number of extinct species. These extinct species are acknowledged and the lineage is recognized to have transformed through several different species. The ESC was further altered by Wiley (1978) to define species as a single lineage that maintains its identity from other lineages; both are useful concepts and widely applied in a paleontological context.

Process-based concepts

Biological species concept (BSC)

Defined by Mayr (1942) as "groups of [actually or potentially] interbreeding natural populations which are reproductively isolated from other such groups". Still the most widely applied of species concepts, and as the definition highlights reproductive isolation it is widely testable. The basis of the concept is Dobzansky's (1937) observation that the process responsible for species formation is the development of pre- and/or postzygotic reproductive isolating barriers. Most criticisms of this concept are aimed at its taxonomic context. Application of the BSC differentiates large numbers of taxa that are sexually isolated but difficult to tell apart on morphological grounds (Grant 1957), and thus it is a difficult concept to work with in the field. Moreover, asexual species (including those undergoing apomixis or parthenogenesis) are not included in this definition. Sokal and Crovello (1970) observed that if two species hybridized in areas of sympatry and formed interfertile backcrosses, then strict application of the BSC would imply they were one species. This is clearly not a satisfactory state of affairs, and highlights the difficulty with the phenomenon of hybridization in a species concept context. In general, however, the BSC is not strictly applied in cases of limited hybridization, and a low level of interbreeding is tolerated (King 1993, Mayr 1992).

Recognition species concept (RSC)

Defined as "the most inclusive population of biparental organisms which share a common fertilization system" (Patterson 1985). The emphasis of the definition is on a common mate recognition system. However, it implies that all species capable of producing hybrid progeny, whether fertile or sterile, are one species.

Cohesive species concept

Defined as "the most inclusive population of individuals having the potential for cohesion through intrinsic cohesion mechanisms" (Templeton 1989). The concept concentrates on common reproductive mechanisms, and the BSC is altered to stress the importance of mating mechanisms which facilitate reproduction (i.e., produce fertile progeny) rather than the negative aspects of isolation. Templeton argues that it is the former processes, not the latter, that can be selected for and thus drive speciation events. In practice, extremely similar to RSC (above) and both address similar shortcomings of the BSC (King 1993).

Ecological species concept

Van Valen (1976) recognized the importance of ecological niches and their influence on an individual's development and genotypic selection. Whilst an interesting focus to the concept, it has several shortcomings which have limited its application, including the implications that two different species cannot occupy the same niche, and that species forced to extinction through competition could not have been real species (Wiley 1978).

Ecogenetic species concept

Defined as "each species has a unique way of living in and relating to the environment and has a unique genetic system – that is, that which governs the intercrossibility and interfertility of individuals and populations" (Levin 2000). Being the most recent definition, the ecogenetic species concept has had little time for testing. However, it tries to combine the importance of ecological niches and genetic differentiation and their roles in promoting reproductive contact within a species unit and reproductive isolation between other such species units. This is potentially a powerful species concept that tries to intergrate positive aspects of the biological, cohesive, and ecological species concepts, whilst making specific reference to a genetic context.

as their name suggests, rely on the identity of a pattern of morphological or genetic variation that can be used to distinguish one group of species from another (i.e., taxonomic, numerical taxonomic, phylogenetic, evolutionary species concepts; Box 6.1). Process-based definitions concentrate on demonstrating the occurrence or evolution of processes (e.g., reproductive isolation or ecological differentiation) that isolate one group of species from another (i.e., biological, recognition, cohesive, ecological, ecogenetic species concepts; Box 6.1). However, there are elements of both concepts in some definitions, and the presence of characters that differentiate two species (a pattern) may imply genetic isolation (a process). For most purposes, consideration of a combination of concepts is required to identify/prove speciation beyond doubt, that is, that they are morphologically or genetically differentiated and that they are reproductively isolated and/or ecologically differentiated from other taxa. However, often just the former principles are used (morphological or genetic differences), as such differences are generally more easily quantified.

6.2.2 Reproductive isolation and ecological differentiation

If a species concept that relies on reproductive isolation is accepted, then tests of species integrity necessarily involve examination of such mechanisms that may be acting between taxa (Box 6.2).

Box
6.2

Main reproductive isolating barriers (after Levin 1978)

Prezygotic isolating mechanisms include ecological separation, where taxa grow or live in different habitats or niches, despite being sympatric. One example would be insect species that live on different hosts within the same ecosystem. Temporal separation may also prevent reproductive contact between taxa. For example, plants or animals become reproductively receptive at different times of the day (diurnal) or year (seasonal). The sensory organs mediate many animal species barriers, including visual (e.g., color, courtship dances), olfactory and taste (e.g., hormonal), auditory (e.g., frequency and pattern of mating signals), and tactile cues (Butlin and Richie 1994). For plants, a pollinator's perceptive abilities and behavior also offer a range of reproductive isolation possibilities, including mechanical (potential pollinators may be unable to access flower structures or pollen may be deposited on specific body parts) and ethological (i.e., aspects of a pollinator's

sensory capacity may isolate flowers of different species, for example, scent, color or nutritive differences; Grant 1994).

Following a successful mating event, there are reproductive isolating mechanisms that operate before gametes fuse (i.e., **prezygotic** but **postmating**), for example, a destructive incompatibility between the male gamete (sperm or pollen) and the reproductive tract leading to the female gamete.

Postzygotic isolating mechanisms operate following the fusion of gametes. The zygote may be inviable and fail to develop or be rejected or neutralized by the reproductive system housing the female gamete. Even if the product of an intertaxon cross develops to produce a new individual, a whole host of mechanisms may serve to prevent this individual successfully producing fertile gametes, including reproductive inviability, sterility, weakness, or breakdown of the hybrid. These postzygotic effects can be substantial and form effective breeding barriers between taxa.

Table 6a

Premating	Prezygotic
Spatial	
1 Ecological, microsite or host	
Reproductive	
2 Temporal divergence (diurnal or seasonal)	
3 Attraction divergence	
(a) Floral (ethological or mechanical)	
(b) Sense organs (visual, acoustic, olfactory, hormonal, tactile)	
Postmating	
4 Reproductive mode	
5 Cross-incompatibility	
(a) Pollen or sperm	Postzygotic
(b) Seed or zygote	
6 Hybrid inviability or weakness	
7 Isolation of hybrid due to novel reproductive character(s)	
8 Hybrid sterility	
9 Hybrid breakdown	

The classification of these mechanisms is split into those acting before fertilization (i.e., prezygotic) and those acting after fertilization (i.e., postzygotic). Prezygotic mechanisms are those that prevent two gametes from successfully fusing to form a zygote (including ecological and temporal separation, and sensory differences). Following fusion of the gametes, postzygotic mechanisms may operate (including zygote inviability, and hybrid inviability, sterility, weakness or breakdown). These postzygotic effects can be substantial and form effective breeding barriers between taxa.

Ecological shifts can serve to effectively isolate new taxa. Selection appears to play a major role in defining new ecological variants. Ecological shifts often occur following hybridization and polyploidy events when new variation is subject to selection. Such shifts also often occur in profusion when competition is limited and allow new variants to take advantage of unoccupied niches, such as on oceanic islands. The variation necessary for selection to act may arise from a number of sources, including mutation and developmental instability (Levin 2000).

6.3 Speciation

6.3.1 Basis of speciation

Genetic differentiation between species

Evolutionary theory suggests that once two populations or taxa are reproductively isolated then they should start to diverge genetically. De Vries (1905) believed that differences between species could be expressed in units of single gene mutations. However, Morgan (1932) pointed out that even closely related species might differ from each other by many gene changes. Genetic divergence caused by the accumulation of independent base substitutions at loci across the genome differentiates between isolated populations or taxa and causes a loss of fitness when combined in hybrids. These Dobzansky–Muller incompatibilities, as they are called, are known to accumulate at least as fast as the square of the number of substitutions separating the two taxa (Orr and Turelli 2001; a phenomenon also known as the snowball effect).

To test some of these theoretical predictions, crossing studies have been undertaken to examine the relationship between genetic distance and the level of reproductive isolation between closely related taxa (e.g., subspecies of *Layia glandulosa*, Clausen 1951; monkey flower complex, *Mimulus guttatus*, Vickery 1964). Initial analysis broadly indicated that closely related taxa had fewer breeding barriers between them than did more distantly related taxa. In *Drosophila*, reproductive isolation of species is positively correlated with the degree of allozyme divergence (Coyne and Orr 1989). However, there can be major problems with such studies. In *Drosophila*, the nature of the isolating mechanisms was not specified and premating isolation was not included. In addition, the relevance of genetic divergence of certain, supposedly neutral, gene markers to reproductive isolation is clearly questionable (King 1993). Thus, the relationship between reproductive isolation and genetic divergence is not simple and may be complicated by the period of isolation and definitions of the strength of reproductive isolation.

Genetic processes and adaptation

It was the accumulation of beneficial genetic changes that Wright (1932) envisioned as the force that drove evolution through an adaptive landscape. In this evolutionary landscape (Fig. 6.1), the location of an individual or group of individuals (e.g., species) is determined by the particular character combination it possesses, and the vertical height of mountains or valleys represent the fitness of particular character combinations (where the higher the peak the higher the adaptive fitness, Fig. 6.1). Adopting this geographic and topographic analogy, Fisher (1930, 1941)

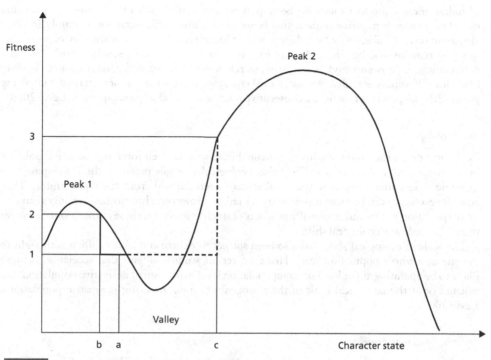

Fig. 6.1

Topography of character fitness. An individual species with character state a which has a fitness of 1 may increase its fitness by moving further up peak 1 to character state b with fitness 2 simply through mutation and selection. However, for the species to cross the maladaptive valley to achieve a character state within the domain of peak 2 (here symbolized as c with a fitness 3) other processes such as major gene mutations or considerable drift following a population crash may be involved.

speculated that mutations of small effect would allow evolutionary fine-tuning, where, through selection, species could ascend the summit of adaptive peaks. However, this does not address how a species could move across a valley of maladaptation to climb a higher adaptive peak (Fig. 6.1). Wright's shifting-balance hypothesis (1932) attempted to explain this difficulty by considering the events following a population crash, where considerable genetic drift occurs, which may not be adaptive but may allow a species to cross to more adaptive peaks (Wright 1977). The importance of this process is still under debate (Coyne, Barton, and Turelli 2000, Goodnight and Wade 2000, Peck, Ellner, and Gould 2000) and its relevance to the speciation debate remains unresolved. Indeed, other processes, such as mutation in major developmental genes, are also expected to contribute to the ability of species to easily cross maladaptive valleys to adaptive peaks (individuals possessing such major gene mutations have been termed "hopeful monsters" by Goldschmidt 1940, Templeton 1981).

Accumulation of major gene and QTL differences

The genetic basis of character differences between species (i.e., due to major genes or several quantitative trait loci, QTLs) needs to be established for individual cases, however, in some species pairs of the same genus, common genes are involved (Levin 2000). Pleiotropic interactions of independent gene loci may also be important and whilst such interactions are generally expected to have a negative impact, this is not always the case. Speciation certainly involves change in several traits, and whilst the number of loci controlling these traits may be small, the question remains whether these changes have occurred together (i.e., synchronously) or independently (i.e., consecutively). A synchronous relationship between character changes is more likely for differences encoded by a functional character suite, whereas for differences involving genes with independent function, a consecutive relationship of change is expected (Levin 2000).

Role of ecology

Ecological divergence often results in genetic differences that reinforce reproductive isolation between species. Therefore it is likely that ecological changes promote the development of genomic differentiation of new species that have been derived from their progenitor. Thus adaptive evolution can be viewed as driving speciation. However, chromosomal changes can also drive speciation events and are usually products of stochastic events, these chromosomal changes usually precede any ecological shift.

    The scale of ecological shifts has also been subject to debate and historically was thought to operate at a whole population level. However, recent studies suggest that speciation is more likely to be operating at the local or metapopulation level, where small derivative populations are isolated or at the ecological limit of their progenitors (i.e., involving parapatric populations, Levin 2000).

Inferring speciation processes

The basis of speciation, whether genetic or ecological, is particularly difficult to discern as the events that have been involved are usually obscured by time (Harrison 1991).

    Despite this inherent difficulty several major theories have been offered to explain the origin of species (Box 6.3), and can be grouped into gradual changes (mostly involving polygenic systems) and sudden speciation events (including those involving major genes).

Box 6.3

Summary of modes of speciation

Several modes of speciation have been postulated and are well studied. The frequency of each mode varies between species groups and some are more prevalent amongst animals, whilst others amongst plants. For ease of classification, the processes are split between gradual and sudden events.

Gradual speciation

Allopatric speciation

This is the classic concept of speciation mode, where populations which have been spatially isolated become genetically differentiated due to the combined effects of mutation, genetic drift, and selection. If

after a period of isolation, two populations come into secondary contact, speciation will be complete if the gene differences are so great that they are reproductively isolated. Although originally thought to be a common mode of speciation it cannot be used to explain the origin of all species, and confirmed examples are lacking.

Speciation on islands

The colonization of unoccupied island systems (whether they be oceanic islands, inland lakes, or mountain tops) and ensuing adaptive radiation into available niches is thought to be responsible for the evolution of many endemic species. Several theories of speciation in island systems have been modeled, including the founder, flush, and transilience models (Carson and Templeton 1984, Giddings, Kaneshiro, and Anderson

1989, Barton 1989). However, whilst some consider island speciation as an ideal situation to study the processes of evolution, others question whether it is only a special case of speciation due to the availability of an unusually large number of unoccupied niches.

Sympatric speciation, reinforcement, and divergence

The theory of reinforcement originally developed by A.R. Wallace is one of the most important concepts of gradual sympatric speciation. Dobzansky (1951), who reiterated the concept, postulated that reproductive isolation can develop between portions of a sympatric population as a byproduct of genetic divergence and is reinforced by selection against unfit hybrid progeny. There are theoretical problems with the model, mainly concerned with the ability to affect selection through different generations, and many cases previously assigned to this process should be more correctly reclassified as micro-allopatric speciation. However, recent models have demonstrated that there is at least a theoretical likelihood of sympatric speciation.

Sudden speciation

Saltational

Lewis' (1966) concept concerns a restricted number of individuals that became isolated from their original population in a new habitat. Multiple, simultaneous, chromosome rearrangements are generated by enforced inbreeding, and strong selection pressures within the new environment favor only a few individuals with particular genome arrangements. The survivors of this drastic selection are interfertile but produce partially or wholly sterile progeny with individuals of the original source population. Several speciation cases are attributed to this process, although its frequency is probably low.

Mating system change

Selfing is one of the most prominent evolutionary pathways leading to species formation in herbaceous plants, and can evolve by mutations in just a few genes. A self-fertilizing lineage is at an advantage in populations occurring at low density, in pioneer environments, or that frequently experience bottlenecks. However, inbreeding depression (the expression of deleterious genes) and the lack of gene recombination must be overcome to favor the maintenance of selfing lineages in the long term.

Hybridization

Speciation can occur following hybridization between two well-established species, and is much more common in plants and lower animals (invertebrates, fish, and amphibians) than in higher animal taxa. Probably due to this taxonomic basis, hybrid speciation was initially considered of little importance, although it is now recognized as a major route to the evolution of new species. Recombination speciation, a process where chromosome segments that distinguish the parental species recombine in the hybrid progeny producing new recombinant types that are fertile *inter se*, but at least partially sterile with both parents, is perhaps the most common mode of speciation following hybridization. The process of hybridization may also affect morphological character coherence, ecological tolerance, and normal fertilization processes may be disrupted causing immediate prezygotic reproductive isolation.

Polyploidy

Polyploidy is the situation when cells of an organism contain multiples of the basic chromosome set. Whilst estimates of polyploidy vary amongst species groups, the highest levels are found in plants and lower animals. In general, polyploids are immediately reproductively isolated from their progenitor species, due to chromosome mispairing and associated infertility in backcross progeny. Polyploidy can also contribute characters that are intrinsically important to the organism, and frequent multiple origins help establish diversity within newly arisen polyploid species. The two types of polyploid are described in more detail in Box 6.4.

6.3.2 Gradual speciation

Allopatry

The classic idea of speciation occurs in populations that have been spatially isolated and which become genetically differentiated due to independent evolutionary processes and lack of gene flow. If, after a period of isolation, two populations come into secondary contact, speciation will be complete if gene differences are so great that they cannot interbreed (i.e., they are reproductively isolated). This process is termed allopatric speciation and until recently was thought to be the predominant driver of speciation. However, this model requires an impossibly large number of range-splitting events to account for the origin of all species and there is little direct evidence for such a model. Thus other modes of speciation, both gradual and sudden, must also be acting.

Speciation in island systems

The colonization of oceanic islands and ensuing adaptive radiation into available niches is thought to be responsible for the evolution of many endemic species (Carlquist 1995). For example, approximately 800 of the world's 2000 species in the family Drosophilidae can be found on the Hawaiian Islands. These species are grouped in two genera, *Drosophila* and *Scaptomyza*, and molecular evidence suggests that these two lineages separated up to 24 million years ago. These lineages may have colonized the islands from two independent continental ancestors or as a single ancestor onto the older islands (Kaneshiro, Gillespie, and Carson 1995). Another example of island speciation is the Galapagos Islands finches, made famous by Charles Darwin. Bill shape is tremendously variable between the species and allows them to exploit different food sources, for example, insects or fruits. The availability of new food niches in the Galapagos system is believed to be responsible for driving finch speciation, even in sympatric populations (Grant and Grant 1989). For plants, variation in growth form has allowed the silver sword alliance of the Hawaiian Islands to exploit many different niches. Of the 28 endemic species in the genera *Argyroxiphium*, *Dubautia*, and *Wilkesia*, growth forms vary from spectacular rosette plants with 2 m tall flowering stalks, to trees, shrubs, cushion plants, and even lianas, which can also be found in a wide range of habitats. Despite their range of habits, the group exhibits low genetic variation and is thought to have originated only 2–5 million years ago from a limited stock of ancestors closely related to the Californian tarweeds, *Madia* and *Raillardiopsis* (Baldwin and Robichaux 1995). Speciation within several other "island" situations is thought to operate under similar constraints, for example, the African rift valley lake cichlids (Mayer, Tichy, and Klein 1998) and the giant lobelias and groundsel of the Eastern Arc Mountains of Africa (Knox and Palmer 1995, 1998).

Speciation on islands has been viewed as an ideal situation in which to pursue studies of the processes of evolution (Carlquist 1995), and it was Darwin's visits to oceanic islands, particularly the Galapagos, which were arguably the most influential in catalyzing his theories of species evolution and natural selection. Several theories of island speciation have been modeled, including the founder, flush, and transilience models (Carson and Templeton 1984, Giddings, Kaneshiro, and Anderson 1989, Barton 1989). Whilst differing in tempo and mechanisms, both models rely on the large number of niches available in island systems for diversification. Indeed, niche availability is sufficient to facilitate stable polymorphisms, even in a single randomly mating population, and is thus the first step toward sympatric speciation (Maynard Smith 1966). However, despite some high profile examples, the fact that a large number of unoccupied niches are available has led some to question whether evolution in island systems is a special, rather than general case of speciation (King 1993).

Reinforcement and divergence

Reinforcement, one of the most important concepts in gradual sympatric speciation, describes the development of reproductive isolation between populations as a byproduct of genetic divergence, that is, reinforced by selection against unfit progenitor–derivative hybrids (Grant 1966). Howard (1993) refined this definition, as the evolution of prezygotic isolating barriers in hybridization zones as a response to hybridization. The highly developed sense organs of animals are the main vectors of reinforcement, either through the preferential selection of mates or the preferential selection of flowers by plant pollinators (see Howard 1993, Grant 1949). Grant (1949) noted that bee species generally prefer a single type and/or color of flower (known as flower constancy). He postulated that for plants such flower preference could form the basis of speciation by reinforcement. For example, the diverse color forms of Phlox species have been linked with prezygotic isolation and divergence due to pollinator preference (Levin and Kerster 1967).

Hewitt (1988) has highlighted the theoretical problems with reinforcement, which are mainly concerned with the ability to affect selection through different generations. Indeed, Butlin (1989) has argued that there are no unequivocal natural examples of reinforcement. Butlin and Ritchie (1994) observed that mammalian mating behavior that contributes to genetic isolation evolves as a result of processes occurring within species. The resulting barriers to gene exchange are therefore incidental consequences, rather than the function, of the characters involved. Butlin (1989) argues that selection probably acts to maximize ecological fitness as a whole rather than minimizing unfit hybrid production, and for which phenological separation may be a byproduct of earlier flowering and fruiting times that maximize seed set. For such cases, for example, phenological divergence in Pinus muricata (Millar 1983), Butlin (1989) suggests that the term "micro-allopatric phenological separation," not reinforcement, should be used.

The influence of diversifying natural selection on resource utilization has also been shown to cause insect species to change their natural host and lead to speciation (e.g., the leaf beetle genus Oreina, Dobler et al. 1996, Tauber and Tauber 1989). Such processes are, however, mainly confined to phytophagous and zoophagous parasites and parasitoids (Bush 1975). This process has been termed micro-allopatric speciation in view of the niche separation that accompanies a host shift.

Evidence for speciation by reinforcement has been difficult to find in natural systems (Butlin 1989, Butlin and Ritchie 1994), but reproductive character displacement (Brown and Wilson 1956) does occur. Butlin (1989) defines reproductive character displacement as the selection for character divergence in already genetically isolated species that maximizes assortative mating. The diversity of calls exhibited by species of Hawaiian crickets, Laupala species (Otte 1989), is most probably due to this process (Butlin and Ritchie 1994). Recent models of the interaction of trait and preference characters at the genomic loci have demonstrated at least a theoretical likelihood of sympatric speciation (Dieckmann and Doebeli 1999, Kondrashov and Kondrashov 1999, Tregenza and Butlin 1999).

Selection mediated by animal sense organs and expressed by preference in females, can also lead to exaggerated ornamentation in male animals. Some of these features can even be so cumbersome that they are detrimental to survival, for example, the tails of peacocks and swordtail fishes (Meyer 1997). Sexual selection has also been implicated in the reproductive isolation of several neospecies examples, including two recently diverged, sympatric stickleback species (Hatfield and Schluter 1996) and the explosive radiation of the haplochromine cichlid fishes of the African great lakes (Deutsch 1997).

6.3.3 Sudden speciation

Four main processes are responsible for sudden speciation events: saltational speciation, breeding system change, hybridization, and polyploidy. For species that have arisen by these means there are some ecological and genetic issues that need to be considered.

Population genetic theory predicts that neospecies that occur in limited number will suffer a minority type disadvantage if they are sympatric, randomly mating with parental taxa, and produce progeny of lower fitness or fertility (Felber 1991, Gardner and Macnair 2000, Husband 2000, Levin 1978). This problem can be overcome if prezygotic isolating mechanisms are in place that reduce intertype and increase assortative mating (e.g., Arnold and Hodges 1995,

Ramsey and Schemske 1998). Modeling approaches have shown that, at least for plants, even relatively minor increases in ecological differentiation, clumped distribution, or a change in mating behavior can favor the sympatric establishment of a new lineage via the range of sudden speciation mechanisms (Buerkle et al. 2000, Gardner and Macnair 2000, McCarthy, Asmussen, and Anderson 1995, Van Dijk and Bijlsma 1994).

Saltational speciation

Lewis (1966) considered a special case of sudden speciation, termed saltational speciation, which occurs when a single or a few individuals become isolated from their original population, into a new habitat. Enforced inbreeding within the daughter population generates multiple simultaneous chromosome rearrangements, and strong selection pressures within the new environment favor a few individuals with certain genome arrangements. Under such conditions, selection produces individuals within the new environment that are interfertile but produce partially or wholly sterile progeny with the progenitor.

   Two predictions arise from the saltational speciation hypothesis. First, in groups undergoing active speciation, imperfect sterility barriers will exist between some populations of the same species, and this has been observed in the plant genus *Clarkia* (Lewis 1966). Second, certain chromosome or genome patterns within different populations of a species should be found more often in certain ecological regions, and this has been observed in *Trillium* species (Stebbins 1971). Good examples of saltational speciation concern the origin of several plant species, for example, *Coreopsis nuecensoides* from *C. nuecensis* (Crawford and Smith 1982), *Stephanomeria malheurensis* from *S. exigua* ssp. *coronaria* (Brauner and Gottlieb 1987), and in *Clarkia* (Lewis 1966). The role of single-gene mutations of large effect may also be important in saltational speciation events, and pleiotropic effects may cause additional reproductive isolation due to morphological changes (Hilu 1983).

Autogamy and asexuality

The evolution of selfing (autogamy) is itself one of the most prominent evolutionary pathways leading to species formation in herbaceous plants (Barrett 1989, Stebbins 1957). A plant species that is able to adopt a self-fertilizing breeding strategy is at an advantage in low density populations and is favored in pioneer environments or populations that frequently experience bottlenecks (Barrett 1989). Indeed it has been a frequent observation that selfing species tend to occur at the geographic/ecological margins of their outcrossing progenitors (Jain 1976). Shifts to autogamy have been recorded both in sibling species and in groups for which most species are self-incompatible, for example, *Holocarpa* and *Layia* (Clausen 1951). However, for autogamous lineages to persist, problems associated with inbreeding depression and lack of recombination need to be overcome. Deleterious recessive genes can be effectively purged from a population or species following several generations of inbreeding, and many autogamous species suffer few effects of inbreeding depression (Barrett and Charlesworth 1991). Autogamy allows fixation of environmentally advantageous chromosomal rearrangements (which can also contribute towards reproductive isolation in saltational speciation, see above), but under environmental change scenarios the lack of recombination of allelic variation may not allow an autogamous species to produce new variation fast enough to adapt to new conditions.

   Asexual reproduction, including agamospermy and vegetative propagation, may also lead to new taxonomic lineages, and modeling approaches have indicated that these can arise very easily (e.g., following single-gene mutations, van Baarlen, Verduijn, and van Dijk 1999). Apomixis has allowed propagation of species in the genus *Sorbus* and a proliferation of dandelions, *Taraxacum* "micro species" (Stace 1991). However, whilst ensuring reproductive isolation, in evolutionary terms taxa adopting this breeding system are viewed as "dead ends," with little or no chance to recombine adaptive genes amongst individuals of a population or species.

Hybridization as a speciation mechanism

Many species definitions and processes imply that there is a lack of genetic contact between diverging populations. However, speciation can occur following hybridization between two separate species. In the past, hybridization has been considered of minor importance (Mayr 1963, Wagner 1970), but it is now recognized as a major route in the evolution of new species (Anderson 1953, Grant 1981, Levin 1979, Rieseberg 1998, Stebbins 1959). Hybridization is much more common in plants and lower animals (invertebrates, fish, and amphibians) than in higher animal taxa (Arnold 1997, Avise 1994, Mayr 1963, Stace 1975). For example, Stace (1991) states that 715 of the 2834 species in the British flora are hybrids.

New species can arise via "recombination speciation" (Grant 1981), a process where chromosomal segments that distinguish the parental species recombine in hybrid progeny producing new recombinant types that are fertile *inter se*, and at least partially sterile with both parents. Recombination speciation has been shown to occur experimentally in crosses between *Gilia malior* and *G. modocensis* (Lewis 1966). Templeton (1981) suggested a rationale for this form of speciation, that hybridization might induce mutator activity, resulting in high levels of chromosome breakage and gene divergence; and some evidence for such processes has been found (Barton, Halliday, and Hewitt 1983). Examples of species believed to have originated via recombination speciation include the plants *Stephanomeria diegensis* (Gallez and Gottlieb 1982), *Iris nelsonii* (Arnold 1993), *Helianthus paradoxus* (Rieseberg, Beckstrom-Sternberg, and Doan 1990), the salamander *Plethodon teyahalee* (Highton, Maha, and Maxson 1989), and the fish *Gila seminuda* (DeMarais et al. 1992).

Detailed studies of hybrid chromosome recombination and repatterning suggest that the process may not be entirely random. In artificial crosses between *Helianthus annuus* and *H. petiolaris* to regenerate their hybrid *H. anomalus* (Rieseberg 1995, Rieseberg, Fossen, and Desrochers 1995, Rieseberg et al. 1995, 1996), many chromosome arrangements were repeated in separate resynthesis lines and the natural hybrid. Rieseberg and co-workers speculate that the genomic architecture of this hybrid lineage is being constrained possibly by fertility factors that in certain combinations produce sterile individuals and are selected against (see Box 6.11).

Hybridization also often leads to the expression of transgressive characters (extreme compared to parental morphologies), due to the interaction of complementary genes, and may provide the raw material for rapid adaptation leading to niche divergence (Rieseberg, Archer, and Wayne 1999). Thus, hybridization may also affect normal fertilization processes causing prezygotic reproductive isolation. Hybridization is known to affect morphological character coherence (Rieseberg 1995, Rieseberg and Ellstrand 1993) and may disrupt floral structure in plants or patterns of mating calls in animals (Grant 1949; e.g., *Penstemon*, Straw 1955).

Ellstrand, Whitkus, and Rieseberg (1996) have noted that most plant groups in which hybridization occurs are principally outcrossing perennials with reproductive modes that stabilize hybridization, such as agamospermy, vegetative spread, or permanent odd ploidy. Some hybridization events may result in sterile entities that propagate only by asexual modes of reproduction. For example, clonal propagation is responsible for the spread of *Spartina* x *townsendii* (Gray, Marshall, and Raybould 1991) and *Potamogeton* hybrids (Hollingsworth, Preston, and Gornall 1995, 1996).

Polyploidy

Polyploidy is the situation when cells of an organism contain multiples of the basic chromosome set. Whilst estimates of polyploidy vary amongst species groups, high levels are commonly found in plants (Soltis, Soltis, and Milligan 1992). For example, by taking a basic chromosome number (x) greater than 13 to indicate polyploidy, Grant (1981) estimated that 47% of angiosperm plants are of polyploid origin. Lewis (1980) argued that chromosome numbers of $x = 9$, 10, and 11 are also probably due to polyploidy and he obtained an estimate of 70–80% for angiosperms

(Masterson 1994). The proportion of polyploids is even higher in ferns, for which it has been estimated by Grant (1981) that about 95% are derived following polyploidy ($x > 13$). A remarkable example of a high level of polyploidy is found in *Ophioglossum reticulatum*, which has 1260 chromosomes, making it an 84-ploid (by taking $x = 15$, Briggs and Walters 1997). Polyploidy is also found amongst several lower animals groups (e.g., invertebrates, fish, and amphibians) and at least two episodes of polyploidy are hypothesized during the evolution of vertebrates (Thompson and Lumaret 1992).

In general, polyploids are immediately reproductively isolated from their progenitors due to chromosome mispairing and associated infertility in backcross progeny. There are two types of polyploidy, autopolyploidy and allopolyploidy. Autopolyploids are species in which multiple chromosome sets are derived from a single species, and allopolyploids are those in which two or more species have made genomic contributions (the main characters of the two types of polyploid are described in Box 6.4). However, the difference between the two categories of polyploid may sometimes be difficult to distinguish and there is somewhat of a continuum between them.

The production of polyploids appears to be mediated by two main processes: the production of unreduced gametes and triploid hybrids. Unreduced gamete production by diploid taxa (i.e., with same ploidy as the parent) has played an important role in the evolution of many auto- and allopolyploid species (Bretagnolle and Thompson 1995, Harlan and deWet 1975, Ramsey and Schemske 1998, Thompson and Lumaret 1992). Whilst the frequency of unreduced gametes may be low, there is evidence that unreduced gametes may perform better under competitive conditions in the reproductive tract than reduced gametes, leading to higher fertilization success (Mulinix and Lezzoni 1988). Production of unreduced gametes may also increase markedly under some conditions in the wild. For example, Bretagnolle and Thompson (1995) noted that low temperatures regularly induced production of unreduced gametes. Therefore, depending on environmental conditions, the incidence of polyploidy may not be such a rare event. Triploid hybrids on the other hand are produced following the fusion of a reduced and unreduced gamete. These intermediate hybrids act as a bridge between ploidy levels, through their ability to produce genomically stable diploid and polyploid gametes (Bretagnolle and Thompson 1995, Ingram 1978, Ratter 1972, 1973a, 1973b). Such a strategy may allow an easy two-step process to producing new polyploid species (e.g., Ingram, Weir, and Abbott 1980). The formation of such

Box 6.4 Autopolyploids versus allopolyploids

There are two types of polyploidy: autopolyploidy and allopolyploidy. Autopolyploids have multiple chromosome sets that are derived from a single species, and allopolyploids (also known as amphidiploids) have genomic contributions from two or more species. Whilst these are clear categories, in nature there is a continuum between the two types of polyploid. For example, if individuals from two very different populations of the same species form an interpopulation polyploid hybrid, it may behave cytologically like an allopolyploid, even though it is an autopolyploid; such a case is usually referred to as segmental allopolyploidy (Stebbins 1947).

The salient features of the two polyploidy extremes are compared in Table 6b.

Polyploidy can contribute characters that are intrinsically important to the organism (Stebbins 1980). For example, plants can have larger flowers, seeds, and firmer textured leaves, and these characters are products of what de Vries and Gates termed "gigas" effects (Briggs and Walters 1997). Many allopolyploids also exhibit novel morphological and molecular characters associated with their hybrid origin, but autopolyploids are not always more vigorous than their progenitors (Rieseberg 1995, Rieseberg and Ellstrand 1993, Soltis and Soltis 1995, Song et al. 1995). The majority of chromosomes in allopolyploids usually form bivalents and their chromosomes segregate normally during mitosis and meiosis (Jenkins and Rees 1991).

Table 6b

Characteristic	Autopolyploids	Allopolyploids
Vigor and morphological expression	Not always more vigorous than parents	Exhibit hybrid vigor compared to progenitors and "gigas" effect of polyploidy. May also exhibit novel morphological characters
Breeding system (plants only)	Tend to be outcrossing	Often autogamous and reproductive success is assured. Reduced problems of inbreeding depression
Gene inheritance	Polysomic, allows increased heterozygosity and allele diversity to be maintained within individuals, compared to diploid progenitors	Gene multiplication and lack of genetic transfer between genomes from different origins results in fixed heterozygosity
Chromosome pairing behavior and fertility	Form multivalents, but suffer fertility consequences due to meiotic difficulties. Genomes diploidize over time	Most chromosomes in allopolyploids form bivalents and segregate normally during mitosis and meiosis
Formation frequency and polytopic origin	Usually arise on multiple occasions locally and within limited periods. Allows crossing between independent lineages	Polytopic origins have been recorded for many allopolyploid taxa, although frequency may not be as high as for autopolyploids
Establishment frequency of new polyploid lineage	Form less frequently than allopolyploids, but still viable mode of speciation	May be a slightly easier and more productive speciation mechanism in nature than autopolyploidy

However, autopolyploids tend to form multivalents and can experience reduced fertility due to chromosome segregation difficulties at meiosis (King 1993). There is some evidence that autopolyploid genomes extensively diploidize over time due to gross gene silencing, facilitating normal chromosome segregation (Gastony 1991, Soltis and Soltis 1988, 1993). Allopolyploid plants are often autogamous (selfing), whereas autopolyploid plants tend to remain outcrossing like their diploid progenitors (Barrett 1989, Thompson and Lumaret 1992). Autopolyploids tend to occur in natural populations less frequently than allopolyploids and have been viewed as maladaptive (Soltis and Soltis 1993). However, autopolyploids usually arise on multiple occasions and within limited periods (Soltis and Soltis 1993), which allows crossing between individuals of independent origin. In addition, the polysomic nature and multiplicity of gene inheritance allow increased heterozygosity

and allelic diversity relative to diploid progenitors (Soltis and Soltis 1993). These genetic attributes of autopolyploids should allow them to be successful in nature, although in plants this is dependent on the successful establishment of outcrossing to maintain diversity (Barrett 1989). In contrast, heterozygosity in allopolyploids results from gene multiplication (fixed heterozygosity, Soltis and Soltis 1993), and so in plants reproductive success can be assured by selfing (Barrett 1989). Thus, whilst autopolyploids appears to have some inherent limitations, in nature both allo- and autopolyploidy have produced many new taxa. Notable products of polyploid evolution include four plant species known to have arisen within the last century: *Senecio cambrensis* (Abbott and Lowe 1996, Ashton and Abbott 1992); *Spartina anglica* (Gray, Marshall, and Raybould 1991); and *Tragopogon mirus* and *T. miscellus* (Norvak, Soltis, and Soltis 1991, Ownbey 1950, Roose and Gottlieb 1976, Soltis and Soltis 1989).

triploids also allows species that comprise more than one ploidal type to maintain species unity (e.g., *Galax urceolata*, Burton and Husband 1999, Petit, Bretagnolle, and Felber 1999).

Models of polyploid evolution often require multiple independent origins (polytopic) to generate successful evolutionary lineages. Indeed, the majority of studied cases of polyploidy have documented recurrent formation (e.g., *Arabis holboellii*, Sharbel and Mitchell-Olds 2001; *Senecio cambrensis*, Ashton and Abbott 1992; *Tragopogon mirus* and *T. miscellus*, Cook et al. 1998), and multiple independent origins appear to be the rule rather than the exception (Leitch and Bennett 1997, Soltis and Soltis 1993). Whilst the independent lineages of some polyploid taxa appear morphologically identical, origins involving parent species from different geographic ranges can give rise to differentiated lineages of the same taxon (e.g., *Senecio cambrensis*, Lowe and Abbott 1996) and in extreme cases different species (e.g., *Saxifraga opdalensis* and *S. svalbardensis* are both derived from a *S. cernua* x *rivularis* cross, Steen et al. 2000).

6.4 Hybridization

In addition to the origin of new species at the homoploid and polyploid level (see above), hybridization can result in several other outcomes (Box 6.5).

6.4.1 Introgression

Introgression is the process of transferring a portion of genetic material from one species into the genetic background of another (and sometimes involves significant genomic quantities, Anderson 1949, Anderson and Hubricht 1938). Genetic transfer occurs via a partially fertile interspecific hybrid, which hybridizes with one or both of the parental species (backcrossing). In the past, many cases of introgression have been inferred but not investigated in detail. However, the process is undoubtedly common, and well-documented examples include ants (Shoemaker, Ross, and Arnold 1996), beetles, butterflies, crickets, grasshoppers, moths, newts, toads, snakes, pocket gopher, ground squirrel, shrews, crows, mice (Hewitt 1988), a range of fish species (Hubbs 1955), and many plant lineages, including irises (Arnold, Bennett, and Zimmer 1990), sunflowers (Heiser 1951, Rieseberg and Soltis 1988, Rieseberg, Beckstrom-Sternberg, and Doan 1990, Rieseberg, Choi, and Ham 1991), ragwort (Abbott, Ashton, and Forbes 1992), and *Clarkia* (Bloom 1976). A review of introgression by Rieseberg and Wendel (1993) lists 165 plant examples, many of which are supported by molecular evidence (Rieseberg and Brunsfeld 1992).

Introgression of genes may help hybrid derivatives to colonize new habitats (Abbott 1992). For example, introgression of genes from *Dacus neohumeralis* into *D. tryoni* resulted in increased heat tolerance in introgressant derivatives (Lewontin and Birch 1966). Introgression can also lead to the origin of stabilized introgressants that may be recognized at a subspecific taxonomic level, for example, *Helianthus annuus* ssp. *texanus* (Heiser 1951, Rieseberg, Beckstrom-Sternberg, and Doan 1990) and *Senecio vulgaris* var. *hibernicus* (Abbott, Ashton, and Forbes 1992).

6.4.2 Hybrid zones

Hybrid zones may be formed following localized hybridization of two formerly allopatric or parapatric taxa that have come into secondary contact (Hewitt 1988). Hybrid zones may be either linear (Hewitt 1989) or mosaic (Harrison and Rand 1989), depending on the distribution of habitats or niches. For hybrid zones that are not short-lived (ephemeral), several models have been put forward to explain their persistence (see Box 6.6). Each of the models considers dispersal and selection as critical factors, but places different emphases on these mechanisms. Almost all models predict that if hybrid zones persist then the result may be permanent introgression of genes across species barriers. For example, in the European mouse, nuclear genes from *Mus domesticus* have introgressed into *M. musculus* across a linear hybrid zone that splits Europe, whereas the reciprocal transfer of genes has not occurred (Hunt and Selander 1973).

Box
6.5

Summary of consequences of hybridization

Hybridization between species can have consequences besides inviable zygotes or infertile progeny: some are ephemeral and some longer lasting.

Hybrid zones – ephemeral

Hybrid zones may be formed following localized hybridization of two formerly allopatric or parapatric taxa that have come into secondary contact. Hybrid zones are often ephemeral events, where the products of hybridization may be removed by selection and cause no further consequences. In such cases selection pressures may exist to reduce the number of gametes being wasted in unfit hybrid progeny and reproductive character displacement may occur.

Hybrid zones – long-lasting

Hybrid zones may also be longer lived, indeed some have persisted for many years. In such cases, a balance between hybridization frequency, dispersal, and selection is acting to maintain hybrid zones. Several models have been put forward to explain the persistence of hybrid zones, for example, the bounded hybrid superiority hypothesis, the mosaic model, tension zones, and the evolutionary novelty model (Box 6.6). Whilst the future fate of a hybrid zone is unknown such observations as

the superiority of some hybrid forms in intermediate or novel habitats, allow the predicted establishment of long-lived lineages.

Introgression

Introgression is the process of transfer of genetic material from one species into the genetic background of another, and follows the successful formation of an interspecific hybrid. The process is common in both animals and plants and may involve significant portions of the genome. Genetic transfer occurs via hybridization (often across hybrid zones) between one or both of the parental species and involves backcrossing with partially or fully fertile hybrid progeny. The level of introgression of genes from a nuclear or organelle origin may also vary, either due to Haldane's rule in animals or dispersal differences of the male and female gametes. The process of introgression can cause several outcomes: from gene pool enrichment that may help the colonization of new habitats, to the origin of stabilized taxa that are recognized at a low taxonomic level, for example, subspecies or variety.

Polyploid or homoploid reticulate speciation

The origin of new species at the homoploid or polyploid level can result following hybridization, and these mechanisms are summarized in Box 6.3.

6.4.3 Differential introgression of organelle and nuclear genomes

The level of introgression of genes from a nuclear or organelle origin may also vary. For example, the mtDNA of *M. domesticus* has transferred into Scandinavian populations of *M. musculus* without the associated transfer of nuclear DNA markers (Gyllensten and Wilson 1987). Several special circumstances may account for differential introgression of plastid versus nuclear genes. In animal species, which possess sex chromosomes of unequal size, the phenomenon where the heterogametic sex is absent, rare or sterile in F1 offspring between two species is known as Haldane's rule. Under these circumstances, the organelle genome (mtDNA) can then be introgressed across species boundaries whilst the nuclear genome is effectively diluted due to backcrossing, and has been observed in a number of taxa (Avise 1994, Coyne and Orr 1989). For plants, chloroplast DNA capture is also commonly observed in some genera, for example, *Gossypium* (Fig. 6.8; Wendel 1989, Wendel, Stewart, and Rettig 1991) and in *Quercus* (Ferris et al. 1993). One reason

---

Box 6.6

Types of hybrid zone

Bounded hybrid superiority hypothesis

Also known as "hybrid-superiority", predicts that hybrid fitness will be superior to parental populations in certain habitats (Moore 1977).

The mosaic model

Predicts that hybrids will exhibit lower fitness than parental taxa regardless of habitat, and are selected against, but that a hybrid zone is maintained by recurrent hybrid formation (Barton and Hewitt 1985). Parental forms are distributed according to environmental adaptations and are interspersed by a mosaic of recurrently forming hybrids, rather than occurring as a linear cline (Harrison 1986, 1990, Howard 1986).

The tension zone

Assumes that hybrid clines are maintained by a balance between dispersal and selection against hybrids (developed by Barton and Hewitt 1985, 1989 from an earlier idea by Key 1968), and where hybrids exhibit reduced viability and

fecundity and are less fit than parental genotypes regardless of habitat (Barton and Hewitt 1985). However, Arnold (1997) found that the ecological and genetic patterns expected under the tension zone model are often not met in natural populations, and Arnold and Hodges (1995) found that many hybrids were not uniformly unfit and that environmental-dependent selection may be important in maintaining hybrid zones. For example, Anderson (1948, 1949, 1953) noted that habitat disturbance often produced a "hybridized habitat" in which a range of hybrid progeny performed better than parental types and could survive for long periods (e.g., Louisiana Iris hybrid swarms, Emms and Arnold 1997).

Evolutionary novelty model

A refinement of the bounded hybrid superiority model, and considers the rarity of F1 formation, endogenous and/or exogenous selection pressures for or against certain hybrid genotypes, and the possibility of invasion of parental or novel habitats by superior hybrid individuals (Arnold 1997). According to Arnold, this model predicts the establishment of geologically long-lived evolutionary lineages.

---

for such cytonuclear disequilibrium in plants is the differential rate of gamete exchange due to the dispersal properties of pollen and seed (Arnold 1992, Ennos 1994).

6.5 Analysis of speciation and hybridization

Even if a speciation or hybridization event is suspected, how does one go about proving it? Lewontin and Birch (1966) observe that, "it is a fundamental difficulty of a historical science like evolution that one can never establish the cause of a past event. It is only possible to show that certain causes are plausible or at most likely, but because each species is a unique historical event we cannot say for certain what its genetic history was." Studies that demonstrate speciation or hybridization events are inevitably complex applications of ecological genetic principles and usually combine both traditional and contemporary methods. For example, in the event of speciation, a pattern-based species concept would require proof of stable and significant morphological and/or genetic differentiation from ancestral/related taxa. The inheritance of plastid genome variation (e.g., mtDNA) may give essential information on differential parental contributions to hybrids or identify the most closely related and possibly ancestral taxa of derived species (also see chapter 5). For process-based species concepts, it may be necessary to demonstrate reproductive isolation and/or ecological differentiation of a derived species from ancestral/related taxa. In the case where hybridization is suspected, interspecific hybrids or hybrid zones can be identified by examining the additive or segregating nature of species-specific molecular markers. For a putative hybrid, the resynthesis of morphologically similar taxa would add strong supporting evidence for a hybridization theory. For both speciation and hybridization cases, the study of an organism's chromosome complement and pairing behavior at meiosis gives information on potential fertility consequences and ploidy level of natural taxa or artificial crosses. In addition, the application of genetic mapping techniques provides insights into the genomic consequences of speciation and hybridization processes.

The study of speciation and hybridization events can also be complicated or masked by other population genetic processes. Avise (1994) observes that careful application of methods studying biochemical and genetic variation may allow an insight into historical processes that lead to speciation, but that recent events are more likely to be correctly identified. It is certainly easier to identify recent hybridization and speciation events before additional drift and mutation processes have obscured all trace of the action of the processes responsible. The causes of speciation events arising via sudden mechanisms (e.g., hybridization or polyploidy) are also usually easier to identify than the products of gradual processes.

The common analytical techniques used to identify neospecies and hybrids have been outlined in the following five sections. The first section covers techniques necessary for testing the reproductive isolation of taxa and for the resynthesis of neospecies and hybrids. The second section describes the utility of chromosome investigation for examining the complementarity of intertaxon crosses and for establishing ploidy level. The third section covers methods to assess morphological and ecological differentiation between taxa and in cases of hybridization, to demonstrate the character additivity and novelty of such individuals. Section four describes the use of molecular markers (from isozymes to AFLPs) for demonstrating genetic differentiation and relationships between species and hybrids. Finally, section five examines the use of genetic mapping in speciation and hybridization studies, particularly for understanding the genomic consequences of these processes.

It is difficult to provide absolute diagnostic criteria. Depending on the case, only a single line of inquiry may be necessary but more likely a combination of the following methods will be required to provide support for a speciation or hybridization theory. The selection of which techniques to apply will depend on such factors as time, resources, and expertise. A discussion of such influences during the planning of an ecological genetic study is presented in chapter 2 and should be consulted before embarking on such a study.

6.5.1 Intertaxon crossing studies and observations

Reproductive isolation of species

According to several species concepts (but most notably the biological species concept), a taxon can be classified as a new species if it is reproductively isolated from all other taxa. Such reproductive isolation can take many different forms (Box 6.2). Testing the reproductive limits of a new species can also, therefore, take many different forms, for example, from observation of mating preferences amongst animals to observing pollen tube growth in plants, but the principles of studying reproductive isolation are outlined below.

The level of reproductive isolation has been tested for taxonomic classifications of *Drosophila* (i.e., species, sibling species, subspecies, and geographic populations; Coyne and Orr 1989, Orr and Orr 1996). To emphasize the difference in pre- and postzygotic reproductive mechanisms, two indices are typically used:

Prezygotic isolation index
Expressed as:

$$1 - \frac{\text{(frequency of heterotypic matings)}}{\text{(frequency of homotypic matings)}}$$

Postzygotic isolation index
Expressed as: a measure of hybrid inviability and hybrid sterility scaled from zero to one.

Similar indices can also be applied to the study of plant pollination studies. For example, in the Louisiana iris complex, pollen tube growth, which can be quantified as a proportion of successful fertilization events, appears to be an important factor limiting the amount of hybridization between *Iris fulva* and *I. hexagona* (Carney, Cruzan, and Arnold 1994, Carney, Hodges, and Arnold 1996, Emms, Hodges, and Arnold 1996).

Where species are sympatric or form hybrid zones, observations may also support laboratory studies and a variety of fitness and reproductive characters can be surveyed across a range of life stages. For example, studies of the insect species pair, *Aquarius remigis* and *A. remigoides* (Hemipteran), found that interspecific crosses exhibited reduced fertility, hatching success, survival, and percentage males (Gallant and Fairbairn 1997). These postmating isolation mechanisms appear at least in part to be related to Haldane's rule. The level of interspecific hybridization between different species of Darwin's finches on Isla Daphne Major, Galapagos, was examined by Grant (1993) over a 16-year period, who observed breeding behavior that increased assortative matings. In the case of *Penstemon clevelandii*, the proven homoploid hybrid derivative of *P. spectabilis* and *P. centranthifolius* (Wolfe et al. 1998a, b), speciation has been accompanied by demonstrable pollinator shifts by carpenter bees, wasps, and hummingbirds in the species complex (Straw 1955).

Resynthesis of hybrid and polyploid taxa and other evolution experiments

The most effective way to test the proposed origin of a hybrid or polyploid taxa is resynthesis by experimental crosses (Rieseberg and Carney 1998). For plant taxa, generating artificial crosses is generally straightforward, although self-incompatibility or selfing mechanisms may have to be overcome (e.g., by a combination of emasculation and pollen donation) before successful intertaxon crosses can be made. Thus, a researcher should have a good knowledge of the study organism's breeding system and developmental ontogeny. Even with this knowledge, some experimentation will be required. In addition to hybrid formation, polyploidization events can be simulated in plants by treating developing floral primordial cells or germinating seed with colchicine. At low concentrations (<1%), colchicine inhibits the formation of spindle fibers and so at mitosis the chromosomes of a cell double in number but do not migrate or segregate into different cells. Depending on the period of treatment, an organism's chromosome complement can be doubled.

For animal taxa, *Drosophila* hybrid lineages have been generated using forced matings in closed systems (Coyne and Orr 1989), but resynthesizing studies may prove difficult due to welfare and husbandry issues for higher taxa. Observation of the frequency and range of variation expressed by natural hybrids may be an alternative option for both animal and plant taxa.

Previous studies that have resynthesized fertile hybrids have provided good tests for the evolutionary potential of hybridization events. In his classic work of 1930, Müntzing (in Briggs and Walters 1997) tried to resynthesize the plant *Galeopsis tetrahit* ($2n = 4x = 32$) by crossing two closely related diploid ($2n = 2x = 16$) species, *G. pubescens* and *G. speciosa*. Sterile diploid F1 hybrids were produced and selfed, and amongst the F2 progeny was one triploid plant ($2n = 3x = 24$), presumably produced by the fusion of a reduced and unreduced gamete. Backcrossing the highly sterile triploid to *G. pubescens* produced a fertile tetraploid ($2n = 4x = 32$) that was morphologically similar to, and interfertile with, *G. tetrahit*. The tetraploid backcross was presumably produced following the fusion of an unreduced gamete of the triploid with a reduced gamete from *G. pubescens*. Other resynthesis studies have been undertaken to demonstrate the evolutionary potential of hybridization, even though no fertile hybrid derivatives have been recorded in the wild. Work by Grant (1965, 1966) showed that fertile hybrid progeny could be obtained in later generations of a cross between *Gilia malior* and *G. modocensis*, both tetraploids in the *Gilia inconspicua* complex. The F1 hybrid of this cross was almost completely sterile; however, some F2 seed was collected and raised to the F10. One F10 plant was fertile and tetraploid; this hybrid was intermediate in morphology between its parents and produced sterile progeny when backcrossed to them, thus demonstrating a classic example of evolution by hybrid speciation (Grant 1981). See chapter 7 for a case study involving *Senecio*.

6.5.2 Chromosome and ploidy analysis

Chromosome number provides crucial evidence of ploidy, whilst examination of chromosome pairing at meiosis helps provide an understanding of the fertility exhibited by a hybrid. Descriptions of the methods used to generate chromosome preparations are provided amongst others by the excellent texts and protocols of Darlington and La Cour (1962), Dyer (1979), and Sessions (1996).

Chromosome studies of the allohexaploid, *Senecio cambrensis*, show that it exhibits no or only slightly irregular meiotic chromosomal pairing in wild and artificially synthesized individuals, and helps explain the plant's high fertility (Harland 1954, Weir and Ingram 1980). In contrast, the triploid hybrid between *S. vulgaris* ($2n = 4x = 40$) and *S. squalidus* ($2n = 2x = 20$), *S.* x *baxteri* ($2n = 3x = 30$), is nearly completely sterile and artificially synthesized triploids show great meiotic imbalance, usually forming 10 bivalents and 10 univalents (Ingram 1977, 1978, Ingram, Weir, and Abbott 1980).

Chromosome painting

A further refinement of chromosome observation can be found with the use of genomic *in situ* hybridization (GISH) techniques to "paint" chromosomes with different fluorescent dyes, and offers an exciting new tool for testing genome relationships between species and hybrids (Stace and Bailey 1999). In one study using GISH, strong evidence was found to support the assertion that *Milium montianum* was of allotetraploid origin and that *M. vernale* was the contributor of one of its genomes (Bennett, Kenton, and Bennett 1992). GISH has also been used to follow the introgression of drought-resistance genes from *Festuca arundinacea* into the long arm of chromosome 2 of the artificially generated *Lolium multiflorum*-like hybrid derivative (Humphreys and Pasakinskiene 1996).

6.5.3 Morphological and ecological variation

Traditionally, morphological characters have been the mainstay of species delimitation and have been used extensively to demonstrate hybridization or introgression (Heiser 1949, 1951, Stace 1975). If a new taxon is reproductively isolated from ancestral/related taxa then over time the action of mutation, drift, and possibly selection may act to differentiate them morphologically. Similarly, neospecies may be isolated ecologically from their progenitor taxa or interspecific hybrids may express a range of ecological tolerances between or beyond their parental extremes.

There are, however, a number of qualifications to these statements, some connected with the selective consequences of morphological and ecological changes, and such patterns need to be borne in mind when interpreting such data.

Speciation considerations

• The morphology of an organism may be constrained by selection pressures. There are cases of sister species (e.g., *Drosophila*, Ayala 1975) which show very little morphological variation but are reproductively isolated and are differentiated at molecular genetic loci.
• Whilst morphology can be very informative, there are often only limited numbers of such

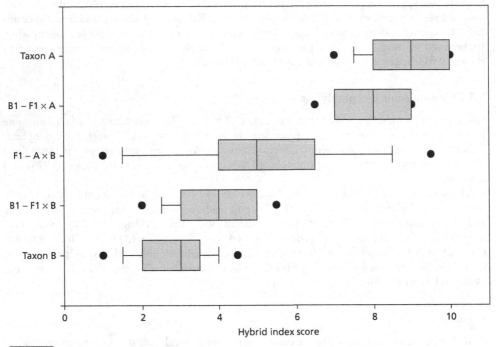

Example of hybrid character score presentation. Those characters that distinguish the parental taxa are standardized to ensure equal contribution. Characters are then weighted, such that character scores for one taxon take a low value, whereas those for the other taxon take high values. The cumulative contribution of standardized scores is then plotted. The box plot used here shows the median for the group as the line within the box, the limits of the box represent the 25th and 75th percentile, the limits of the whisker plot the 10th and 90th percentile and the maximum and minimum values are indicated by the points. Whilst such plots may provide good evidence for hybridization or introgression, and demonstrate the variable nature of F1 products, individual character expression is masked.

characters. In addition, shared characters may result from convergent evolution, rather than shared derived ancestry (Rosen 1979).
• Reproductive isolation of a species may be accompanied by changes in morphology or ecological tolerance. In such cases, differentiation from ancestral taxa may appear to be greater than would be expected given the period of isolation.
• Morphological variation may be the result of a plastic environmental response. For example, several ecotypes may exist that are not reproductively isolated from one another, yet, environmental influence has enhanced morphological differentiation. Where possible, taxa to be examined should be reared under common environmental conditions before undergoing morphological examination.

Hybridization considerations

The processes of hybridization and polyploidy are known to affect morphological variation and ecological tolerances. Polyploids tend to exhibit a "gigas" effect, where individuals are more robust and larger than their parental taxa.

Hybrid species also often exhibit transgressive characters, those which are outside of the range of variation exhibited by either parent. In a review of hybrid studies, Rieseberg, Archer, and Wayne (1999) found that in 91% of 171 cases, hybrid derivatives expressed at least one transgressive character. Indeed, the authors state that "hybridization may provide the raw material for adaptation and provide a simple explanation for niche divergence and phenotypic novelty often associated with hybrid lineages."

Indices, scatter plots, distance measures, and significance testing

Hybrids have traditionally been identified by their overall morphological or ecological intermediacy (Wilson 1992). The use of a hybrid index, where morphological or ecological characters are scored can be used to provide evidence of hybridization. Zero or low scores are given to character magnitudes similar to one parent, and high scores for character magnitudes like the other parent. Truly intermediate hybrids should exhibit an intermediate hybrid index (Fig. 6.2; Anderson 1949). However, little information on the actual character of a hybrid is given using this method. An improvement is to use scatter diagrams, where two quantitative characters are plotted on the x and y axes and several discrete characters can be represented as informative symbols (Fig. 6.3; Anderson 1949, Briggs and Walters 1997). Such methods can also be used to demonstrate morphological or ecological differences between speciating taxa, but the number of characters that can be represented on such plots is ultimately limited.

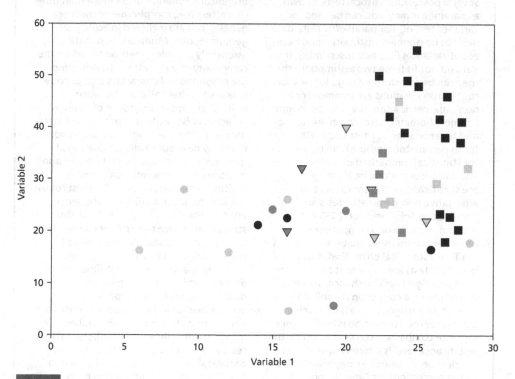

Fig. 6.3

Multiple characters can be represented on a single plot. In this example two continuous characters are plotted on the x and y axes, and two discrete characters are coded for within the plot according to shape (circle, triangle, square) and shading (black, dark gray, light gray) of the plot point. Whilst such a plot provides a good visual representation of hybridization, the number of characters that can be presented in such a fashion is limited.

Instead of plotting character scores, distance methods, which use standardized variables, can also be applied to morphometric and ecological data. Two commonly used distance estimates are Euclidean and Mahalanobis distance. Distances derived from such methods can either be presented graphically using dendrogram building methods or distance plots (chapter 6 and Box 6.7). To test morphological or ecological differences between taxa more rigorously, significance tests can be applied to individual or multiple characters. Such tests rely on comparing the amount of variation present within groups to that amongst groups, and examples are the t-test and analysis of variance (ANOVA). However, before such methods can be applied it is necessary to test whether the data can be used with such sensitive tests, termed parametric methods. There are three assumptions of the input data for parametric tests: (i) normally distributed; (ii) equal variance (heteroscedastic); and (iii) independent. If any of these tests are violated, then data can

---

Box 6.7

Multiple comparisons and multivariate statistics

Data assumptions, variance tests and multiple comparison corrections

Several powerful statistical tests, known as parametric methods, can be used for variance testing. For parametric tests, data need to be normally distributed and have equal variances (i.e., heteroscedastic). If data do not fit these two assumptions then they can be transformed (e.g., log or square root transformations are common) or a non-parametric alternative can be sought. For morphometric data, which are normal and heteroscedastic, t-test and ANOVA (both parametric methods) can be used for statistical comparison of means and variances to examine if groups are significantly different (these and alternative non-parametric tests are discussed by Sokal and Rohlf 1994). One problem with making comparisons between groups using multiple characters is a Type I statistical error. That is, so many statistical tests are made that a result turns out to be significant by chance. To avoid this problem, a correction is applied to the level at which significance is determined. A commonly used correction is Bonferroni's multiple comparison correction, where the significance level is altered depending on the number of pairwise comparisons made (other corrections for Type I errors and methods to tackle Type II errors, where a test is accepted when it is false, are covered in Sokal and Rohlf 1994).

For multiple character data sets, multivariate statistics are available for analysis without incurring problems of Type I or II errors. If an unlimited number of axes could be plotted and visualized in space then an entity's characteristics could be reflected in the character scores at the intersection of all axes. However, in practice two (or at most three) axes are plottable graphically, whereas often more than three characters (i.e., morphometric traits or genetic loci) are included in ecological genetic studies. Multivariate statistics essentially provide a method to reduce the complexity of data sets whilst reflecting the magnitude of characters that provide maximum differentiation between individual samples or groups of samples (whichever is specified). For non-parametric data (e.g., some ecological data and non-normally distributed allelic variation at genomic loci, although such tests can also be applied to parametric data types), multidimensional scaling is the most robust method for plotting distances between entities. For data with presence or absence scores (e.g., ecosystem species composition descriptions and multilocus dominant molecular data, i.e., AFLPs), principal coordinate and discriminant function analysis can be used. For parametric data (including most morphometric and environmental data, and genetic data that approximate to a normal distribution, e.g., microsatellite allelic variation), powerful tests such as principal components and canonical variate analysis can be used to highlight statistical differences between individuals, populations, and species.

Multidimensional scaling

Multidimensional scaling is a technique designed to construct a plot that reflects relationships between samples from precalculated distances between them (Manly 1992). Typically Euclidean distances are calculated between all pairwise comparisons of objects using all character information. The method plots the objects for which distance estimates are available in multidimensional space, using coordinates assumed from an initial configuration setup. A regression is then calculated between the assumed and actual pairwise distance matrices and the disparity between the two distance matrices estimated and used to alter the original coordinates of the objects in multidimensional space to minimize the disparity between assumed and actual matrices.

Principal components analysis

Plots of principal components are especially valuable tools in exploratory data analysis. Principal component analysis is a multivariate technique for examining relationships among several quantitative variables. The method finds linear combinations of variates that maximize the variation contained within them, thereby displaying most of the original variability in a smaller number of dimensions. Principal components are linear combinations of the original variables, with coefficients equal to the eigenvectors of the correlation or covariance matrix. They are sorted by descending order of eigen values, which are equal to the variances of the components. In addition, eigen values are orthogonal (i.e., at right angles) and principal component scores are jointly uncorrelated, although these two properties are quite distinct (Manly 1992). See Box 6.8 for an example.

Canonical variate and discriminant function analysis

Canonical variate analysis (CVA) is a powerful parametric test to distinguish between entities that fall into natural groupings. The within-group sums of squares and products are pooled over all groups and from the group means and sizes the between-group sums of squares and products are calculated. The derivative finds linear combinations of the original variables that maximize the ratio of between-group to within-group variation,

thereby giving functions of the original variables that can be used to discriminate between the groups. The squared distances between group means are Mahalanobis D^2 statistics when all dimensions are used; otherwise they are approximations. Mahalanobis' distances are calculated using group means and variances, and a dendrogram can be constructed from the distance matrix. Discriminant function analysis can be further used to assign units to groups that have previously been discriminated by CVA following comparison of all unit canonical axes to group variances (Manly 1992).

Principal coordinate analysis

Principal coordinate analysis (PCO), or metric scaling, is an ordination method which operates with data in the form of a symmetric matrix of associations, for example, pairwise distances. Principal coordinates analysis, following Gower (1966), attempts to find a set of points for the n units in a multidimensional space and places similar units close together and dissimilar units further apart. The coordinates of the points are arranged so that their centroid is at the origin. Furthermore they are arranged relative to their principal axes, so that the first dimension of the solution gives the best one-dimensional fit to the full set of points, and so on for subsequent dimensions. Associated with each dimension of the set of coordinates is a latent root, which is the sum of squares of the coordinates of all the points in that dimension. For n units, if there is an exact solution it will be in at most $n - 1$ dimensions. If an incomplete solution results, either because the Euclidean property does not hold or because not all dimensions are used, then a residual can be calculated from the point to the centroid and to the original data. Eigenroots are computed by the Householder–Oretoga–Wilkenson method and coordinate axes assessed for significance by comparison to the average value of the diagonal elements. The robustness of clusters defined by PCO is examined by comparison of the minimum similarity between individuals within a cluster to the average similarity between all individuals.

be transformed or less sensitive non-parametric tests can be used (Sokal and Rohlf 1994). For an introduction and excellent detailed discussion of these significance tests, the reader is referred to texts such as Sokal and Rohlf (1994).

Multivariate statistics

As increasingly powerful computer facilities have become more widely available, various multivariate statistical procedures have been applied to the analysis of morphology and ecological tolerance of taxa (Cooley and Lohnes 1971, Manly 1992). These tests allow the examination of multiple combinations of characters in a single analysis, which can be presented graphically and can be used to estimate the characters that most differentiate particular taxonomic groupings. Examples of multivariate tests that can be applied are principal component analysis and canonical variate analysis (Manly 1992), the basis of which is described in Box 6.7, and see Box 6.8 for an example of its use.

For the homoploid hybrid species, *Helianthus anomalus*, prezygotic isolation from its parent taxa, *H. annuus* and *H. petiolaris*, appears to have been achieved by ecological separation (Schwarzbach, Donovan, and Rieseberg 2001). This evidence was gathered by examining a range of morphological and ecological characters for the three species and indicates that ecological separation appears to help reduce competitive interaction with parental taxa.

Landmark analysis

More recently, mathematical analyses of morphological variation have been developed with emphasis on the examination of biological shape and form. In particular, changes in specific homologous developmental structures or "landmarks" have been used to demonstrate shape changes between closely related organisms (Bookstein 1978). In these analyses, the coordinates of several comparable anatomical features describe the outline of a structure and quantification of the relative distortion of the coordinate positions between the landmarks of biological entities provides an informative description of shape change (Bookstein 1978, 1991, Gower 1971, Rohlf and Slice 1990, Sneath 1967). Further statistical developments of such methods have been used in biological contexts (Box 6.8).

6.5.4 Molecular genetic analyses

Considerations for speciation studies

Following reproductive isolation, population genetic theory predicts that a new species will become genetically differentiated from its progenitor(s) over time as mutations accumulate within separate lineages. Unlike morphological characters, which may be subject to the action of selection, neutral molecular markers are more likely to adhere to this prediction. Thus, the level of reproductive isolation between sympatric and/or potentially interbreeding species can be tested at a population genetic level by examining differentiation or estimating gene flow between lineages. However, for molecular genetic variation, several qualifications need to be borne in mind:

• For recent speciation events, there may not have been sufficient time for enough mutations to accumulate in the separate lineages to lead to significant differentiation. Such cases will also be typified by low genetic variation, associated with a speciation bottleneck, and will be independent of marker type used.

• For nuclear-encoded loci, neospecies derived via hybridization (either homoploid or polyploid) are likely to exhibit a genotype that is the additive product of its parents. Recently formulated polyploid and hybrid taxa will have low genetic diversity due to the restricted nature of their route of origin. However, diversity can be quickly recovered by polytopic formation involving divergent parents.

Box 6.8

Analysis of biological shapes

Morphometric analysis, which applies geometric information of biological features to study shape similarity and change between entities, was first pioneered by D'Arcy Thompson (1961), and can be used to great effect to demonstrate morphological variation between individuals, populations, or species. One method is to define homologous landmarks using reliable anatomical features, for a set of individuals and to examine how the position of landmarks changes between samples. Such information can be very informative to the nature of morphological change that accompanies speciation and hybridization events. A range of advanced statistical methods can be applied to examine aspects of shape change by describing the relative landmark positional changes as simply as possible by applying standardizing statistics and matrix transformations. For example, plotting Procrustes coordinates standardizes the size and rotation of individuals to allow better comparison of relative shape change between individuals. Shape variation can be visualized using various warping techniques (including thin-plate spline transformation) to Procrustes coordinates. Using this method the first specimen is taken as a reference, and a thin-plate transformation is applied to assess change relative to all other specimens, and is viewed as an *x* and *y*

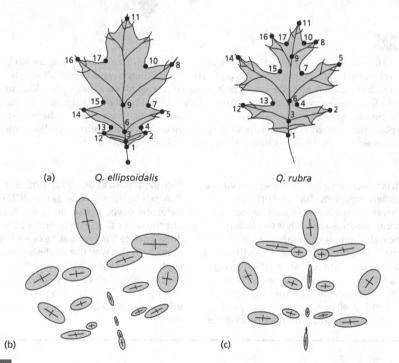

(a)   *Q. ellipsoidalis*

Q. rubra

(b)

(c)

Fig. 6a

(a) The typical leaf shapes of the two species are indicated together with the relative position of 17 landmarks recorded on a sample of leaves and trees along the hybrid transect. (b) and (c) By using a generalized fit analysis of landmark variation (Hill 1980, Jensen 1990, Rohlf and Slice 1990), the average position and a two-dimensional representation of standard variation (cross) and limit of variation (ellipse) are plotted. The two plots represent the averaged landmark location from different sample localities, the first where *Q. ellipsoidalis*-like individuals grow and the second are *Q. rubra*-like individuals.

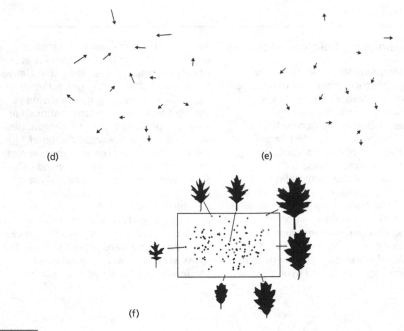

(d)

(e)

(f)

Fig. 6a

(d) and (e) For the two sample localities (in b and c), the consensus landmark configuration from each site is compared to the overall consensus landmark configuration averaged across all sample sites. The residual vectors from a generalized resistant fit are plotted for the *Q. ellipsoidalis*-like individuals and the *Q. rubra*-like trees, respectively. Arrows indicate the direction and magnitude of change that the respective landmarks have undergone from an "average" intermediate leaf to the two extreme morphologies depicted. (f) Finally a principal component analysis is performed on linear and angular measurements derived from the landmark coordinates (e.g., leaf length).

matrix which defaults from a square where landmark positions change. Further multivariate statistics can also be used to complement such analyses including principal components analysis, cluster analysis, and discriminant analysis (Box 6.7).

By way of example leaf variation along a transect of hybridization between two North American oak tree species, *Quercus rubra* and *Q. ellipsoidalis*, was analyzed using some standard landmark procedures

(Fig. 6a; Jensen et al. 1993). Whilst the first PCA axis (*x*) mainly reflects size differences between leaves, the second axis (*y*) picks out the other shape differences that differentiate the two oak taxa and their hybrids. In this way the analysis provides evidence that there has been a history of hybridization between the two oak species, mainly in the form of introgression from *Q. ellipsoidalis* into *Q. rubra* (Jensen et al. 1993).

• The relationship between genetic differentiation and classification is not straightforward. Therefore although some species concepts promote genetic differentiation criteria for taxonomy (e.g., numerical taxonomy species concept, see Box 6.1), it is not the case that there is an "expected" level of genetic differentiation for two taxa separated at the specific taxonomic levels (e.g., species, genus, family).

Despite these shortfallings there are several criteria available for defining species boundaries solely using genetic distance/differentiation methods (Box 6.9).

Box
6.9

Delimiting species using molecular markers

Several methods have been developed to delimit species using molecular genetic variation with objective criteria, and most of the methods use analytical techniques covered in previous chapters of this book (e.g. using a minimum limit of genetic distance or separation based on dendrogram topology). Sites and Marshall (2003), who provide a recent review, group the methods into non-tree based and tree-based approaches, and is the basis of the following outline and description.

Non-tree based methods

Hybrid zone barrier

Indirect estimates of gene flow based on Wright's population structure statistics (number of migrants per generation, Nm, chapters 3 and 4) have been used as a basis for examining the strength of species integrity across hybrid zones (Porter 1990). For sister taxa the value of Nm is calculated among populations within species and between sympatric populations of the different species. These calculations estimate the level of random mating between populations of each species and the background genetic similarity between species, not due to gene flow, respectively. If the Nm calculated between species components of a hybrid zone is not significantly greater than that found for the range of sympatric estimates then the hybridizing populations can be considered as two species. However, if some gene flow is recorded, but is weak (Nm between 0.5 and 1), then there may be exchange of selectively favored alleles and the two groups should not be viewed as separate species.

Genetic and geographic distance correlations

To disassociate isolation by distance effects from genetic differentiation between species (chapters 3 and 4) a correlation between pairwise genetic and geographic distance can be performed (Good and Wake 1992). A regression is performed between these pairwise estimates for populations within individual assigned groups (which can correspond to species or even geographic regions of a single species), and the regression line is expected to pass through the origin. The test is applied when looking at the regression relationship between populations from the two groups/species. If the line passes through the origin then there is an isolation-by-distance effect and these groups/species are not genetically isolated, however, if the two groups are genetically differentiated (and would be the expected situation for two different species), then the line will not pass through the origin. Furthermore if there is zero correlation between the genetic and geographic pairwise distances then genetic differentiation between the two groups is independent of geographic distance.

Absolute genetic distance

Specific values of Nei's genetic distance based on allozyme variation (Nei D, chapter 3) have been suggested to be used to delimit species. As 97% of well-defined vertebrate species have a Nei identity (Nei I) of >0.85 (approx. equal to Nei D of 0.16) and most values within species are greater than 0.85 (Thorpe 1982), Highton (1990) proposed that a cut off of $D = 0.15$–0.16 should be used to delimit species, as this is the value which generally correlates with complete speciation for sister taxa. The method assumes a regular mutation rate for loci scattered around the genome, and is correlated with the appearance of the reproductive isolation between species groups.

Field for recombination

For independent assorting nuclear loci, a field for recombination (Doyle 1995) can be identified by grouping individuals according to heterozygous allelic combinations. For each locus individuals are grouped if they share alleles that are combined in the heterozygous state in the sample. Individuals possessing alleles that do not form heterozygotes or form exclusive allelic heterozygotous states (i.e., non-overlapping sets of heterozygous individuals), are classified as separate groups, as heterozygotes provide evidence for recombination within a gene pool (chapter 3). The method cumulatively combines these fields for recombination over loci to derive a multilocus estimate that is used to define species. The method is intentionally non-topographic and attempts to disassociate sequence level processes (e.g., mutation and recombination), which are often complicating factors of allele trees, from whole organism demographic processes (e.g., gene flow and natural selection), which are usually more relevant to species boundary delimitation.

Population aggregation analysis

The importance of multicharacter/locus genetic differentiation for groups of potential species can be examined using a population aggregation analysis (Davis and Nixon 1992). A population sample, comprising several individuals, from each potential group, is scored for state at a number of attributes (loci or characters). A population profile is then defined for each group by assigning those attributes which have a fixed state (the same in all individuals of the sample) as "characters", and those that are polymorphic as "traits". The latter are deemed unimportant in species comparisons, unless polymorphisms do not overlap. The population profiles are then compared sequentially between groups and two groups are defined as separate species if they have fixed "character" differences between them, otherwise they are amalgamated into a larger single species group.

Tree-based methods

Cladistic haplotype aggregation

As with the population aggregation analysis, the presence of DNA haplotypes within samples is used to define population profiles and populations are grouped if they possess identical profiles. The phylogeny of unaggregated groups of haplotypes is then calculated (chapter 5), and separate species are defined from topographically contiguous groups of populations (Brower 1999).

Exclusivity criterion

Using genealogies from unlinked loci for the same individuals, a strict consensus tree is derived (chapter 5), and species are defined according to those with exclusive nodes (Baum and Shaw 1995). Under this method species must be basal, i.e., they do not contain other taxa, and the method assumes that unlinked genes have similar evolutionary histories.

Genealogical difinition

Phylogenetic trees from morphology or DNA (which should be non-recombining, chapter 5) for individual samples from geographically located populations can be used to define species (Wiens and Penkrot 2002). The method is used in conjunction with nested clade analysis and assumes that failure of DNA haplotypes from a single locality to cluster together on the tree is evidence of gene flow (chapter 5). The method is population based and defines species as those populations that are strongly supported, exclusive, and concordant with geography. A range of trees and outcomes are possible and Wiens and Penkrit (2002) use easy-to-follow flow diagrams to illustrate population grouping and species definition.

Test of cohesion

Templeton (1998) has proposed a test of species boundaries that considers both historical and ongoing demographic influences on population genetic structure. The method tests two hypotheses: (H_1) organisms sampled are derived from a single evolutionary lineage, and (H_2) populations of lineages identified by rejection of H_1 exchange genes (genetically exchangeable) and/or possess similar adaptations or environmental tolerances (ecologically interchangeable). Species are recognized only after rejection of both hypotheses at the same levels of divergence. H_1 is tested by calculating the significance of association between nested clade networks and geographic locations (chapter 5). An inference key is provided to help discriminate the biological processes that lead to statistical significance and allow assessment of the rejection of H_1. Even if two genetic lineages are separate there is still the possibility that they constitute a single species and so H_2 tests whether such fragmented lineage exchange genes or are ecologically interchangeable. This test is performed using nested clade analysis to test association of traits for genetic exchangeability (e.g., mating isolation) or ecological interchangeability (e.g., habitat requirements or environmental tolerances) with the evolutionary lineages defined in the rejection of H_1. H_2 is rejected if either set of traits are significantly associated with evolutionary lineage, and the separate evolutionary lineages are presumed to be separate cohesive species if H_2 is rejected at the same clade level at which H_1 was rejected.

Box
6.10

Measuring an introgression ratio

The introgression ratio (*IG*) reflects the proportion of haplotypes shared between two species in a sympatric area or population. The value of *IG* is expected to be one when there is no difference between the species and zero when they are totally different and can be calculated according to Belahbib et al. 2001.

A global measure can be taken by comparing actual with expected introgression ratio, calculated assuming that haplotype distribution is not geographically structured. One locus may be considered (e.g., cytoplasmic genomes)

or for nuclear loci an average could be taken.

Alternatively individual populations and species pairs can be considered separately and calculated according to the formula:

$$IGR = \frac{IG(x_i \cdot x_j)}{IG(x_i \cdot y_i)}$$

where $IG(x_i \cdot x_j)$ is the introgression ratio among populations of species x (referred to as the focal species) in population I and in forest j, and $IG(x_i \cdot y_i)$ is the introgression ratio between species x and y in forest i. If the ratio is larger than one then individuals of species x in forest i are more similar to individuals of the same species in forest j than to individuals of species y.

Considerations for hybridization studies

It should now be obvious that demonstrating differentiation between species against a background of hybridization can be a difficult task. However, hybridization between two genetically differentiated taxa can also result in several diagnostic patterns of variation:

• First-generation hybrids should exhibit an additive genotype between parental types for nuclear-encoded molecular markers. However, in later generations or backcross products, markers will segregate or recombine. Thus, across a hybrid zone a range of genotypes may be observed between parental extremes. Selection pressures may serve to maintain the limits of species and their hybrids.

• The outcome of introgressive hybridization may be that only a small portion of the genome of one species may be present in the genomic background of another species. Such cases can be easily misidentified during cursory surveys.

• For cytoplasmically encoded genomes, inheritance tends to be uniparental. In most animals and plants, the mitochondrial genome is inherited through the maternal line. In angiosperms, the chloroplast genome is predominantly inherited through the maternal line, whilst in gymnosperms it is predominantly inherited through the paternal line (see chapter 2). These modes of inheritance can be useful for working out the maternal and paternal parents of particular hybrid products, including new species.

• Where two species are thought to have been hybridizing in a zone of contact, the amount of introgression of genes and allele sharing can be assessed using an introgression ratio (Box 6.10), whilst commonly applied to cytoplasmic markers it can also be used for nuclear genes.

Examples of application of molecular markers

For a review of the characteristics of molecular markers that are useful for examining speciation and hybridization, see Box 6.11 and chapter 2. Following are some examples of how a range of molecular markers have been successfully applied to such studies.

Nuclear genome – single locus markers. Allozyme and microsatellite variation have been widely used to demonstrate neospecies and detect hybridization or introgression. Isozyme analysis of the progenitor–derivative species pair, *Corepsis nuecensoides* and *C. nuecensis*, demonstrated that whilst there was close genetic similarity between the two species the former exhibited

Box
6.11

Behavior of molecular markers for speciation and hybridization studies

Variation at characterized DNA loci including the mitochondrial, chloroplast, and nuclear genomes, can be analyzed using a variety of methods, including restriction fragment length polymorphism (RFLP), single strand conformation polymorphism (SSCP), sequencing, and even single nucleotide polymorphism (SNPs). Whilst markers from the uniparentally inherited organelle genomes can be used to demonstrate closest ancestor or differential parental contributions within hybrids, markers from biparentally inherited genomes can be used to examine genetic differentiation of neospecies or show that hybrids exhibit a genetic profile which combines the specific markers of its putative parents (i.e., is additive), also see chapter 2.

Organelle-encoded loci

The rapid rate of phylogenetically interpretable base substitution and its predominant maternal inheritance make the mtDNA genome in animals an ideal marker for establishing progenitor species relationships and hybrid maternal lineage (Wolfe, Li, and Sharp 1987). Whilst the mutation rate of plant chloroplast DNA is lower than that of animal mtDNA, useful levels of intraspecific variation exist providing a comparable organelle marker for plants (Harris and Ingram 1991, Soltis, Doyle, and Soltis 1992, Wolfe, Li, and Sharp 1987). In addition, the discovery of variable mononucleotide microsatellite repeat motifs in the genome offers good future potential for variable markers, although phylogenetic interpretation may be limited (Provan, Powell, and Hollingsworth 2001, Provan et al. 1999). The cpDNA molecule is predominantly maternally inherited in angiosperms but paternally inherited in gymnosperms (Wagner et al. 1991), offering the potential for a pollen marker in the latter group. Despite its low rate of mutation and high rate of internal recombination, which prevent useful reliable interpretation (Palmer 1992), the maternally inherited mitochondrial genome of plants does harbor some intraspecific variation that can be used to complement cpDNA studies by highlighting extra organelle variation or by providing a complementary seed-only dispersed marker for studying gymnosperms. Large sets of universal PCR primers are available which cover many species groups, organelles, and genomic loci.

Nuclear-encoded loci

Allozyme proteins are predominantly encoded by nuclear loci and are codominantly expressed (Crawford and Ornduff 1989, Gallez and Gottlieb 1982), and despite having a slow rate of mutation that is not phylogenetically interpretable, they have been used extensively to demonstrate genetic differentiation, founder bottlenecking, and hybrid additivity. Microsatellite markers which are DNA-encoded and codominantly expressed, offer faster mutation rates, increasing their potential application to trace neospeciation and hybridization events.

Anonymous loci can be isolated from genomic libraries and variation surveyed using RFLP analysis or sequencing. The neutrality or adaptive potential of loci can be assessed by using database searches (such methods have been used for animals and plants, e.g., Karl, Bowen, and Avise 1992, Lowe et al. 1998). A number of potentially neutral markers are also available for a wide range of organisms using universal PCR primers for loci located in known gene introns and are likely to be used in future analyses of speciation and hybridization (e.g., Strand, LeebensMack, and Milligan 1997). Nuclear-encoded genes for developmental characters that may have adaptive and breeding system significance will also undoubtedly be of increased importance in the future for furthering our understanding of adaptive changes during speciation. For example the PER and VOX genes of insects and in plants the MADS box and *cycloidea* loci (Gustafsson and Albert 1999, Gillies et al. 2002, Möller et al. 1999).

Multigene families of loci-encoding nuclear proteins are also potential target markers. However, gene duplication issues provide difficulties in ensuring that the equivalent gene locus is targeted (orthologous) in different taxa, rather than a non-coding duplicate (paralogous). This is also a problem for ribosomal DNA loci and several related families of rDNA-like sequences can be found within single genomes (Hershkovitz, Zimmer, and Hahn 1999). However, the role of concerted evolution (Elder and Turner 1995) serves to homogenize multiple copy DNA regions (e.g., rDNA) in most species and may also be the case for other low copy number loci (Doyle and Doyle 1999, Zimmer et al. 1980). The rate of substitution of ribosomal DNA is intermediate between that of plant chloroplast and animal mitochondrial genomes and variation is phylogenetically informative (Baldwin 1992). In cases of hybridization, the rDNA profile often exhibits an additive pattern combining parental types during the first generations but with time becomes homogenized in the direction of one or other of the parental types (see Fig. 6.3). It may be necessary to undertake considerable exploratory analysis when applying genomic regions that have had relatively little characterization to the study of new species, and these issues are discussed in more detail in chapter 5.

Whole genome approaches

Techniques that "randomly target" genomic loci, such as RAPDs, inter-SSRs, and AFLPs, offer great potential for identifying genetic differentiation amongst taxa and highlighting the additive genomic profiles that accompany hybridization (Weising et al. 1995). There are problems with the use of these markers, including genomic origins (which can be from nuclear or organelle loci), homology of comigrating bands, and their independence (chapter 2, Harris 1999, Rieseberg 1996, Robinson and Harris 1999), therefore a suitably robust method must be used in order that the findings of a study are not compromised.

Table 6c

Characteristics of molecular markers for studies of speciation and hybridization.

Marker	Mutation rate	Phylogenetically interpretable	Inheritance	Differentiation	Diversity/ Bottlenecks	Hybrid additivity
MtDNA animals	V. fast	Yes	Maternal	****	****	No
MtDNA plants	V. slow	No	Maternal	*	*	No
CpDNA plants	Slow	Yes	Maternal to paternal	**	**	No
Isozymes	Slow	No	Biparental	**	**	Yes
Microsatellites	Fast	No	Biparental	***	****	Yes
Specific nuclear loci	Slow to v. fast	Yes	Biparental	***	**	Probably
Multigene families/loci	Fast	Sometimes	Biparental	***	**	Depends
Multiloci whole genome markers	Slow to v. fast	No	Biparental	****	****	Yes

The number of stars rates the use of the method for a particular application, where **** is very useful and * is of limited use.

significantly reduced diversity compared to the latter and is concordant with their presumed species relationship (Crawford and Smith 1982). Isozyme analysis has shown that the recently arisen allotetraploid, *Spartina anglica*, combines the diagnostic genotypes of *S. alterniflora* and *S. maritima* and most probably arose by chromosome doubling of the sterile hybrid, *S.* x *townsendii* (Gray, Marshall, and Raybould 1991, Raybould et al. 1991).

Nuclear genome – multigene familes. The rDNA multigene family, encoded at several nuclear loci, has been successfully applied to study hybridization between *Iris* species in the southeast of the USA. Arnold, Bennett, and Zimmer (1990) showed that individuals in allopatric populations of *I. fulva* and *I. hexagona* possessed species-specific rDNA profiles, and that the distribution of rDNA markers among plants in parapatric populations indicated that bidirectional backcrossing and introgression were common (Nason, Ellstrand, and Arnold 1992). Nuclear rDNA analysis (Arnold 1993) also confirmed that the diploid species *I. nelsonii* combined genetic markers found in allopatric populations of *I. fulva*, *I. hexagona*, and *I. brevicaulis* and may, therefore, be considered to be a derivative of hybridization between these three species.

Cytoplasmic genome markers – mtDNA. The highly variable mtDNA genome of animals has been used extensively to demonstrate cases of speciation and hybridization. For example, an ancient (postglacial) introgression event between Arctic charr (*Salvelinus alpinus*) and lake trout (*S. namaycush*) was inferred using mtDNA analysis (Wilson and Bernatchez 1998). It was found that Arctic charr mtDNA had reached complete fixation in a population of lake trout from southern Quebec despite the population appearing to have lake trout morphology and microsatellites. Fixation of the Arctic charr mtDNA could have occurred through several mechanisms, including drift and selection of the mitochondrial type and some associated nuclear genes, although retention of ancestral polymorphism was ruled out.

Despite its low rate of mutation (Wolfe, Li, and Sharp 1987) and high rate of internal recombination (Palmer 1992, see chapter 2) the mitochondrial genome of plants has also been used to study neospecies and hybrids. For example, a screen of variation in the mitochondrial genome of oaks, together with cpDNA variation, was used to examine the rate of plastid genome transfer in sympatric and hybridizing populations of oak (Dumolin-Lepèque, Kremer, and Petit 1999, Dumolin-Lepèque, Pemonge, and Petit 1999).

Cytoplasmic genome markers – cpDNA (plants). For plants, analysis of cpDNA variation of the *Tragopogon* allotetraploids, *T. mirus* and *T. miscellus*, and their respective diploid parents, revealed that the maternal parent in three independent origins of *T. mirus* was *T. porrifolius*; whereas both *T. dubius* and *T. pratensis* had acted as maternal parents in the recurrent origin of *T. miscellus* (Soltis and Soltis 1989). This work supported much earlier work on this species complex using isozymes (Roose and Gottlieb 1976). See also the classic hybrid speciation events for *Senecio* taxa from the British Isles in chapter 7.

Total genome approaches. RAPD analysis has been successfully used to confirm an intergeneric hybrid speciation event between *Margyricarpus digynus* and *Acaena argentea*, which had been previously suspected from morphological evidence. The hybrid *Margyracaena skottsbergii* is endemic to the Juan Fernandez Islands and was found to exhibit an additive RAPD profile which combined all 18 fragments specific to *Acaena argentea* and all 23 unique fragments observed in *Margyricarpus digynus* (Crawford et al. 1993). Using AFLP analysis, Beismann et al. (1997) found evidence to support morphological characterization that *Salix* x *rubens* is the interspecific hybrid of white willow (*S. alba*) and crack willow (*S. fragilis*). They also identified a clonal colony in crack willow due to the sharing of an identical 335-band profile amongst several individuals.

Analyzing and presenting molecular genetic data

Phenograms and tables. One of the simplest methods of demonstrating a diagnostic or additive genotype is to represent pictorially the observed genotype/phenotype (e.g., Fig. 6.4). However,

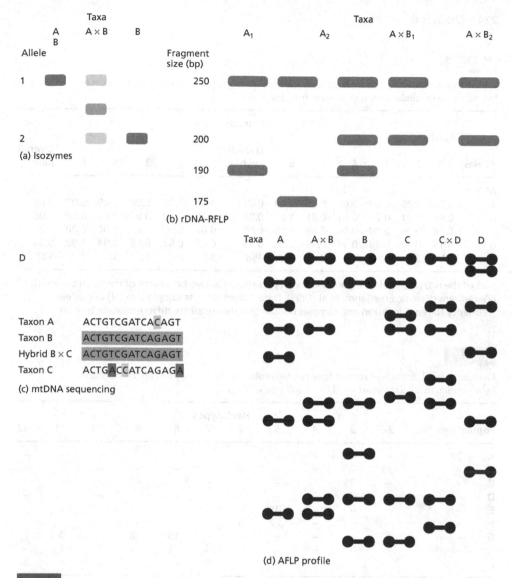

Taxa

| | A | A × B | B | | | A₁ | A₂ | A × B₁ | A × B₂ |

(a) Isozymes

| Allele | | | | Fragment size (bp) | | | | | |

Taxon A   ACTGTCGATCACAGT
Taxon B   ACTGTCGATCAGAGT
Hybrid B × C   ACTGTCGATCAGAGT
Taxon C   ACTGACCATCAGAGA

(c) mtDNA sequencing

(b) rDNA-RFLP

(d) AFLP profile

Diagrammatic presentation of typical behavior and expected profiles of molecular markers demonstrating species delimitation and interspecific hybrids. (a) Isozyme profile of allelic variation at dimeric enzyme for diploid taxa. The individual of taxon A expresses allele 1 and that of taxon B allele 2. The interspecific hybrid combines alleles 1 and 2 from the parent taxa and also exhibits the heterodimeric band, which migrates midway between the two homodimeric enzyme molecules. (b) Following restriction digestion of the rDNA gene, all taxa have a 250 bp fragment. Taxon B is monomorphic for a 200 bp fragment, whereas taxon A is polymorphic for two fragments (190 bp and 175 bp). The first interspecific hybrid (A × B₁) combines the molecular profile of taxa A₁ and B, whereas hybrid A × B₂ only expresses the profile of taxon B. The latter pattern of expression in a hybrid may be due to concerted evolution or maternal effects (see main text). (c) The hybrid (B × C) has inherited the mtDNA type of taxon B, which for most animals and angiosperms, is expected to be from the maternal parent due to known inheritance patterns. (d) The hybrid A× B combines the AFLP molecular profile of its two putative parents, however, some fragments present in the parent may not be present in the hybrid due to the dominant expression of heterozygous bands. The hybrid C × D represents an individual that expresses a combination of additive and novel fragments compared with its putative parent taxa (C and D). Such a profile may arise due to rapid genome reorganization following hybridization or accumulation of mutations in an established hybrid species that has been reproductively isolated from its parent taxa for some time.

Table 6.1

Commonly used methods of representing raw molecular data supporting cases of speciation or hybridization in tabular format – allozyme frequency data.

	Larch							Pine					
							Overall						Overall
Alleles	1	2	3	4	5	6	larch	1	2	3	4	5	pine
Mdh-S													
a	0.00	0.00	0.03	0.01	0.00	0.00	0.01	0.00	0.00	0.00	0.00	0.00	0.00
b	0.86	0.81	0.74	0.81	0.81	1.00	0.80	0.10	0.33	0.01	0.06	0.00	0.05
c	0.00	0.00	0.00	0.01	0.00	0.00	0.00	0.00	0.00	0.01	0.00	0.00	0.01
d	0.13	0.19	0.23	0.17	0.19	0.00	0.18	0.89	0.67	0.98	0.94	1.00	0.94
N	37	31	48	42	8	2	168	34	3	65	18	7	127

Part of the isozyme allele frequency data set presented for two host races of the larch budmoth (*Zeiraphera diniana*; Emelianov et al. 1995). Data shown reflects sample size (*N*) and allele frequency for each location and samples pooled that demonstrate differentiation by host.

Table 6.2

Commonly used methods of representing raw molecular data supporting cases of speciation or hybridization in tabular format – mtDNA haplotype scoring.

	Haplotypes											
Populations	1	2	3	4	5	6	7	8	9	10	11	12
A	27	–	–	–	–	–	–	–	–	–	–	–
B	–	35	–	–	–	–	–	–	–	–	–	–
C	–	–	36	–	–	–	–	–	–	–	–	–
D	–	–	–	27	–	–	–	–	–	–	–	–
E	–	–	–	–	10	–	–	–	–	–	–	–
F	–	–	–	–	2	3	4	–	–	–	–	–
G	–	–	–	–	–	–	–	15	18	1	3	–
H	–	–	–	–	–	–	4	1	8	7	1	–
I	–	–	–	–	–	–	6	–	–	–	–	8

Variation in a mtDNA RFLP data set presented for Australian populations of the ghost bat (*Macroderma gigas*; Wilmer et al. 1999). Haplotype frequencies should be presented together with a table describing the number of mutations, distance estimates for haplotypes, or a dendrogram.

representation of complex patterns of variation or multiple loci is too cumbersome using this method. Tables may be used to present this type of data, where instead of a graphical rendering of the variation, the genotype is represented by a number or letter (Tables 6.1, 6.2, 6.3). Several other criteria from delimiting species are described in Box 6.9. In the case of hybridization, the proportion of a genome that is shared between species within a contact zone can be assessed relative to the frequency of species-specific markers outside the contact zone (a measure called an introgression ratio, Box 6.10, which can be quoted as a single figure for a species pair).

Scatter plots and multivariate statistics. To summarize genotype/phenotype information, a genetic distance algorithm may be applied. There are many different distance algorithms and choice of

Table 6.3

Commonly used methods of representing raw molecular data supporting cases of speciation or hybridization in tabular format – RAPD fragment frequencies.

Species	S.vulgarls var. *vulgaris*	S. vulgaris var. *hibernicus*	S. eboracensis	S. squalidus
RAPD product size (bp)				
OPH01				
750	0.92	0.80	0.54	0.20
650	1.00	1.00	1.00	0.80
620	0.00	0.00	1.00	0.60
550	0.00	0.00	1.00	1.00
N (individuals)	13	5	11	5

Part of frequency data set of RAPD fragments for common groundsel (*Senecio vulgaris* var. *vulgaris*) and Oxford ragwort (*S. squalidus*) and two of their hybrid derivatives (Lowe 1996). *Senecio eboracensis* combines the RAPD profile of its two parent taxa, whereas the introgressant, *S. vulgaris* var. *hibernicus*, is indistinguishable from var. *vulgaris*.

the most suitable depends on the behavior of the marker system that is being employed (Box 6.11 and Appendix B). For many molecular genetic data sets, one of the simplest methods of presenting appropriate genetic distances graphically is a multidimensional scaling plot (Box 6.7). Other multivariate statistics can also be applied. For some isozyme and microsatellite data it may be possible to assume a normal distribution of alleles and apply a parametric differentiation to data, such as principal component analysis. For the simple presence and absence data derived from dominant multilocus markers (i.e., RAPD and AFLP), a principal coordinate analysis offers an alternative solution to dendrogram plotting (Box 6.7, e.g., Fig. 6.5).

Dendrograms. Dendrograms provide visual representation for a large set of entities (whether it be characters, individuals, or species). Almost any distance measure can be presented in this way. Tree-building methods used to represent simple distance relationships between taxa or individuals, include UPGMA and neighbor-joining methods (see chapter 5, e.g., Fig. 6.6). Certain data types (e.g., RAPD and AFLP) may be more appropriately presented using neighbor-joining methods, which allow different mutation rates for each branch. Phylogenetic methods (e.g., parsimony analysis) can also be used, and are invaluable for reconstructing the evolutionary order of speciation events (e.g., using mtDNA, Fig. 6.7). However, such concepts are part of a large body of research on phylogenetics and cladistics, and are beyond the scope of this chapter but see chapter 5 for an introduction to these methods.

Hybrid taxa, which are normally detected by the possession of additive phenotypes, may also be identified due to their intermediate position between parental taxa in a dendrogram constructed from phenetic distance measures based on allele or marker frequencies. For example, in a UPGMA dendrogram of allozyme allelic frequencies, hybrids and introgressants between *Colochortus minimus* and *C. nudus* were placed in an intermediate position relative to the parental taxa (Ness, Soltis, and Soltis 1990).

Discrepancies between phylogenies based on nuclear and organelle DNA may also indicate past hybridization events (Fig. 6.8). For example, phylogenetic analysis of restriction-site variation of nuclear rDNA genes in the 21 taxa comprising *Helianthus* sect. *Helianthus* revealed major discrepancies when compared with the cpDNA phylogeny (Rieseberg 1991). The rDNA evidence strongly suggested both recent and ancient introgression and provided compelling evidence that *H. anomalus*, *H. deserticola*, and *H. paradoxus* were diploid species, and originated as hybrids

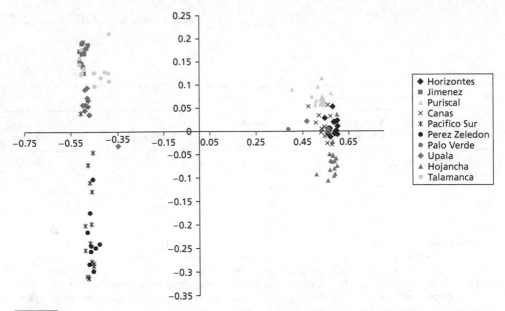

Fig. 6.5

Commonly used methods for graphical representation of speciation and hybridization data – PCO plot of AFLP data. For Costa Rican populations of Spanish cedar, *Cedrela odorata*, the first two principal coordinates are calculated from a 145 AFLP fragments data set (Cavers, Navarro, and Lowe 2003). Moist forest trees (right) are highly differentiated from dry forest types (left), and the authors argue that these two ecotypes may actually represent separate species.

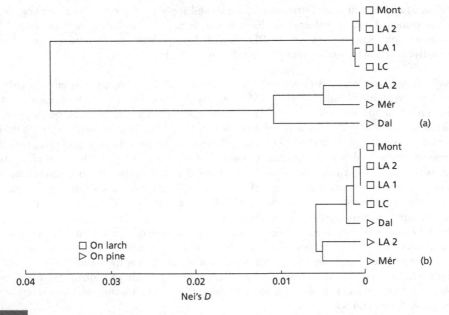

Fig. 6.6

Commonly used methods for graphical representation of speciation and hybridization data – UPGMA dendrogram for isozyme data. Isozyme variation within populations of the larch budmoth from larch and pine hosts sampled from paired locations is demonstrated using UPGMA clustering based on Nei's (1978) unbiased genetic distance of isozyme variation (Emelianov et al. 1995). Plot (a) shows genetic divergence between host races, although this relationship collapses when three differentiating loci are removed (Plot (b)).

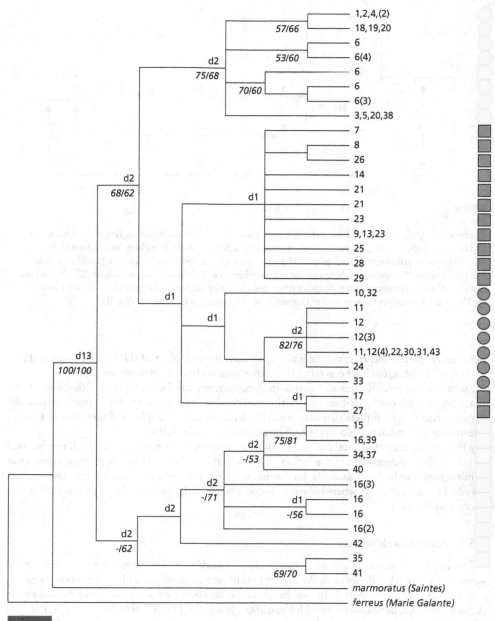

Fig. 6.7

Commonly used methods for graphical representation of speciation and hybridization data – dendrogram of mtDNA variation. A strict consensus tree, indicating support values (above) and bootstrap values (below), was constructed from mtDNA variation in the Dominican anole (Malhotra and Thorpe 2000). The symbols identify four major genetic clades, which are also morphologically and geographically differentiated. The authors use this evidence to support a vicariant speciation hypothesis on this Lesser Antillean island.

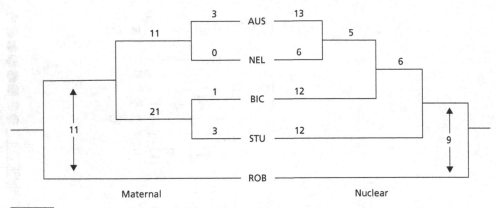

Examples of hybridization producing tree discrepancies – discordant phylogenies. Discordant maternal and nuclear phylogenetic trees for Australian cotton species (*Gossypium*; Rieseberg and Wendel 1993). A monophyletic group contains *G. bickii* (BIC), *G. australe* (AUS) and *G. nelsonii* (NEL) for the nuclear markers, but cpDNA indicates shared recent ancestry between *G. bickii* and *G. sturtianum* (STU), the latter from a different taxonomic group. Such a pattern could arise by ancient introgression of *G. bickii* ancestral cpDNA into the nuclear background of *G. sturtianum*. Outgroup is *G. robinsonii* (ROB).

between *H. annuus* and *H. petiolaris*. Taken together, the cpDNA and rDNA data suggested that evolution in *Helianthus* has been reticulate rather than exclusively dichotomous.

Sometimes the resolution of hybrids in dendrograms can be ambiguous (Rieseberg 1995; Fig. 6.9). In such cases, split decomposition is a dendrogram-based method, which graphically presents relational conflicts that may occur in data sets containing hybrid taxa, and is a useful alternative to standard dendrogram-building algorithms (Fig. 6.10).

Multivariate statistical methods are a robust alternative to dendrograms, which can be used to great effect to demonstrate speciation or hybridization events. They are recommended over dendrogram methods of analysis for initial investigations where evolutionary assumptions should be minimized or when hybridization is suspected (Rieseberg and Ellstrand 1993, see above and Box 6.7).

6.5.5 Genomic architecture

Following development of a suite of morphological and/or molecular markers that distinguish between closely related species or document hybridization, knowledge of their genomic location and arrangement can shed light on the genomic outcomes of speciation and hybridization (Charlesworth, Charlesworth, and McVean 2001, Jasny and Hines 1998). In particular, the field of comparative genomics, which concentrates on examining gross and fine-scale changes between gene location and sequence offers tremendous insights into the genomic changes accompanying speciation and hybridization (Eppig 1996). The techniques used to construct and analyze genomic maps are described in detail by a number of other workers (e.g., Weir 1996), however, it is useful to review here a number of general principles and the findings from some specific case studies.

The role of chromosomal repatterning and genomic recombination is key to some speciation models, for example, saltational speciation and polyploidy, and so the study of genomic change following these processes is an essential aspect of speciation hypothesis testing. *Arabidopsis* is the best genetically characterized plant species, and advantage has been taken of this vast background of sequence and mapping information to examine the genomic consequences of speciation and

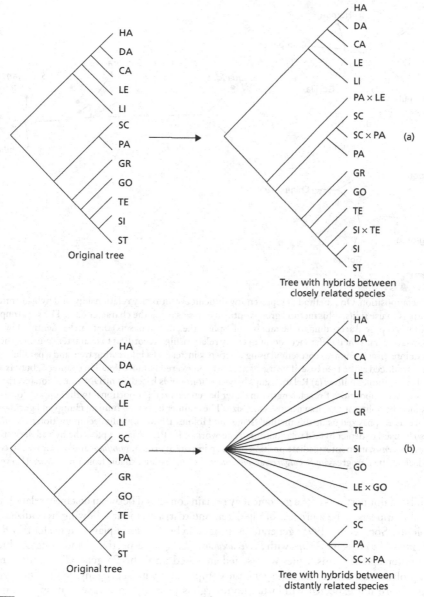

Fig. 6.9

Examples of hybridization producing tree discrepancies – collapsing nodes. Cladograms of Central American *Aphelandra* species (McDade 1992). Hybrids between closely related taxa (plot (a)) are resolved in an intermediate position; however, those between distantly related taxa (plot (b)) disrupt tree topology.

hybridization within the closely related Brassicaceae. Phylogenetic comparison of mapped loci reveals that the crop genus *Brassica* contains approximately three times the genome complement of *Arabidopsis thaliana* and has probably arisen via ancient polyploidization (Lydiate et al. 1993). Examination of gene and sequence order also reveals that major duplication and rearrangements have occurred, but that there are regions of the *Brassica* and *Arabidopsis* genomes that have been conserved (Cavell et al. 1998, Kowalski et al. 1994, Largercrantz and Lydiate 1996).

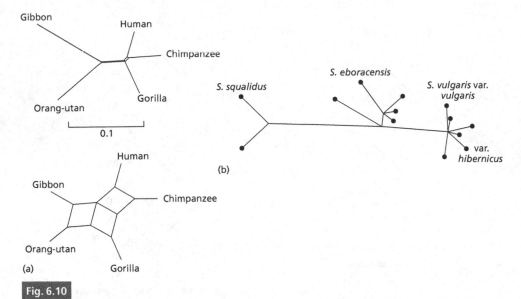

Fig. 6.10

Split decomposition. One method of representing difficult evolutionary relationships and hybrid entities in tree form is to use splits. This method groups entities into sets using the character data. The set grouping of entities is repeated according to the number of "splits" (i.e., different sets) that can be identified by the character data. When differences occur in entity relationships according to character splitting, these different interpretations are represented using a network instead of a bifurcating tree, and a parallelogram can be introduced at the tree base. The use of networks is covered in more detail in chapter 5, but taxonomic and hybrid examples follow. (a) Relationships between hominoids based on mtDNA variation interpreted using split decomposition. Branch length can either be represented proportional to the support for the corresponding split (top) or drawn at the same size. The conflict involving human, chimp and gorilla sequences is emphasized using this method (Page and Holmes 1998). (b) Split decomposition network can also be used to depict hybrid relationships. A network of RAPD variation plots the hybrid derivative, *Senecio eboracensis*, in an intermediate position between its parent taxa, *S. squalidus* and *S. vulgaris*. In contrast, individuals of the introgressant *S. vulgaris* var. *hibernicus*, are not distinguishable from var. *vulgaris* (Lowe 1996).

Whilst major portions of the genome may remain conserved between distantly related species, there also appears to be a process of rapid genome rearrangement following hybridization and polyploidy. Song et al. (1995) generated synthetic hybrid *Brassica* plants equivalent to *B. napus* and *B. juncea* by crossing *B. rapa* with *B. oleracea* and *B. nigra*, respectively, and using colchicine to generate tetraploid plants. Lines were selfed and bred until the F5 generation and 89 nuclear DNA probes were used to examine genomic composition within progeny lines. Whilst genome changes were not expected within the homozygous progeny, extensive rearrangements were detected. The magnitude of change was greatest in the progeny of the cross between the more divergent parental taxa. In addition, the maternally contributed nuclear genome exhibited less change than that of paternal origin. The results indicate that even within a relatively short time period, polyploid hybrids can generate novel genomic diversity, which may contribute to their successful establishment (Soltis and Soltis 1995). Such observations have also been noted in diploid hybrid genomes (Box 6.12).

Rapid genome reorganizations have also been noted in the later generations of other synthesized hybrids and polyploids (e.g., cotton, maize, and wheat; Gale and Devos 1998, Leitch and Bennett 1997, Stephens 1949). Other common features of polyploid genomes have been observed, including gene silencing, which eventually leads to a diploidization of the genome, and gene diversification, which may lead to differentiation in the role of duplicated genes (Soltis and

Box
6.12

Genetic maps and genomic reorganization during sunflower speciation

In order to investigate the role of genome reorganization in diploid hybrid speciation, Rieseberg and coworkers generated genetic maps for the sunflower, *Helianthus anomalus*, a diploid hybrid derivative of *H. annuus* and *H. petiolaris* (Rieseberg 1995, Rieseberg, Fossen, and Desrochers 1995, Rieseberg et al. 1996). Parental genomic contributions were found to be combined and mixed in the hybrid line (Fig. 6b(a)), indicating that polyploids may have tremendous flexibility in optimizing their genomes and allow them to occupy new ecological niches. The genomic compositions of three synthetic hybrid lineages, equivalent to *H. anomalus*, were compared with individuals of the ancient hybrid species from natural populations (Fig 6b(b)). Statistically significant concordance was found between the natural and resynthesized lines, and large blocks of genes appeared to have been inherited and protected from recombination following hybrid speciation. From this evidence, Rieseberg and coworkers suggest that selection may largely govern genomic organization in hybrid speciation and that significant associations among unlinked genes imply that interactions between coadapted parental genes constrain the genomic composition of hybrid species.

Fig. 6b

(a) Genetic maps were generated for the hybrid species *H. anomalus* and its two parent taxa *H. annuus* and *H. petiolaris* using RAPDs. Comparative genetic mapping techniques were used to compare the affinity of genomic segments of the hybrid to its parents. The letters A to W indicate chromosomal segments, where a matching letter in the different species indicates gross homology, and shaded areas indicate significant translocations. Whilst some chromosomes are unchanged between the species, it is clear that *H. anomalus* recombines a large amount of material from both parents *H. annuus* and *H. petiolaris* in its genomic complement (Rieseberg 1995).

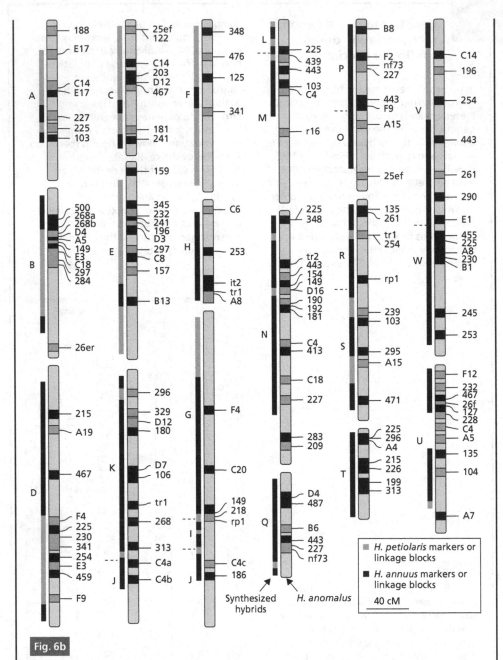

Fig. 6b

(b) The genomic composition of experimentally generated (left arms) and wild lines (right arms) of the sunflower homoploid hybrid, *H. anomalus*, are examined using comparative genetic mapping techniques (letters A to W indicate linkage groups). The distribution of parental markers (*H. petiolaris* – gray, and *H. annuus* – black blocks) within natural populations of the ancient hybrid are strikingly similar (and significantly different from random expectation) to three recently and independently synthesized lines of *H. anomalus* (Rieseberg et al. 1996).

Soltis 1993). The role of transposable elements and methylation as a driving force in the evolution of global gene-silencing mechanisms and genome restructuring has been proposed (Soltis and Soltis 1999). As is the case for diploid genomes, there may be a period of transilience within polyploids following speciation, when the genome is more tolerant of change (Box 6.12). Indeed, similar patterns of chromosomal change have been detected throughout a range of polyploid species (e.g., wheat and tobacco), and suggest that genome change occurs at the time of or shortly after speciation events. The extent to which these genomic changes are similar across independent lines and origins remains a crucial question and in need of further testing in the future (Soltis and Soltis 1999).

6.6 Future developments

Important advances have been made in statistical techniques for analyzing patterns of morphological and genetic variation resulting from speciation and hybridization events. However, it is likely that future advances will come from gene sequence and genomic organization investigations, which will improve our understanding of the processes acting behind these events.

Investigation of function, adaptation, and change of genes involved in promoting reproductive isolation and ecological tolerance will doubtless increase as loci become available that can be screened at a population level. The use of markers for genes with evolutionary development importance (so-called "evodev" markers) could provide evidence for major gene effects or changes that drive speciation events, and can be used to establish whether gene changes following speciation have occurred simultaneously or consecutively.

The application of gross comparative genomic and microsynteny approaches will improve our understanding of the generality of transilience and translocation processes that have so far been observed to accompany hybridization and polyploidization events. In addition, these methods can be used to test hypotheses on the influence of ecologically and environmentally induced genomic changes contributing to speciation and to establish the genetic basis of transgressive characters for ecological tolerance that appear to follow many cases of hybridization and polyploidization.

6.7 Further reading

A detailed description of a range of methods for studying reproductive isolation in plants and animals is outlined in amongst others Levin (2002). Description of the methods used to generate chromosome preparations are provided amongst others by the excellent texts and protocols of Darlington and La Cour (1962), Dyer (1979), and Sessions (1996).

For morphological comparisons, an introduction and excellent detailed discussion of significance tests is provided by Sokal and Rohlf (1994), but see Box 6.10 for an introduction. The use of landmark analysis and other shape-based methods are provided by Bookstein (1989, 1991), Rohlf and Slice (1990), and Sneath (1967).

The basis of multivariate statistics, including principal component analysis and canonical variate analysis is given in detail by Manly (1992), but an introduction is provided in Box 6.7. Chapter 5 provides further reading suggestions for dendrogram methods.

The techniques used to construct and analyze genomic maps are covered in detail by excellent texts such as Weir (1996).

REFERENCES

Abbott, R.J. 1992. Plant invasions, interspecific hybridization and the evolution of new plant taxa. *Trends in Ecology and Evolution*, **7**: 401–5.

Abbott, R.J., Ashton, P.A., and Forbes, D.G. 1992. Introgressive origin of the radiate groundsel, *Senecio vulgaris* L. var. *hibernicus* Syme: Aat-3 evidence. *Heredity*, **68**: 425–35.

Abbott, R.J. and Lowe, A.J. 1996. A review of hybridization and evolution in British *Senecio*. In D.J.N. Hind and H.J. Beetje, eds. *Compositae: Systematics. Proceedings of the International Compositae Conference, Kew, 1994*. Kew, London: Royal Botanic Gardens, pp. 679–89.

Anderson, E. 1948. Hybridization of the habitat. *Evolution*, **2**: 1–9.

Anderson, E. 1949. *Introgressive Hybridization*. New York: Wiley.

Anderson, E. 1953. Introgressive hybridisation. *Biological Review*, **28**: 280–307.

Anderson, E. and Hubricht, L. 1938. The evidence for introgressive hybridization. *American Journal of Botany*, **25**: 396–402.

Arnold, M.L. 1992. Hatural hybridization as an evolutionary process. *Annual Review of Ecology and Systematics*, **23**: 237–61.

Arnold, M.L. 1993. *Iris nelsonii* (Iridaceae): origin and genetic composition of a homoploid hybrid species. *American Journal of Botany*, **80**: 577–83.

Arnold, M.L. 1997. *Natural Hybridization and Evolution* (Oxford Series in Ecology and Evolution). Oxford: Oxford University Press.

Arnold, M.L., Bennett, B.D., and Zimmer, E.A. 1990. Natural hybridization between *Iris fulva* and *Iris hexagona*: pattern of ribosomal DNA variation. *Evolution*, **44**: 1512–21.

Arnold, M.L. and Hodges, S.A. 1995. Are natural hybrids fit or unfit relative to their parents? *Trends in Ecology and Evolution*, **10**: 67–71.

Ashton, P.A. and Abbott, R.J. 1992. Multiple origins and genetic diversity in the newly arisen allopolyploid species, *Senecio cambrensis* Rosser (Compositae). *Heredity*, **68**: 25–32.

Avise, J.C. 1994. *Molecular Markers, Natural History and Evolution*. New York: Chapman and Hall.

Ayala, F.J. 1975. Genetic differentiation during the speciation process. *Evolutionary Biology*, **8**: 1–78.

Baldwin, B.G. 1992. Phylogenetic utility of the internal transcribed spacers of nuclear ribosomal DNA in plants: an example from the Compositae. *Molecular Phylogenetics and Evolution*, **1**: 3–16.

Baldwin, B.G. and Robichaux, R.H. 1995. Historical biogeography and ecology of the Hawaiian silversword alliance (Asteraceae): new molecular phylogenetic perspectives. In W.L. Wagner and V.A. Funk, eds. *Hawaiian Biogeography: Evolution on a Hot Spot Archipelago*. Washington, DC: Smithsonian Institute Press, pp. 259–87.

Barrett, S.C.H. 1989. Mating system evolution and speciation in heterostylous plants. In D. Otte and J.A. Endler, eds. *Speciation and its Consequences*. Sunderland, MA: Sinauer, pp. 257–83.

Barrett, S.C.H. and Charlesworth, D. 1991. Effects of a change in the level of inbreeding on genetic load. *Nature*, **352**: 522–4.

Barton, N.H. 1989. Founder effect speciation. In D. Otte and J.A. Endler, eds. *Speciation and its Consequences*. Sunderland, MA: Sinauer, pp. 229–56.

Barton, N.H., Halliday, R.B., and Hewitt, G.M. 1983. Rare electrophoresis variants in a hybrid zone. *Heredity*, **50**: 139–46.

Barton, N.H. and Hewitt, G.M. 1985. Analysis of hybrid zones. *Annual Review of Ecology and Systematics*, **16**: 113–48.

Barton, N.H. and Hewitt, G.M. 1989. Adaptation, speciation and hybrid zones. *Nature*, **341**: 497–503.

Baum, D.A. and Shaw, K.L. 1995. Genealogical perspectives on the species problem. In P.C. Hoch and A.G. Stephenson, eds. *Experimental and Molecular Approaches to Plant Biosystematics*. Missouri: Missouri Botanic Garden, pp. 289–303.

Beismann, H., Barker, J.H.A., Karp, A., and Speck, T. 1997. AFLP analysis sheds light on distribution of two *Salix* species and their hybrid along a natural gradient. *Molecular Ecology*, **6**: 989–93.

Belahbib, N., Ouassou, A., Sbay, H., Kremer, A., and Petit, R.J. 2001. Frequent cytoplasmic exchange between oak species that are not closely related: *Quercus suber* and *Q. ilex* in Morocco. *Molecular Ecology*, **10**: 2003–12.

Bennett, S.T., Kenton, A.Y., and Bennett, M.D. 1992. Genomic in *situ* hybridization reveals the allopolyploid nature of *Milium montianum* (Graminaea). *Chromosoma*, **101**: 420–4.

Bloom, W.L. 1976. Multivariate analysis of the introgressive replacement of *Clarkia nitens* by *Clarkia speciosa* subsp. *Polyantha*. *Evolution*, **30**: 412–24.

Bookstein, F.L. 1978. *The Measurement of Biological Shape and Shape Change*. Lecture Notes in Biomathematics. Berlin: Springer-Verlag.

Bookstein, F.L. 1989. Principal warps: thin-plate splines and the decomposition of deformations. *IEEE Transactions on Pattern Analysis and Machine Intelligence*, **11**: 567–85.

Bookstein, F.L. 1991. *Morphometric Tools for Landmark Data*. Cambridge: Cambridge University Press.

Brauner, S. and Gottlieb, L.D. 1987. A self-compatible plant of *Stephanomeria exigua* subsp. *coronaria* (Asteraceae) and its relevance to the origin of its self-pollinating derivative *S. malheurensis*. *Systematic Botany*, **12**: 299–304.

Bretagnolle, F. and Thompson, J.D. 1995. Gametes with the somatic chromosome number: mechanisms of their formation and role in the evolution of autopolyploid plants. Tansley Review No. 78. *New Phytologist*, **129**: 1–22.

Briggs, D. and Walters, S.M. 1997. *Plant Variation and Evolution*. 3rd edn. Cambridge: Cambridge University Press.

Brower, A.V.Z. 1999. Delimitation of phylogenetic species with DNA sequences: a critique of Davis and Nixon's population aggregation analysis. *Systematic Biology*, 48: 199–213.

Brown, W.L. and Wilson, E.O. 1956. Character displacement. *Systematic Zoology*, 5: 49–64.

Buerkle, C.A., Morris, R.J., Asmussen, M.A., and Rieseberg, L.H. 2000. The likelihood of homoploid hybrid speciation. *Heredity*, 84: 441–51.

Burton, T.L. and Husband, B.C. 1999. Population cytotype structure in the polyploid *Galax urceolata* (Diapensiaceae). *Heredity*, 82: 381–90.

Bush, G.L. 1975. Modes of animal speciation. *Annual Review of Ecology and Systematics*, 6: 339–64.

Butlin, R.K. 1989. Reinforcement of premating isolation. In D. Otte and J.A. Endler, eds. *Speciation and its Consequences*. Sunderland, MA: Sinauer, pp. 158–79.

Butlin, R.K. and Ritchie, M.G. 1994. Behaviour and speciation. In P.J.B. Slater and T.R. Halliday, eds. *Behaviour and Evolution*. Cambridge: Cambridge University Press, pp. 43–79.

Carlquist, S. 1995. Introduction. In W.L. Wagner and V.A. Funk, eds. *Hawaiian Biogeography: Evolution on a Hot Spot Archipelago*. Washington, DC: Smithsonian Institute Press, pp. 1–13.

Carney, S.E., Cruzan, M.B., and Arnold, M.L. 1994. Reproductive interactions between hybridising irises: analyses of pollen-tube growth and fertilization success. *American Journal of Botany*, 81: 1169–75.

Carney, S.E., Hodges, S.A., and Arnold, M.L. 1996. Effects of differential pollen-tube growth on hybridization in the Louisiana irises. *Evolution*, 50: 1871–8.

Carson, H.L. and Templeton, A.R. 1984. Genetic revolutions in relation to speciation phenomena: the founding of new populations. *Annual Review of Ecology and Systematics*, 15: 97–131.

Cavell, A.C., Lydiate, D.J., Parkin, I.A.P., Dean, C., and Trick, M. 1998. Collinearity between a 30-centimorgan segment of *Arabidopsis thaliana* chromosome 4 and duplicated regions within the *Brassica napus* genome. *Genome*, 41: 62–9.

Cavers, S., Navarro, C., and Lowe, A.J. 2003. A combination of molecular markers (cpDNA PCR-RFLP, AFLP) identifies evolutionarily significant units in *Cedrela odorata* L. (Meliaceae) in Costa Rica. *Conservation Genetics*, In press.

Charlesworth, D., Charlesworth, B., and McVean, G.A.T. 2001. Genome sequences and evolutionary biology, a two-way interaction. *Trends in Ecology and Evolution*, 16: 235–42.

Clausen, J. 1951. *Stages in the Evolution of Plant Species*. Oxford: Oxford University Press.

Cook, L.M., Soltis, P.S., Brunsfeld, S.J., and Soltis, D.E. 1998. Multiple independent formations of *Tragopogon* tetraploids (Asteraceae): evidence from RAPD markers. *Molecular Ecology*, 7: 1293–302.

Cooley, W.W. and Lohnes, P.R. 1971. *Multivariate Data Analysis*. New York: Wiley.

Coyne, J.A., Barton, N.H., and Turelli, M. 2000. Is Wright's shifting balance process important in evolution? *Evolution*, 54: 306–17.

Coyne, J.A. and Orr, H.A. 1989. Two rules of speciation. In D. Otte and J.A. Endler, eds. *Speciation and its Consequences*. Sunderland, MA: Sinauer, pp. 180–207.

Cracraft, J. 1983. Species concepts and speciation analysis. *Current Ornithology*, 1: 159–87.

Crawford, D.J., Braumer, S., Cosner, M.B., and Stuessy, T.F. 1993. Use of RAPD markers to document the origin of the intergeneric hybrid x *Margyracaena skottsbergii* (Rosaceae) on the Juan Fernandez Islands. *American Journal of Botany*, 80: 89–92.

Crawford, D.J. and Ornduff, R. 1989. Enzyme electrophoresis and evolutionary relationships among three species of *Lasthenia* (Asteraceae: Heliantheae). *American Journal of Botany*, 76: 289–96.

Crawford, D.J. and Smith, E.B. 1982. Allozyme variation in *Coreopsis nucensoides* and *C. nuecensis* (Compositae), a progenitor-derivative species pair. *Evolution*, 36: 379–86.

Darlington, C.D. and La Cour, L.F. 1962. *The handling of chromosomes*. London: George Allen and Unwin.

Darwin, C. 1859. *The Origin of Species by Means of Natural Selection*. London: John Murray.

Davis, J.I. and Nixon, K.C. 1992. Populations, genetic variation, and the delimitation of phylogenetic species. *Systematic Biology*, 41: 421–35.

DeMarais, B.D., Dowling, T.E., Douglas, M.E., Minckley, M.L., and Marsh, P.C. 1992. Origin of *Gila seminuda* (Teleostei: Cyprinidae) through introgressive hybridization: implications for evolution and conservation. *Proceedings of the National Academy of Sciences USA*, 89: 2747–51.

Deutsch, J.C. 1997. Colour diversification in Malawi cichlids: evidence for adaptation, reinforcement or sexual selection? *Biological Journal of the Linenan Society*, 62: 1–14.

de Vries, H. 1905. *Species and Varieties, their Origin by Mutation*. Chicago: Open Court.

Dieckmann, U. and Doebeli, M. 1999. On the origin of species by sympatric speciation. *Nature*, 400: 354–7.

Dobler, S., Mardulyn, P., Pasteels, J.M., and Rowell-Rahier, M. 1996. Host-plant switches and the evolution of chemical defense and life history in the leaf beetle genus *Oreina*. *Evolution*, 50: 2373–86.

Dobzansky, T. 1937. *Genetics and the Origin of Species*. New York: Columbia University Press.

Dobzansky, T. 1951. *Genetics and the Origin of Species*. 3rd edn. New York: Columbia University Press.

Dobzansky, T., Ayala, F.J., Stebbins, G.L., and Valentine, J.W. 1977. *Evolution*. San Francisco: Freeman.

Doyle, J. 1995. The irrelevance of allele tree topologies for species delimitation, and a non-topological alternative. *Systematic Botany*, **20**: 574–88.

Doyle, J.J. and Doyle, J.L. 1999. Nuclear protein-coding genes in phylogeny reconstruction and homology assessment: some examples from Leguminosae. In P.M. Hollingsworth, R.M. Bateman, and R.J. Gornall, eds. *Molecular Systematics and Plant Evolution*. London: Taylor and Francis, pp. 229–54.

Dumolin-Lapègue, S., Kremer, A., and Petit, R.J. 1999. Are chloroplast and mitochondrial variation species independent in oaks? *Evolution*, **53**: 1406–14.

Dumolin-Lapègue, S., Pemonge, M.-H., and Petit, R.J. 1999. Association between chloroplast and mitochondrial lineages in oaks. *Molecular Biology and Evolution*, **15**: 1321–31.

Dyer, A.F. 1979. *Investigating Chromosomes*. London: Edward Arnold.

Elder, J.F. and Turner, B.J. 1995. Concerted evolution of repetitive DNA sequences in Eukaryotes. *Quarterly Review of Biology*, **70**: 297–320.

Ellstrand, N.C., Whitkus, R., and Rieseberg, L.H. 1996. Distribution of spontaneous plant hybrids. *Proceedings of the National Academy of Sciences USA*, **93**: 5090–3.

Eppig, J.T. 1996. Comparative maps: adding pieces to the mammalian jigsaw puzzle. *Current Biology*, **6**: 723–30.

Emelianov, I., Mallet, J., and Baltensweiler, W. 1995. Genetic differentiation in *Zeiraphera diniana* (Lepidoptera: Tortricidae, the larch budmoth): polymorphism, host race or sibling species? *Heredity*, **75**: 416–24.

Emms, S.K. and Arnold, M.L. 1996. The effect of habitat on parental and hybrid fitness: transplant experiments with Louisiana Irises. *Evolution*, **51**: 1112–19.

Emms, S.K., Hodges, S.A., and Arnold, M.L. 1996. Pollen-tube competition, siring success, and consistent asymmetric hybridization in Louisiana irises. *Evolution*, **50**: 2201–6.

Ennos, R.A. 1994. Estimating the relative rates of pollen and seed migration among plant populations. *Heredity*, **72**: 250–9.

Felber, F. 1991. Establishment of a tetraploid cytotype in a diploid population: effect of relative fitness of the sytotypes. *Journal of Evolutionary Biology*, **4**: 195–207.

Ferris, C., Oliver, R.P., Davy, A.J., and Hewitt, G.M. 1993. Native oak chloroplasts reveal an ancient divide across Europe. *Molecular Ecology*, **2**: 337–44.

Fisher, R.A. 1930. *The Genetical Theory of Natural Selection*. Oxford: Clarendon Press.

Fischer, R.A. 1941. Average excess and average effect of a gene substitution. *Annals Eugenics*, **11**: 53–63.

Gale, M.D. and Devos, K.M. 1998. Plant comparative genetics after 10 years. *Science*, **282**: 656–9.

Gallant, S.L. and Fairbairn, D.J. 1997. Patterns of postmating reproductive isolation in a newly discovered species pair, *Aquarius remigis* and *A. remigoides* (Hemiptera; Gerridae). *Heredity*, **78**: 571–7.

Gallez, G.P. and Gottlieb, L.D. 1982. Genetic evidence for the hybrid origin of the diploid plant *Stephanomeria diegensis*. *Evolution*, **36**: 1158–67.

Gardner, M. and MacNair, M. 2000. Factors affecting the co-existence of the serpentine endemic *Mimulus nudatus* Curran and its presumed progenitor, *Mimulus guttatus* Fischer ex DC. *Biological Journal of the Linenan Society*, **69**: 443–59.

Gastony, G.J. 1991. Gene silencing in a polyploid homosporous fern: paleopolyploidy revisted. *Proceedings of the National Academy of Sciences USA*, **88**: 1602–5.

Giddings, L.V., Kaneshiro, K.Y., and Anderson, W.W. 1989. *Genetics, Speciation and the Founder Principle*. Oxford: Oxford University Press.

Gillies, A.C.M., Cubas, P., Coen, E.S., and Abbott, R.J. 2002. Making rays in the Asteraceae: genetics and evolution of radiate versus discoid flower heads. In Q.C.B. Cronk, R.M. Bateman, and J.A. Hawkins, eds. *Developmental Genetics and Plant Evolution*. London: Taylor and Francis, pp. 233–46.

Goldschmidt, R.B. 1940. *The Material Basis of Evolution*. New Haven: Yale University Press.

Good, D.A. and Wake, D.B. 1992. Geographic variation and speciation in the torrent salamanders of the genus *Rhyacotriton* (Caudata: Rhyacotritonidae).

Goodnight, C.J. and Wade, M.J. 2000. The ongoing synthesis: a reply to Coyne, Barton and Turelli. *Evolution*, **54**: 317–24.

Gower, J.C. 1966. Some distance properties of latent root and vector methods used in multivariate analysis. *Biometrika*, **53**: 315–28.

Gower, J.C. 1971. Statistical methods of comparing different multivariate analyses of the same data. In R. Hodson, D.G. Kendall, and P. Tatu, eds. *Mathematics in the Archaeological and Historical Sciences*. Edinburgh: Edinburgh University Press, pp. 138–49.

Grant, P.R. 1993. Hybridization of Darwin's finches on Isla Daphne Major, Galapagos. *Philosophical Transactions of the Royal Society, London B*, **340**: 127–39.

Grant, P.R. and Grant, B.R. 1989. Sympatric speciation and Darwin's finches. In D. Otte and J.A. Endler, eds. *Speciation and its Consequences*. Sunderland, MA: Sinauer, pp. 433–57.

Grant, V. 1949. Pollination systems as isolating mechanisms in flowering plants. *Evolution*, **3**: 82–97.

Grant, V. 1957. The plant species in theory and practice. In E. Mayr, ed. *The Species Problem*. American Association for the Advancement of Science, Publication No. 50, pp. 39–80.

Grant, V. 1965. Selection for vigor and fertility in the progeny of a highly sterile species hybid in *Gilia. Genetics*, **53**: 757–75.

Grant, V. 1966. The origin of a new species of *Gilia* in a hybridization experiment. *Genetics*, **54**: 1189–99.

Grant, V. 1981. *Plant Speciation*. New York: Columbia University Press.

Grant, V. 1994. Modes and origins of mechanical and ethological isolation in angiosperms. *Proceedings of the National Academy of Sciences USA*, **91**: 3–10.

Gray, A.J., Marshall, D.F., and Raybould, A.F. 1991. A century of evolution in *Spartina anglica*. *Advances in Ecological Research*, **21**: 1–62.

Gustafsson, M.H.G. and Albert, V.A. 1999. Inferior ovaries and angiosperm diversification. In P.M. Hollingsworth, R.M. Bateman, and R.J. Gornall, eds. *Molecular Systematics and Plant Evolution*. London: Taylor and Francis, pp. 403–31.

Gyllensten, U.B. and Wilson, A.C. 1987. Interspecific mitochondrial DNA transfer and the colonization of Scandinavia by mice. *Genetic Research, Cambridge*, **49**: 25–9.

Harlan, J.R. and de Wet, J.M.J. 1975. On Ö. Winge and a prayer: the origins of polyploidy. *Botanical Review*, **41**: 361–90.

Harland, S.C. 1954. The genus *Senecio* as a subject for cytogenetical investigation. *Proceedings of the Botanical Society of the Biritish Isles*, **1**: 256.

Harris, S.A. 1999. RAPDs in systematics – a useful methodology? In P.M. Hollingsworth, R.M. Bateman, and R.J. Gornall, eds. *Molecular Systematics and Plant Evolution*. London: Taylor and Francis, pp. 211–28.

Harris, S.A. and Ingram, R. 1991. Chloroplast DNA and biosystematics: the effects of intraspecific diversity and plastid transmission. *Taxon*, **40**: 393–412.

Harrison, R.G. 1986. Pattern and process in a narrow hybrid zone. *Heredity*, **56**: 337–49.

Harrison, R.G. 1990. Hybrid zones: windows on evolutionary process. *Oxford Surveys in Evolutionary Biology*, **7**: 69–128.

Harrison, R.G. 1991. Molecular changes at speciation. *Annual Review of Ecology and Systematics*, **22**: 281–308.

Harrison, R.G. and Rand, D.M. 1989. Mosaic hybrid zones and the nature of species boundaries. In D. Otte and J.A. Endler, eds. *Speciation and its Consequences*. Sunderland, MA: Sinauer, pp. 111–33.

Hatfield, T. and Schluter, D. 1996. A test for sexual selection on hybrids of two sympatric sticklebacks. *Evolution*, **50**: 2429–34.

Heiser, C.B. 1949. Study in the evolution of the sunflower species *Helianthus annuus* and *H. bolanderi*. *University of Clifton Publications in Botany*, **23**: 157–96.

Heiser, C.B. 1951. Hybridization in the annual sunflowers: *Heolianthus annuus* x *H. debilis* var. *cucumerifolius*. *Evolution*, **5**: 42–51.

Hershkovitz, M.A., Zimmer, E.A., and Hahn, W.J. 1999. Ribosomal DNA sequences and angiosperm systematics. In P.M. Hollingsworth, R.M. Bateman, and R.J. Gornall, eds. *Molecular Systematics and Plant Evolution*. London: Taylor and Francis, pp. 268–326.

Hewitt, G.M. 1988. Hybrid zones – natural laboratories for evolutionary studies. *Trends in Ecology and Evolution*, **3**: 158–67.

Hewitt, G.M. 1989. The subdivision of species by hybrid zones. In D. Otte and J.A. Endler, eds. *Speciation and its Consequences*. Sunderland, MA: Sinauer, pp. 85–110.

Highton, R. 1990. Taxonomic treatment of genetically differentiated populations. *Herpetologia*, **46**: 113–21.

Highton, R., Maha, G.C., and Maxson, L.R. 1989. Biochemical evolution in the slimy salamanders of the *Plethodon glutinosus* complex in the eastern United States. *Illinois Biologica Monographs*, **57**: 1–153.

Hill, R.S. 1980. A numerical taxonomic approach to the study of angiosperm leaves. *Botanical Gazette*, **141**: 213–29.

Hilu, K.W. 1983. The role of single-gene mutations in the evolution of flowering plants. *Evolutionary Biology*, **16**: 97–128.

Hollingsworth, P.M., Preston, C.D., and Gornall, R.J. 1995. Isozyme evidence for hybridization between *Potamogeton natans* and *P. nodosus* (Potamogetonaceae) in Britain. *Botanical Journal of the Linenan Society*, **117**: 59–69.

Hollingsworth, P.M., Preston, C.D., and Gornall, R.J. 1996. Isozyme evidence for the parentage and multiple origins of *Potamogeton* x *suecicus* (*P. pectinatus* x *P. filiformis*, Potamogetonaceae). *Plant Systematics and Evolution*, **202**: 219–32.

Howard, D.J. 1986. A zone of overlap and hybridization between two ground cricket species. *Evolution*, **40**: 34–43.

Howard, D.J. 1993. Reinforcement: origin, dynamics and fate of an evolutionary hypothesis. In R.G. Harrison, ed. *Hybrid Zones and the Evolutionary Process*. Oxford: Oxford University Press.

Hubbs, C.L. 1955. Hybridization between fish species in nature. *Systematic Zoology*, **4**: 1–20.

Humpreys, M.W. and Pasakinskiene, I. 1996. Chromosome painting to locate genes for drought resistance transferred from *Festuca arundinacea* into *Lolium multiflorum*. *Heredity*, **77**: 530–4.

Hunt, W.G. and Selander, R.K. 1973. Biochemical genetics of hybridisation in European house mice. *Heredity*, **31**: 11–33.

Husband, B.C. 2000. Constraints on polyploid evolution: a test of the minority cytotype exclusion principle. *Proceedings of the Royal Society London B*, **267**: 217–23.

Ingram, R. 1977. Synthesis of the hybrid *Senecio squalidus* L. x *S. vulgaris* L. f. *radiatus* Hegi. *Heredity*, **39**: 171–3.

Ingram, R. 1978. The genomic relationship of *Senecio squalidus* L. and *Senecio vulgaris* and the significance of genomic balance in their hybrid *S.* x *Baxteri* Druce. *Heredity*, **40**: 459–62.

Ingram, R., Weir, J., and Abbott, R.J. 1980. New evidence concerning the origin of inland radiate groundsel, *S. vulgaris* var. *hibernicus* Syme. *New Phytologist*, **84**: 543–6.

Jain, S.K. 1976. The evolution of inbreeding in plants. *Annual Review of Ecology and Systematics*, **10**: 173–200.

Jasny, B.R. and Hines, P.J. 1998. A genome sampler. *Science*, **282**: 651.

Jenkins, G. and Rees, H. 1991. Strategies of bivalent formation in allopolyploid plants. *Proceedings of the Royal Society London B*, **243**: 209–14.

Jensen, R.J. 1990. Detecting shape variation in oak leaf morphology: a comparison of rotational-fit methods. *American Journal of Botany*, **77**: 1279–93.

Jensen, R.J., Hokanson, S.C., Isebrands, J.G., and Hancock, J.F. 1993. Morphometric variation in oaks of the Apostle Islands in Wisconsin: evidence of hybridization between *Quercus rubra* and *Q. ellipsoidalis* (Fagaceae). *American Journal of Botany*, **80**: 1358–66.

Kaneshiro, K.Y., Gillespie, R.G., and Carson, H.L. 1995. Chromosomes and male genitalia of Hawaiian *Drosophila*: tools for interpreting phylogeny and geology. In W.L. Wagner and V.A. Funk, eds. *Hawaiian Biogeography: Evolution on a Hot Spot Archipelago*. Washington, DC: Smithsonian Institute Press, pp. 57–71.

Karl, S.A., Bowen, B.W., and Avise, J.C. 1992. Global population genetic-structure and male-mediated gene flow in the green turtle (*Chelonia mydas*) – RFLP analyses of anonymous nuclear loci. *Genetics*, **131**: 163–73.

Key, K.H.L. 1968. The concept of stasipatric speciation. *Systematic Zoology*, **17**: 14–22.

King, M. 1993. *Species Evolution: The Role of Chromosome Change*. Cambridge: Cambridge University Press.

Knox, E.B. and Palmer, J.D. 1995. Chloroplast DNA variation and the recent radiation of the giant *Senecios* (Asteraceae) on the tall mountains of eastern Africa. *Proceedings of the National Academy of Sciences USA*, **92**: 10,319–53.

Knox, E.B. and Palmer, J.D. 1998. Chloroplast DNA evidence on the origin and radiation of the giant lobelias in eastern Africa. *Systematic Botany*, **23**: 109–49.

Kondrashov, A.S. and Kondrashov, F.A. 1999. Interactions among quantitative traits in the course of sympatric speciation. *Nature*, **400**: 351–4.

Kowalski, S.P., Lan, T.-H., Feldmann, K.A., and Paterson, A.H. 1994. Comparative mapping of *Arabidopsis thaliana* and *Brassica oleracea* chromosomes reveals islands of conserved organization. *Genetics*, **138**: 1–12.

Largercrantz, U. and Lydiate, D.J. 1996. Comparative genome mapping in *Brassica*. *Genetics*, **144**: 1903–10.

Leitch, I.J. and Bennett, M.D. 1997. Polyploidy in angiosperms. *Trends in Plant Sciences*, **2**: 470–6.

Levin, D.A. 1978. The origin of isolating mechanisms in flowering plants. *Evolutionary Biology*, **11**: 185–317.

Levin, D.A. 1979. The nature of plant species. *Science*, **204**: 381–4.

Levin, D.A. 2002. *The Origin, Expansion, and Demise of Plant Species* (Oxford Series in Ecology and Evolution). New York: Oxford University Press.

Levin, D.A. and Kerster, H.W. 1967. Natural selection for reproductive isolation in *Phlox*. *Evolution*, **21**: 679–87.

Lewis, H. 1966. Speciation in flowering plants. *Science*, **152**: 167–72.

Lewis, W. 1980. *Polyploidy: Biological Relevance*. New York: Plenum Press.

Lewontin, R.C. and Birch, L.C. 1966. Hybridization as a source of variation for adaptation to new environments. *Evolution*, **20**: 315–36.

Lowe, A.J. 1996. *The origin and maintenance of a new tetraploid Senecio hybrid in York, England*. PhD thesis, University of St Andrews, UK.

Lowe, A.J. and Abbott, R.J. 1996. Origins of the new allopolyploid species *Senecio cambrensis* (Asteraceae) and its relationship to the Canary Islands endemic *Senecio teneriffae*. *American Journal of Botany*, **83**: 1365–72.

Lowe, A.J., Russell, J.R., Powell, W., and Dawson, I.K. 1998. Identification and characterization of nuclear, cleaved amplified polymorphic sequences (CAPS) loci in *Irvingia gabonensis* and *I. wombolu*, indigenous fruit trees of west and central Africa. *Molecular Ecology*, **7**: 1771–88.

Lydiate, D., Sharpe, A., Largercrantz, U., and Parkin, I. 1993. Mapping the *Brassica* genome. *Outlook on Agriculture*, **22**: 85–92.

McCarthy, E.M., Asmussen, M.A., and Anderson, W.W. 1995. A theoretical assessment of recombinational speciation. *Heredity*, **74**: 502–9.

McDade, L. 1992. Hybrids and phylogenetic systematics. II. The impacts of hybrids on cladistic analysis. *Evolution*, **46**: 1329–46.

Malhotra, A. and Thorpe, R.S. 2000. The dynamics of natural selection and vicariance in the Dominican anole: patterns of within-island molecular and morphological divergence. *Evolution*, **54**: 245–58.

Manly, B.F.J. 1992. *Multivariate Statistical Methods: A Primer*. London: Chapman and Hall.

Masterson, J. 1994. Stomatal size in fossil plants: evidence for polyploidy in majority of angiosperms. *Science*, **264**: 421–4.

Mayer, W.E., Tichy, H., and Klein, J. 1998. Phylogeny of African cichlid fishes as revealed by molecular markers. *Heredity*, **80**: 702–14.

Mayr, E. 1942. *Systematics and the Origin of Species*. New York: Columbia University Press.

Mayr, E. 1963. *Animal Species and Evolution*. Boston: Harvard University Press.

Mayr, E. 1992. A local flora and the biological species concept. *American Journal of Botany*, **79**: 222–38.

Maynard-Smith, J. 1966. Sympatric speciation. *American Naturalist*, **100**: 637–50.

Meyer, A. 1997. The evolution of sexually selected traits in male swordtail fishes (*Xiphophorus*: Poecilidae). *Heredity*, **79**: 329–37.

Millar, C.I. 1983. A steep cline in *Pinus muricata*. *Evolution*, **37**: 311–19.

Möller, M., Clokie, M., Cubas, P., and Cronk, Q.C.B. 1999. Integrating molecular phylogenies and developmental genetics: a Gesneriaceae case study. In P.M. Hollingsworth, R.M. Bateman, and R.J. Gornall, eds. *Molecular Systematics and Plant Evolution*. London: Taylor and Francis, pp. 375–402.

Moore, W. 1977. An evaluation of narrow hybrid zones in vertebrates. *Quaterly Review of Biology*, **52**: 263–77.

Morgan, T.H. 1932. *The Scientific Basis of Evolution*. New York: Norton.

Mulinix, C.A. and Lezzoni, A.F. 1988. Microgametophytic selection in two alfalfa (*Medicago sativa* L.) clones. *Theoretical and Applied Genetics*, **75**: 917–92.

Nason, J.D., Ellstrand, N.C., and Arnold, M.L. 1992. Patterns of hybridization and introgression in populations of oaks, manzanitas and irises. *American Journal of Botany*, **79**: 101–11.

Nei, M. 1978. Estimation of average heterozygosity and genetic distance from small numbers of individuals. *Genetics*, **89**: 583–90.

Ness, B.D., Soltis, D.E., and Soltis, P.S. 1990. An examination of polyploidy and putative introgression in *Calochortus* subsection Nudi (Liliaceae). *American Journal of Botany*, **77**: 1519–31.

Norvak, S.J., Soltis, D.E., and Soltis, P.A. 1991. Ownbey's Tragopogons: 40 years later. *American Journal of Botany*, **78**: 1586–600.

Orr, H.A. and Orr, L.H. 1996. Waiting for speciation: The effect of population subdivision on the time to speciation. *Evolution*, **50**: 1742–9.

Orr, H.A. and Turelli, M. 2001. The evolution of postzygotic isolation: accumulating Dobzhansky–Muller incompatibilities. *Evolution*, **55**: 1085–94.

Otte, D. 1989. Speciation in Hawaiian crickets. In D. Otte and J.A. Endler, eds. *Speciation and its Consequences*. Sunderland, MA: Sinauer, pp. 482–526.

Ownbey, M. 1950. Natural hybridization and amphidiloidy in the genus *Tragapogon*. *American Journal of Botany*, **37**: 487–99.

Page, R.D.M. and Holmes, E.C. 1998. *Molecular Evolution, A Phylogenetic Approach*. Oxford: Blackwell Science, pp. 201–8.

Palmer, J.D. 1992. Mitochondrial DNA in plant systematics: applications and limitations. In P.E. Soltis, D.E. Soltis, and J.J. Doyle, eds. *Plant Molecular Systematics*. New York: Chapman and Hall, pp. 36–49.

Patterson, H. 1985. The recognition species concept. In E.S. Vrba, ed. *Species and Speciation*. Pretoria: Transvaal Museum Monograph No. 4, pp. 21–9.

Peck, S.L., Ellner, S.P., and Gould, F. 2000. Varying migration and deme size and the feasibility of the shifting balance. *Evolution*, **54**: 324–7.

Petit, C., Bretagnolle, F., and Felber, F. 1999. Evolutionary consequences of diploid–polyploid hybrid zones in wild species. *Trends in Ecology and Evolution*, **14**: 306–11.

Porter, A.H. 1990. Testing nominal species boundaries using gene flow statistics: the taxonomy of two hybridizing admiral butterflies (*Limenitis*: Nymphalidae). *Systematic Zoology*, **39**: 131–47.

Provan, J., Powell, W., and Hollingsworth, P.M. 2001. Chloroplast microsatellites: new tools for studies in plant ecology and evolution. *Trends in Ecology and Evolution*, **16**: 142–7.

Provan, J., Soranzo, N., Wilson, N.J., McNicol, J.W., Morgante, M., and Powell, W. 1999. The use of uniparentally inherited simple sequence repeat markers in plant population studies and systematics. In P.M. Hollingsworth, R.M. Bateman, and R.J. Gornall, eds. *Molecular Systematics and Plant Evolution*. London: Taylor and Francis, pp. 35–50.

Ramsey, J. and Schemske, D.W. 1998. Pathways, mechanisms, and rates of polyploid formation in flowering plants. *Annual Review of Ecology and Systematics*, **29**: 467–501.

Ratter, J.A. 1972. Cytogenetic studies in Spergularia VI: the evolution of true breeding, fertile tetraploids from a triploid interspecific hybrid. *Notes of the Royal Botanic Gardens Edinburgh*, **32**: 117–25.

Ratter, J.A. 1973a. Cytogenetic studies in Spergularia VIII: barriers to the production of viable interspecific hybrids. *Notes of the Royal Botanic Gardens Edinburgh*, **32**: 297–301.

Ratter, J.A. 1973b. Cytogenetic studies in Spergularia IX: summary and conclusions. *Notes of the Royal Botanic Gardens Edinburgh*, **32**: 411–28.

Raybould, A.F., Gray, A.J., Lawrence, M.J., and Marshall, D.F. 1991. The evolution of *Spartina anglica* C.E. Hubbard (Gramineae): origin and genetic variability. *Biological Journal of the Linnean Society*, **43**: 111–26.

Rieseberg, L.H. 1991. Homoploid reticulate evolution in *Helianthus* (Asteraceae): evidence from ribosomal genes. *American Journal of Botany*, **78**: 1218–37.

Rieseberg, L.H. 1995. The role of hybridization in evolution: old wine in new skins. *American Journal of Botany*, **82**: 944–53.

Rieseberg, L.H. 1996. Homology among RAPD fragments in interspecific comparisons. *Molecular Ecology*, **5**: 99–105.

Rieseberg, L.H. 1998. Molecular ecology and hybridisation. In G.R. Carvalho, ed. *Advances in Molecular Ecology*. London: IOS Press.

Rieseberg, L.H., Archer, M.A., and Wayne, R.K. 1999. Transgressive segregation, adaptation and speciation. *Heredity*, **83**: 363–72.

Rieseberg, L.H., Beckstrom-Sternberg, S., and Doan, K. 1990. *Helianthus annuus* ssp. *texanus* has chloroplast DNA and nuclear ribosomal RNA genes of *Helianthus debilis* ssp. *cucumerifolius*. *Proceedings of the National Academy of Sciences USA*, **87**: 593–7.

Rieseberg, L.H. and Brunsfeld, S.J. 1992. Molecular evidence and plant introgression. In P.E. Soltis, D.E. Soltis, and J.J. Doyle, eds. *Plant Molecular Systematics*. New York: Chapman and Hall, pp. 151–76.

Rieseberg, L.H. and Carney, S.E. 1998. Plant hybridization, Tansley review no. 102. *New Phytologist*, **140**: 599–624.

Rieseberg, L.H., Choi, H.C., and Ham, D. 1991. Differential cytoplasmic versus nuclear introgression in *Helianthus*. *Journal of Heredity*, **82**: 489–93.

Rieseberg, L.H. and Ellstrand, N.C. 1993. What can molecular and morphological markers tell us about plant hybridization? *Critical Reviews in Plant Science*, **12**: 213–41.

Rieseberg, L.H., Fossen, C.V., and Desrochers, A.M. 1995. Hybrid speciation accompanied by genomic reorganization in wild sunflowers. *Nature*, **375**: 313–16.

Rieseberg, L.H., Sinervo, B., Linder, C.R., Ungerer, M.C., and Arias, D.M. 1996. Role of gene interactions in hybrid speciation: evidence from ancient and experimental hybrids. *Science*, **272**: 741–5.

Rieseberg, L.H., Soltis, D.E., and Palmer, J.D. 1988. A molecular re-examination of introgression between *Helianthus annuus* and *H. bolanderi* (Compositae). *Evolution*, **42**: 227–38.

Rieseberg, L.H. and Wendel, J.F. 1993. Introgression and its consequences in plants. In R. Harrison, ed. *Hybrid Zones and the Evolutionary Process*. New York: Oxford University Press, pp. 70–109.

Roberts, H.F. 1929. *Plant hybridization before Mendel*. Princeton, NJ: Princeton University Press.

Robinson, J.P. and Harris, S.A. 1999. Amplified fragment length polymorphisms and microsatellites: a phylogenetic perspective. In E.M. Gillet, ed. *Which DNA Marker for Which Purpose? Development, optimisation and validation of molecular tools for assessment of biodiversity in forest trees, in EU DGXII Biotechnology FW IV Research Programme Molecular Tools for Biodiversity*. http://webdoc.sub.gwdg.de/ebook/y/1999/whichmarker/index.htm 1

Rohlf, F.J. and Slice, D.E. 1990. Extensions of the procrustes method for the optimal superimposition of landmarks. *Systematic Zoology*, **39**: 40–59.

Roose, M.L. and Gottlieb, L.D. 1976. Genetic and biochemical consequences of polyploidy in *Tragopogon*. *Evolution*, **30**: 818–30.

Rosen, D.E. 1979. Fishes from the uplands and intermontane basins of Guatemala: Revisionary studies and comparative geography. *Bulletin of the American Museum of Natural History*, **162**: 267–376.

Russell, B. 1961. *History of Western Philosophy*. London: George Allen and Unwin.

Schwarzbach, A.E., Donovan, L.A., and Rieseberg, L.H.. 2001. Transgressive character expression in a hybrid sunflower species. *American Journal of Botany*, **88**: 270–7.

Sessions, S.K. 1996. Chromosome: molecular cytogenetics. In D.M. Hillis, C. Moritz, and B.K. Mable, eds. *Molecular Systematics*. Sunderland, MA: Sinauer, pp. 121–68.

Sharbel, T.F. and Mitchell-Olds, T. 2001, T. Recurrent polyploid origins and chloroplast phylogeography in *Arabis holboellii* complex (Brassicaceae). *Heredity*, **87**: 59–68.

Shoemaker, D.D., Ross, K.G., and Arnold, M.L. 1996. Genetic structure and evolution of a fire ant hybrid zone. *Evolution*, **50**: 1958–76.

Simpson, G.G. 1961. *Principles of Animal Taxonomy. The Species and Lower Categories*. New York: Columbia University Press.

Sites, J.W. Jr. and Marshall, J.C. 2003. Delimiting species: Renaissance issue in systematic biology. *Trends in Ecology and Evolution*, **18**: 462–70.

Sneath, P.H.A. 1967. Trend-surface analysis of transformation grids. *Journal of Zoology*, **151**: 65–122.

Sokal, R.R. and Crovello, T.J. 1970. The biological species concept: a critical evaluation. *American Naturalist*, **104**: 127–53.

Sokal, R.R. and Rohlf, F.J. 1994. *Biometry*. 3rd edn. New York: W.H. Freeman.

Soltis, D.E. and Soltis, P.S. 1989. Allopolyploid speciation in *Tragapogon*: insights from chloroplast DNA. *American Journal of Botany*, **76**: 1119–24.

Soltis, D.E. and Soltis, P.S. 1993. Molecular data and the dynamic nature of polyploidy. *Critical Reviews in Plant Science*, **12**: 243–73.

Soltis, D.E. and Soltis, P.S. 1995. The dynamic nature of polyploid genomes, Commentary. *Proceedings of the National Academy of Science USA*, **92**: 8089–91.

Soltis, D.E. and Soltis, P.S. 1999. Polyploidy: recurrent formation and genome evolution. *Trends in Ecology and Evolution*, **14**: 348–52.

Soltis, D.E., Soltis, P.E., and Milligan, B.G. 1992. Intraspecific chloroplast DNA variation: systematic and phylogenetic implications. In P.E. Soltis, D.E. Soltis, and J.J. Doyle, eds. *Molecular Systematics of Plants*. New York: Chapman and Hall, pp. 117–50.

Soltis, P.S., Doyle, J.J., and Soltis, D.E. 1992. Molecular data and polyploid evolution in plants. In P.E. Soltis, D.E. Soltis, and J.J. Doyle, eds. *Molecular Systematics of Plants*. New York: Chapman and Hall, pp. 177–201.

Soltis, P.S. and Soltis, D.E. 1988. Electrophoretic evidence for genetic diploidy in *Psilotum nudum*. *American Journal of Botany*, **75**: 1667–71.

Song, K., Lu, P., Tang, K., and Osborn, T.C. 1995. Rapid genome change in synthetic polyploids of *Brassica* and its implications for polyploid evolution. *Proceedings of the National Academy of Sciences USA*, **92**: 7719–23.

Stace, C.A. 1975. *Hybridization and the Flora of the British Isles*. London: Academic Press.

Stace, C.A. 1991. *New Flora of the British Isles*. Cambridge: Cambridge University Press.

Stace, C.A. and Bailey, J.P. 1999. The value of genomic *in situ* hybridisation (GISH) in platn taxonomic and evolutionary studies. In P.M. Hollingsworth, R.M. Bateman, and R.J. Gornall, eds. *Molecular Systematics and Plant Evolution*. London: Taylor and Francis, pp. 199–210.

Stebbins, G.L. 1947. Types of polyploids: their classification and significance. *Advances in Genetics*, **1**: 403–29.

Stebbins, G.L. 1957. Self-fertilization and population variability in higher plants. *American Naturalist*, **41**: 337–54.

Stebbins, G.L. 1959. The role of hybridization in evolution. *Proceedings of the American Philosophical Society*, **103**: 231–51.

Stebbins, G.L. 1971. *Chromosomal Evolution in Higher Plants*. London: Arnold.

Stebbins, G.L. 1980. Polyploidy in plants: unresolved problems and prospects. In W. Lewis, ed. *Polyploidy: Biological Relevance*. New York: Plenum Press, pp. 495–520.

Steen, S.W., Gielly, L., Taberlet, P., and Brochmann, C. 2000. Same parental species, but different taxa: molecular evidence for hybrid origins of the rare endemics *Saxifraga opdalensis* and *S. svalbardensis* (Saxifragaceae). *Botanical Journal of the Linnean Society*, **132**: 153–64.

Stephens, S.G. 1949. The cytogenetics of speciation in *Gossypium* I. Selective elimination of the donor parent genotype in interspecific backcrosses. *Genetics*, **34**: 627–37.

Strand, A.E., LeebensMack, J., and Milligan, B.G. 1997. Nuclear DNA-based markers for plant evolutionary biology. *Molecular Ecology*, **6**: 113–18.

Straw, R.M. 1955. Hybridization, homogamy and sympatric speciation. *Evolution*, **9**: 441–4.

Tauber, C.A. and Tauber, M.J. 1989. Sympatric speciation in insects: perception and perspective. In D. Otte and J.A. Endler, eds. *Speciation and its Consequences*. Sunderland, MA: Sinauer, pp. 307–44.

Templeton, A.R. 1981. Mechanisms of speciation – A population genetic approach. *Annual Review of Ecology and Systematics*, **12**: 23–48.

Templeton, A.R. 1989. The meaning of species and speciation: a genetic perspective. In D. Otte and J.A. Endler, eds. *Speciation and its Consequences*. Sunderland, MA: Sinauer, pp. 3–27.

Templeton, A.R. 1998. Nested clade analysis of phylogeographic data: testing hypotheses about gene flow and populations history. *Molecular Ecology*, **7**: 381–97.

Thompson, D'A.W. 1961. In J.T. Bonner, ed. *On Growth and Form, abridged edition*. Cambridge: Cambridge University Press.

Thompson, J.D. and Lumaret, R. 1992. The evolutionary dynamics of polyploid plants: origins, establishment and persistence. *Trends in Ecology and Evolution*, **7**: 302–7.

Thorpe, J.P. 1982. The molecular clock hypothesis: biochemical evaluation, genetic differentiation and systematics. *Annual Review of Ecology and Systematics*, **13**: 139–68.

Tregenza, T. and Butlin, R.K. 1999. Speciation without isolation. *Nature*, **400**: 311–12.

Van Baarlen, P., Verduijn, M., and van Dijk, P.J. 1999. What can we learn from natural apomicts? *Trends in Plant Sciences*, **4**: 43–4.

Van Dijk, P. and Bijlsma, R. 1994. Simulations of flowering time displacement between two cytotypes that for inviable hybrids. *Heredity*, **72**: 522–35.

Van Valen, L. 1976. Ecological species, multispecies, and oaks. *Taxon*, **25**: 223–39.

Vickery, R.K. 1964. Barriers to gene exchange between members of the *Mimulus gittatus* complex (Scrophulariaceae). *Evolution*, **18**: 52–69.

Wagner, D.B., Dong, J., Carlson, M.R., and Yanchuk, A.D. 1991. Paternal leakage of mitochondrial DNA in *Pinus*. *Theoretical and Applied Genetics*, **82**: 510–14.

Wagner, W.H. Jr. 1970. Biosystematics and evolutionary noise. *Taxon*, **19**: 146–51.

Weir, B.S. 1996. *Genetic Data Analysis*, 2nd Edn. Sunderland, MA: Sinauer.

Weir, J. and Ingram, R. 1980. Ray morphology and cytological investgations of *Senecio cambrensis* Rosser. *New Phytologist*, **86**: 237–41.

Weising, K., Nybom, H., Wolff, K., and Meyer, W. 1995. *DNA Fingerprinting in Plants and Fungi*. London: CRC Press.

Wendel, J.F. 1989. New world tetraploid cottons contain old world cytoplasm. *Proceedings of the National Academy of Sciences USA*, **86**: 4132–6.

Wendel, J.F., Stewart, J.McD., and Rettig, J.H. 1991. Molecular evidence for homoploid reticulate evolution among Australian species of *Gossypium*. *Evolution*, **45**: 694–711.

Wiens, J.J. and Penkrot, T.A. 2002. Delimiting species using DNA and morphological variation and discordant species limits in spiny lizards (*Sceloporus*). *Systematic Biology*, **51**: 61–91.

Wiley, E.O. 1978. The evolutionary species concept reconsidered. *Systematic Zoology*, **27**: 17–26.

Wilmer, J.W., Hall, L., Barratt, E., and Moritz, C. 1999. Genetic structure and male-mediated gene flow in the ghost bat (*Macroderma gigas*). *Evolution*, **53**: 1582–91.

Wilson, P. 1992. On inferring hybridity from morphological intermediacy. *Taxon*, **41**: 11–23.

Wilson, C.C. and Bernatchez, L. 1998. The ghost of hybrids past: fixation of arctic charr (*Salvelinus alpinus*) mitochondrial DNA in an introgressed population of lake trout (*S. namaycush*). *Molecular Ecology*, **7**: 127–32.

Wolfe, K.H., Li, W.-H., and Sharp, P.M. 1987. Rates of nucleotide substitution vary greatly among plant mitochondrial, chloroplast and nuclear DNA. *Proceedings of the National Academy of Sciences USA*, **84**: 9054–8.

Wolfe, A.D., Xiang, Q.-Y., and Kephart, S.R. 1998a. Assessing hybridisation in natural populations of *Penstemon* (Scrophulariaceae) using hypervariable intersimple sequence repeat (ISSR) bands. *Molecular Ecology*, **7**: 1107–25.

Wolfe, A.D., Xiang, Q.-Y., and Kephart, S.R. 1998b. Diploid hybrid speciation in *Penstemon* (Scrophulariaceae). *Proceedings of the National Academy of Sciences USA*, **95**: 5112–15.

Wolpoff, M.H. 1989. Multiregional evolution: the fossil alternative to Eden. In P. Mellars and C. Stringer, eds. *The Human Revolution: Behavioral and Biological Perspectives on the Origin of Modern Humans*. Princeton, NJ: Princeton University Press, pp. 62–108.

Wright, S. 1932. The roles of mutation, inbreeding, crossbreeding, and selection in evolution. *Proceedings of the VI International Congress on Genetics*, **1**: 356–66.

Wright, S. 1977. *Evolution and the Genetics of Populations*. Vol. 3. Chicago: University of Chicago Press.

Zimmer, E.A., Martin, S.L., Beverley, S.M., Kan, Y.W., and Wilson, A.C. 1980. Rapid duplication and loss of genes coding for the chain of hemoglobin. *Proceedings of the National Academy of Sciences USA*, **77**: 2158–62.

7

Case studies in ecological genetics: Lycaenid butterflies, ragworts, bears, and oaks

". . . Two views? There are a dozen views until the correct answer is known."
C.S. Lewis (1945). *That Hideous Strength*

Summary

1 Allozyme markers, together with detailed fieldwork, have been the major approaches to the investigation of ecological genetics in the Lycaenid butterflies.

2 Indirect estimates of gene flow in Lycaenid butterflies are typically higher than field assessments of adult dispersal distances. This may be due to current gene flow or post-glacial migration patterns.

3 Ant and plant-host associates can impose genetic structuring on populations, although these are typically less important that geographical separation.

4 Detailed natural history observations of wild European ragworts (*Senecio*) have led to hypotheses that can be tested through the integration of ecological, cytological, biochemical, and molecular data.

5 Morphological, crossing, and biochemical data support the polytopic origin of radiate *Senecio vulgaris* by the introgression of genes from *S. squalidus* into non-radiate *S. vulgaris*.

6 Morphological, crossing, cytological, biochemical, and molecular data, support multiple, independent allopolyploid origins of *S. cambrensis* in Wales and Scotland and the recent origin of *S. eboracensis* in York, England.

7 DNA analyses of contemporary and ancient mtDNA patterns in the brown bear have shown distinct, large-scale phylogeographic patterns across its global range.

8 European expansion of brown bears following glaciation has occurred northwards from preglacial refugia in the south, whilst North American populations may have contained more variation preglacially, than at present.

9 The utilization of highly variable nuclear microsatellites for fine-scale gene flow studies, together with ecological investigations reveal low dispersal of female brown bears.

10 Molecular phylogeography, pollen core analysis, and modeling have revealed that since the last ice age, European oaks have colonized central and northern regions from refugia located in the south of the continent.

11 Nuclear differentiation in European oaks indicates efficient dispersal of this wind-pollinated species: pollen is dispersed over several kilometers, whilst seed dispersal is generally more limited with occasional long-distance events.

12 There is evidence of extensive historical and contemporary hybridization between oak species, yet diagnostic morphological characteristics and a range of prezygotic and postzygotic breeding barriers persist.

7.1 Introduction

The theme that has been developed throughout this book is one of the integration of field and laboratory investigations, through the detailed analysis of survey and experimental data. The earliest stage of this process involves experimental design (chapter 2), so that the best use is made of any resources available. A critical part of this procedure is the identification of clear, testable hypotheses and access to suitable material to address these hypotheses. Such hypotheses may well involve an examination of genetic diversity and population differentiation (chapter 3), whilst gene flow (chapter 4) may also be an important issue. If speciation (chapter 6) is the ultimate result of ecological genetic processes, then it becomes important to analyze the data within a phylogenetic context (chapter 5). What we have presented, using numerous examples, are the principles of planning and analyzing ecological genetic projects and data, which bridge a potential knowledge gap between the type of genetic markers to apply and the biological interpretation of the derived data.

The case studies presented in this chapter have been chosen to illustrate the range of issues that can be tackled using ecological genetic approaches. One consistent theme in these investigations is that they are long-term studies that have integrated many disciplines.

One example of this has been researches on European ragworts, which, for at least 90 years, have integrated genetic studies and experimental ecology in order to understand their ecological genetics. As is seen in the case study, the introduction of Oxford ragwort into the United Kingdom in the early eighteenth century had the unpredictable outcome of the origin of two new species and a new variety all within the last 150 years: this is a powerful illustration of evolution in action. However, there is nothing to suggest that Oxford ragwort is unique in this respect. The Lycaenid butterflies typically have limited dispersal abilities, with their larvae feeding on a single host species. In addition, at least half of these butterflies have an association (ranging from parasitism to mutualism) with specific ants. Co-evolution, allied to short dispersal distances, has made the Lycaenid butterflies excellent subjects for the investigation of gene flow and speciation. The brown bear, with its large size, public appeal, and increasing rarity illustrates the role of ecological genetics in conservation planning and resource management. Furthermore, the role of ancient DNA approaches in phylogeographic and gene flow investigations is also illustrated. The social and economic importance of European oaks has led to intense genetic study over the last 50 years, making them an ideal case study of ecological genetic principles. A recent boom in population genetic and gene flow studies has been complemented by an extensive background literature of morphological and adaptive variation in natural populations and provenances. Furthermore, oaks have been pivotal for our understanding of the role of glacial refugia and dynamics of continental recolonization since the last glaciation.

7.2 Lycaenid butterflies

The Lycaenidae, a subfamily of the true butterflies, and commonly called the gossamer winged butterflies or blues, coppers, and hairstreaks, comprises approximately 4000 species worldwide. As with the majority of insects, Lycaenids have very limited dispersal abilities. Moreover, they are typically monophagous (restricted to one larval food plant) and the larvae of at least half of the subfamily have specific ant (subfamily Myrmicidae) associations, which range from parasitism to mutualism.

The aesthetic appeal of butterflies, particularly the Lycaenidae, makes them an obvious group for scientific investigation. Furthermore, the increasing rarity of many Lycaenids has imparted a greater urgency to such investigation. This is enhanced by the Lycaenid's highly specific requirements, so they are often viewed as good indicators of general habitat quality. Given the conservation significance of the Lycaenids, much genetic work has focused on gene flow, as the capacity for dispersal directly influences a species' conservation biology. The specificity of a particular species to a host plant and/or ant associate, and the potential reproductive isolation that this confers, has also made the Lycaenidae excellent subjects for the study of speciation.

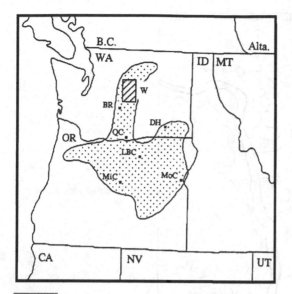

Fig. 7.1

Range of *Euphilotes enoptes* (shown by stippling) in Washington (WA) and Oregon (OR) states, USA. The hatched area represents the sampled area within the Columbia River basin (see Fig. 7.2). The black squares represent six additional, more distant populations. These are from Bethel Ridge (BR), Davis Hollow (DH), Little Butler Creek (LBC), Mill Creek (MiC), Morgan Creek (MoC), and Quarte Creek (QC) (Peterson 1996).

7.2.1 *Euphilotes enoptes* (The Pacific Dotted Blue)

Peterson (1995, 1996) investigated gene flow between populations in the sedentary species *Euphilotes enoptes* over a variety of distances using allozymes. This widespread taxon is found in the Pacific Northwest of the United States, and the larvae are monophagous, feeding on the long-lived prostrate shrub, *Eriogonum compositam* (Polygonaceae). The host plant has a wide altitudinal range from sagebrush steppe (250 m above sea level) to clearings in subalpine forest (1750 m), where it has a patchy distribution of varying density, ranging from a few plants within a few square meters to thousands of individuals in colonies covering 10 or more hectares; individual patches are between 10 and 1000 meters apart.

The initial investigation (Peterson 1995) focused on the degree of gene flow between populations separated by distances up to 30 km within the Columbia basin (Figs. 7.1 and 7.2). Twelve populations, comprising four samples along three altitudinal transects, within the Wenatchee region of Washington State, USA were assayed at six variable allozyme loci, using 6–30 individuals from each population. The three transects varied in maximum and minimum altitude but generally ranged from 1000–1300 m. No relationship was found between geographic distance and the degree of genetic relatedness, using either Nei's (1972) genetic distances or linear regression of log transformed gene flow estimated against log transformed geographic distances (Slatkin 1993). This is surprising since mark–release–recapture studies show that the majority of adults move less than 500 m. It appears that the movement of a few ovipositing females over 1 km is sufficient to eliminate genetic differentiation up to 30 km. Patches of the host plant at low elevations flower earlier than those at higher elevations (Fig. 7.3). The butterflies track the sequential flowering, moving uphill from patch to patch, although it is highly unlikely that many butterflies will travel the full altitudinal distance.

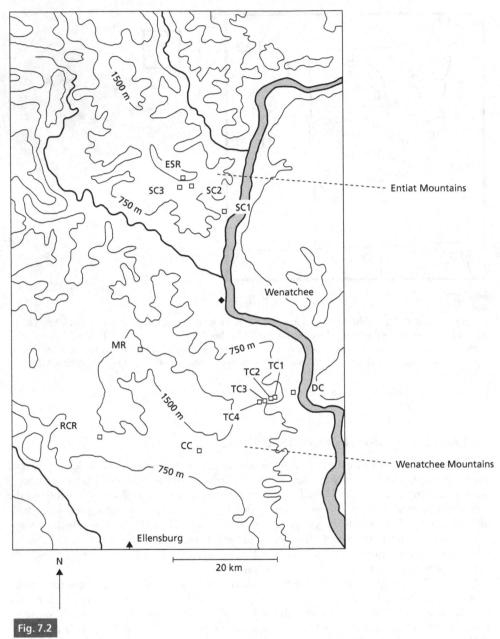

Fig. 7.2

Sample sites of *Euphilotes enoptes* in Wenatchee and Entiat mountain region (Columbia River basin) Washington state, USA (from Peterson 1996). The Columbia River is shaded. 750 m and 1500 m contour lines are also shown. Samples were collected from Cooke Cyn (CC), Davies Cyn (DC), Entiat Summit Road (ESR), Mission Ridge (MR), Reecer Creek (RCR), Swahave Creek (SC; 3 populations), and Tarpiscan Creek (TC; 4 populations).

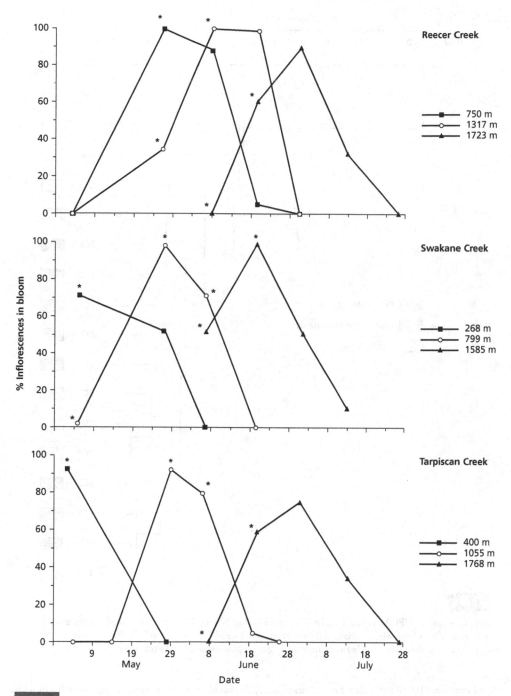

Euphilotes enoptes larval host plant (*Eriogonum compositum*) flowering phenology at low, medium, and high elevations, along three elevational transects. Asterisks indicate dates on which adult *Euphilotes enoptes* were observed (from Peterson 1995).

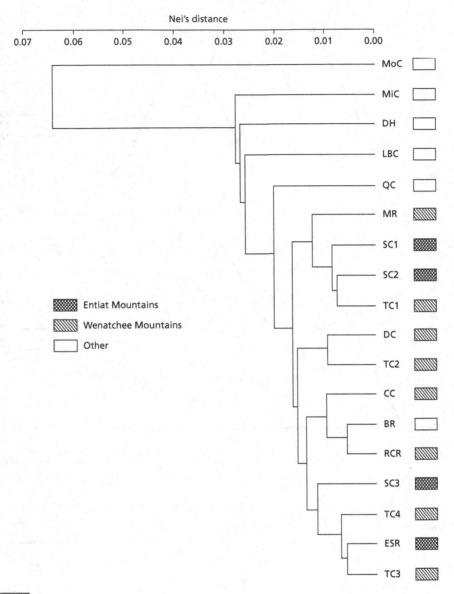

Nei's distance

Fig. 7.4

Euphilotes enoptes UPGMA cluster analysis, clustered using Nei's (1972) distances. Shading designates the geographic area from which each population originated (see also Figs 7.1 and 7.2) (from Peterson 1996). There is a broad relationship between increasing geographic distance and greater genetic distance.

Peterson (1996) extended his previous study by taking six more samples up to 400 km away from the initial samples in the Entiat/Wenatchee area (Fig. 7.1). Some of the data from the original survey were utilized plus additional Entiat/Wenatchee samples to give 18 samples in all. Sites coupled with the Entiat/Wenatchee region were all within 30 km of each other (Fig. 7.2). Gene flow was calculated indirectly and Nei's (1972) genetic distances were calculated to determine if there was a positive correlation between genetic distance and geographic distance. When the distant populations were considered, genetic distance was positively correlated with geographic distance from the Entiat/Wenatchee area. For instance, the two populations most distant from the Wenatchee area (Mill Creek and Morgan Creek; MiC and MoC, respectively in Fig. 7.4) were the most differentiated on the UPGMA phenogram. The main exception to this relationship was the Davis Hollow population, which had a greater degree of genetic separation from Wenatchee/Entiat than direct distance between the two areas would suggest. However, the shortest distance between Davis Hollow and Entiat/Wenatchee is not the route that butterfly movements would follow. Due to the absence of suitable habitats within the Columbia Basin, gene flow would most likely occur around the periphery of the basin, where the host plant is found.

7.2.2 *Mitoura* (hairstreak) species

Gene flow between different host races (i.e., populations that exhibit a preference for particular plant or ant hosts) within the genus *Mitoura* was investigated by Nice and Shapiro (2001). Host races have been considered to be species in the making (Bush 1994), since fidelity to a particular host renders taxa at least partially reproductively isolated and assortative mating of individuals within the same host, allied to different selection pressures, may lead to speciation. However, speciation will only occur if there is insufficient gene flow to offset the developing genetic differentiation between the various host races.

Mitoura is a large, complex genus of temperate North American butterflies whose larvae feed on cypresses and cedars (Cupressaceae). The three closely related hairstreak species (*M. nelsoni*, *M. muiri*, *M. siva*), which were the subject of the study by Rice and Shapiro (2001), are found on the western side of the United States, primarily in California. They are all univoltine (a single flight period within one year) and flight period timing varies within a species, depending on the altitude and habitat characteristics of the particular location where a population is found. Typically, populations found at higher altitudes have later flight periods.

Perhaps unsurprisingly given the reproductive isolation processes that may be occurring and the parapatric and interdigitating nature of their populations there has been taxonomic dispute regarding the status of these three taxa. Scott (1986) treated them as three subspecies of *M. grynea*, whilst Tilden and Smith (1986) considered them to be three separate species (an approach maintained here). Genitalia differences, which are key prezygotic isolating mechanisms in butterflies, are inconsistent between the taxa, showing differences between taxa in some areas but not in others. However, the adults can be separated based on wing markings, flight period, and host-plant association.

Mitoura nelsoni and *M. siva* larvae each have single host plants (*Calocedrus decurrens* and *Juniperus occidentalis*, respectively), while *M. muiri* larvae are found on three different host plants (the plant host varying across the range of this species). At the northern end of its range *Cupressus sargentii* and *C. macnabiana* are the larval food plants, whilst further south, where *C. sargentii* is absent and *C. macnabiana* is rare, larvae are found on *Juniperus californica*. *Juniperus californica* also has a different ecology to the other two host plants of *M. muiri*, being found on calcicolous, rather than ultramafic, soils. In *M. muiri*, wing phenotype differences are found between populations on different host plants (Table 7.1).

Within the study area, a region of the Sierra Nevada (California), *M. nelsoni* and *M. muiri* occur within 6–7 km of each other, although they are phenotypically and altitudinally separated. *Mitoura nelsoni* occurs at approximately 1300 m in mesic forest, flying in June–July and with *Calocedrus decurrens* as the larval host plant. *Mitoura muiri* is found at lower altitudes (500 m), flying from April to May and with the larvae feeding on *Cupressus sargentii*. *Mitoura muiri* was also sampled from the Californian coastal mountain range (Fig. 7.5).

The third species, *M. siva*, is found within 10 km of *M. nelsoni* in the northern Sierra Nevada and southern Cascade mountains, although the two species are separated by altitude. Nice and Shapiro (2001) used allozymes and mtDNA to ascertain whether genetic separation of *M. muiri*

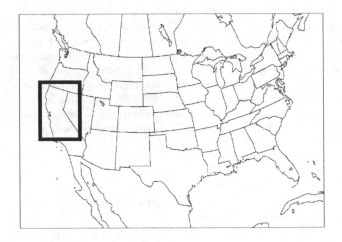

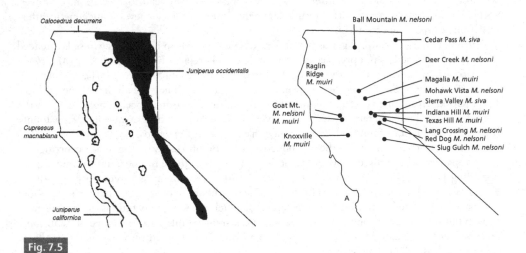

Location of *Mitoura* species sample sites and host plant distributions in California (from Nice and Shapiro 2001).

was due to host-plant fidelity and hence whether host-race formation is occurring. Alternatively, given the large geographic range of the three species, genetic variation may be geographically influenced.

Genetic analysis was carried out using 13 allozyme loci (10 of which were polymorphic), plus sequencing of the mtDNA-encoded cytochrome oxidase subunit. For the allozyme study, 287 individuals were assayed and sample sizes varied between 12 and 30 per population from 12 populations (seven populations of *M. nelsoni*; two populations of *M. muiri*; and three populations of *M. siva*, Fig. 7.5). Fewer individuals (54) were surveyed for the mtDNA study though samples were taken over a similar geographic range and many of the same populations were used.

Table 7.1

Distribution of larval host plant and flight time (phenology) of three closely related *Mitoura* species in the Eastern USA. All species are univoltine and adult fly period varies between locality and elevation. These three species differ in larval host plant, adult flight phenology, and wing markings.

Species	Distribution	Larval host plant	Phenology
M. nelsoni	British Columbia to Southern California coast ranges, plus mid-elevations West Sierra Nevada	*Calocedrus decurrens*	April–July
M. muiri	Within range of *M. nelsoni*, in north of range inner coast range of California	*Cupressus sargentii* and *C. macnabiana*. In south of range *Juniperus californica*	March–June
M. siva	Great Basin, East Sierra Nevada to Oregon and Washington	*Juniperus occidentalis*	April–July

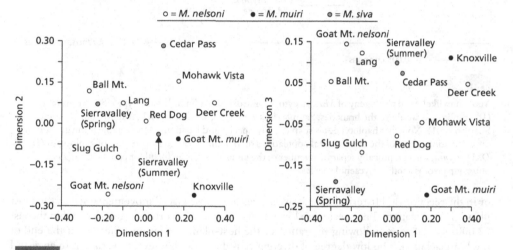

Fig. 7.6

Multidimensional scaling of allozyme data for *Mitoura nelsoni*, *M. muiri*, and *M. siva* (from Nice and Shapiro 2001). There are no obvious groupings linking species together or geographically close areas.

Using multidimensional scaling (Fig. 7.6), the allozyme data revealed no geographic or taxonomic distinction between the three species. The mtDNA results were largely in accord with the allozyme data. However, some geographic patterns were observed. All three taxa possessed a common mtDNA haplotype (A), whilst a single *M. siva* individual had a unique haplotype (B). More significantly, three of the four coastal Californian *M. muiri* populations possessed novel haplotypes (C and D; Fig. 7.7). Analysis of these data indicates that gene flow is more restricted between coastal populations than elsewhere.

The extremely low level of genetic differentiation recorded between populations and species suggests a correspondingly high level of gene flow between and within species, either currently

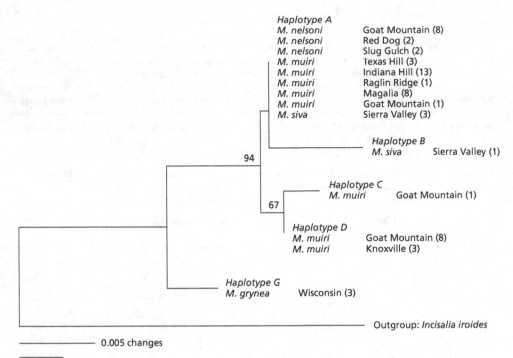

Fig. 7.7

Maximum likelihood phylogeny of *Mitoura* cytochrome oxidase subunit II haplotypes. Bootstrap values (100 replicates) are above the branches. Sample sizes at each population are in brackets (from Nice and Shapiro 2001). Note that haplotype *A* is shared across species and locations. Also notable is the genetic different ation of two of the three coastal populations of *M. muiri* revealed by presence of haplotypes *C* and *D*. *M. grynea*, a geographically separate member of the same genus, is included for comparison. The outgroup taxon is another Lycaenid species.

or in the recent past. Historical gene flow, as a result of post-glacial movements, seems the most plausible explanation. The species are likely to have colonized northern America within the last 12,000–15,000 years, following migration of the host-plant species into the area at the end of the last glaciation. The low degree of divergence is due to a high level of variation in ancestral populations followed by a relatively short time for lineage sorting.

There is genetic differentiation of, but ecological similarity between, the coastal *M. muiri* populations and those of the Sierra Nevada. Within the Sierra Nevada, the three species show close genetic similarity but ecological separation. Despite the absence of genetic differences between populations and species the apparent specificity of taxa to their host plants, and the phenological and morphological differences suggest that incipient speciation of broadly sympatric populations is occurring. Niegel and Avise (1986) have identified this as one of the potential routes to speciation, and similarly low levels of genetic divergence have been described for other Californian butterflies (Nice and Shapiro 1999).

7.2.3 *Jalmenus evagoras* (The Common Imperial Blue)

Alongside host-plant specificity, the association that many Lycaenids have with Myrmecid ants is another factor that may impose genetic structure within a taxon and hence contribute towards isolation and, ultimately, speciation. This phenomenon was investigated by Costa, McDonald, and Pierce (1996) in the Australian Lycaenid, *Jalmenus evagoras*. This species is tended by several

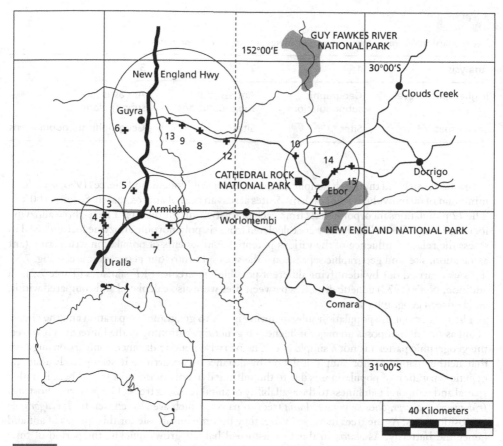

Fig. 7.8

Sampling sites of *Jalmenus evagoras* in New South Wales, Australia. Butterflies at sites 1–7 are tended by the ant species *Iridomyrmex vicinus*, those at sites 8–15 by *I. anceps*. The four geographically defined subpopulations are circled (from Costa, McDonald, and Pierce 1996).

ant species, although in the *Acacia* grassland study area the association is restricted to two ant species (*Iridomyrmex anceps* and *I. vicinus*). An individual butterfly colony is only found with one of the two species. The butterfly–ant association is a mutualistic; the larvae secrete amino acids and carbohydrates for the ants, whilst gaining protection from predators through the action of ant guards. The host plant is *Acacia harpophylla*, where both mating and pupation take place, and is shared with other *Jalmenus* species. However, each tree is host to only one ant species and these ant–tree associations persist from year to year. Butterflies choose host plants by the presence of their specific ant associate on the tree.

Population structure in the butterfly may be influenced by the patchy distribution of the appropriate tree–ant combination allied to the fidelity of the butterfly to the host tree, via the ant association. Individual butterflies appear to spend their entire adult lives around the host tree. Male butterflies emerge first and then immediately fertilize any emergent females. Evidence for population substructuring, via the ant association, comes from observations that larger winged *J. evagoras* adults produce more larvae per cluster when associated with *Iridomyrmex anceps* rather than *I. vicinus*.

Table 7.2

Hierarchical structuring of three separate analyses of *F*-statistics in *Jalmenus evagoras*.

Analysis	1	2	3
Higher spatial level	Geographic subpopulations	Ant associate subpopulations	Ant associate subpopulations
Lower spatial level	Sites	Sites	Geographic subpopulations

In an extensive and thorough sampling regime, Costa, McDonald, and Pierce (1996) sampled a minimum of 60 individuals from each of 15 sites across an area of approximately 50 km × 100 km. The 12 populations incorporated six from each of the two ant mutualists. Thirty-three allozyme loci were assayed and *F*-statistics were calculated from six polymorphic allozyme loci and used to assess the relative influence of the various potential components of population structuring (ant association, site, and geographic separation – the sites split into four geographic areas; Fig. 7.8). This was carried out by identifying different possible hierarchies of *F*-statistics (Table 7.2). In addition, Nei's (1972) genetic distances between sites were also calculated and compared within and between geographic areas.

The majority of the population substructuring is due to geographic separation and the choice of ant association imposes no major influence on genetic partitioning in the butterfly. However, the geographic pattern is not a simple one. The pairwise genetic distance comparison suggests that neither panmixia nor simple isolation-by-distance is occurring. It seems likely that the genetic separation of populations is due to the shifting dynamic of the host tree. There is both a spatial and temporal patchiness to the availability of suitable *Acacia* trees. The preferred host ant, *Iridomyrmex anceps*, choose young *Acacia* trees to patrol, which are also chosen by *J. evagoras* as oviposition sites. As the trees increase in size, they become unsuitable for this species of ant and hence the butterfly. As *Acacia* in this fire-disturbed habitat grows quickly, the period of time when the tree–ant association is suitable is short. The period may be made even shorter by the larvae's ability to defoliate the host plant entirely and consequently starve.

This relationship leads to a shifting pattern of discrete colonies of the butterfly as the ant associate alters with developing age of the *Acacia* and thus the suitability of the tree for *J. evagoras*. The lack of population substructuring is due to the regular availability of new suitable sites and butterfly dispersal to these sites. The absence of inbreeding revealed by the *F*-statistics suggests that butterfly colonies are the result of several pioneer individuals to a new site rather than the product of a single colonization event and this butterfly dispersal may have produced the current connectedness. The lack of marked population substructuring supports Slatkin's (1977) view that regular extinction and recolonization events reduce the opportunity for genetic differentiation.

7.2.4 European *Maculinea* species (Large Blues)

The genetic impact of fragmented populations caused by human disturbance has been considered in the Large Blue butterflies, *Maculinea nausithous* and *M. teleius*. These species are among five European members of the genus, all of which are now listed in the IUCN Red Data Book as vulnerable or rare. Their increasing rarity is due to the specificity of ant association, which in turn leads to highly specialized requirements of the seminatural habitats in which the species are found. Like *J. evagoras*, these two species have an association with a Myrmicid ant. However, unlike *J. evagoras*, which is fed by its ant host, these two *Maculinea* species predate the Myrmicid larvae.

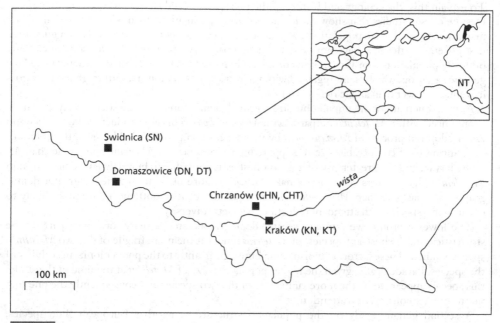

Fig. 7.9

Location of collection sites of *Maculinea* species in southern Poland and Russia (top right map of Eastern Europe and Western Siberia). Abbreviations have 1–2 letters for the town followed by an N or T for *M. nausithous* and *M. teleius* respectively (from Figurny-Puchalska, Gadeberg, and Boomsma 2000).

Both butterfly species are found in wet river valley meadows, and females of both species oviposit on *Sanguisorba officinalis*, although the pattern of egg laying is different in the two species. The host ant species is also different, *Maculinea teleius* depends upon *Myrmica scabrinodis*, whilst *Maculinea nausithous* predates upon *Myrmica rubra*. The ant species differ in their pattern of distribution; *M. rubra* is more patchily distributed but has larger colonies than *M. scabrinodis*. In addition, *Maculinea nausithous* has twice as many pupae per *Myrmica* nest as *Maculinea teleius*. This in turn influences the butterfly distribution pattern. Figurny-Puchalska, Gadeberg, and Boomsma (2000) investigated the genetic variation and structure of four populations of *M. nausithous* and *M. teleius* from Russia and Poland using allozymes, the two species co-existing at three of these populations (Fig. 7.9).

Allelic differentiation between pairs of populations was calculated by Fisher's exact tests or log-likelihoods. Wright's *F*-statistics were calculated to assess the degree of genetic structuring. Nei's (1972) genetic distance was calculated to assess isolation-by-distance of the populations.

The three Polish populations of *M. teleius* sampled were genetically distinct from each other despite all displaying a low level of allozyme diversity. By comparison, the sole Russian population was more diverse. This pattern of diversity may be attributed to postglacial migration from eastern refugia with Russian populations closer to the source of origin than Polish ones. Alternatively, the Polish populations may have a patchy distribution with a small effective population size and corresponding genetic drift. By comparison, the Russian site may have a greater level of variation due to it being part of a well-functioning metapopulation system and hence a large effective population size.

Maculinea nausithous had a higher level of expected heterozygosity and population differentiation than *M. teleius*. Given the common habitat of the two species and the apparent greater abundance of *M. teleius*, the higher level of genetic variation in *M. nausithous* is perhaps surprising.

To explain this, the ecology and history of the two species need to be considered more closely. Despite census studies that show *M. teleius* has larger populations than *M. nausithous*, this data may not reflect effective population sizes. In addition current population size is not a guarantee that genetic bottlenecks have not occurred in the past. Due to annual fluctuations in climate, butterfly populations can vary in size considerably from year to year. Such bottlenecks reduce genetic variation, which is not regained with recovery in population size, although heterozygosity reduces slower than allelic richness (chapter 3).

Annual fluctuations in population size, termed demographic stochasticity, may be more pronounced within *M. teleius* compared with *M. nausithous*. This is a product of the quicker and easier adoption process of *M. nausithous* by its host ant compared with *M. teleius*, the ability of host-ant nests of comparable size to support higher numbers of *M. nausithous* larvae than *M. teleius* larvae, and the greater risk of host-ant nest extinction due to butterfly larvae predation in *M. teleius*. These factors combine to make *M. teleius* more likely to experience greater demographic stochasticity than *M. nausithous*. Consequently, genetic bottlenecks are more likely to occur in this species with attendant reduction in genetic diversity.

The lower reproductive rate and greater demographic stochasticity (driven in part by the stochasticity of the host-ant species) of *M. teleius* render it the more fragile of the two *Maculinea* species studied. These factors may also contribute significantly to the poor colonization ability of the species. Paradoxically, given the higher population size of *M. teleius* at the time of the study, this species appears to be the more vulnerable of the two species and consequently the one that merits greater conservation attention.

A second genetic study on the population structure of another European Blue species, *Maculinea alcon*, has been undertaken by Gadeberg and Boomsma (1997) on populations in Jutland (Denmark). Like the other European *Maculinea* species, *M. alcon* is rare and declining due to its specific habitat requirements being incompatible with modern agriculture. Consequently, it is found in fragmented populations of variable sizes. This species has a single host plant in Jutland, *Gentiana pneumonanthe*, with an obligate mutualism with a Myrmicid ant species. Gadeberg and Boomsma's (1997) study set out to establish whether the geographically isolated populations are also genetically isolated and whether any pattern was due to isolation by distance along a latitudinal gradient or the co-existence of one genetically distinct form.

Populations of *M. alcon* in Jutland were identified and sampled if the population size was deemed large enough to support the destructive sampling necessary for allozyme work. Further threat to the population was reduced by collecting eggs and waiting for caterpillar emergence, rather than removing adults for assay (butterfly populations suffer high larval mortalities, so removing eggs will have minimal effect on the adult population size).

With five polymorphic loci, the populations showed a significant lack of heterozygotes (significantly high F_{IS} values). There is also a geographical pattern to the variation, with northern colonies exhibiting significantly less variation than those in the south. In addition, linkage disequilibrium was identified between two pairs of loci in central Jutland populations. Linkage disequilibrium needs to be taken into account when F-statistics are being calculated as it inflates F_{ST} values (chapter 3). After discounting linkage groups, F_{ST} is still significantly high.

These high positive F_{ST} values suggests that there is a considerable amount of genetic separation between populations. The lack of heterozygotes implies that there is a degree of sib-mating in a small effective population. Kaaber (1964), investigating wing morphology, suggested that within Denmark, *M. alcon* constitutes two species, *M. alcon* and *M. rebeli*. Subsequent work by Thomas et al. (1989) in identifying the different plant and ant hosts of the two butterflies has confirmed that *M. rebeli* is a species distinct from *M. alcon*. Specifically, *M. rebeli* utilizes a different *Myrmica* larval host and a different food plant, *Gentiana cruciata*, from *M. alcon*. This food plant is absent from Denmark. Thus, *M. alcon* is similarly not found in Denmark. However, the two different wing morphologies found in the study area may reflect two different subpopulations within any given colony. There is a close association between the frequency of the morphs and certain allele frequencies, although this is not consistent in all populations. If adults retain fidelity within their wing pattern when choosing mates (positive assortative mating), the lack of heterozy-

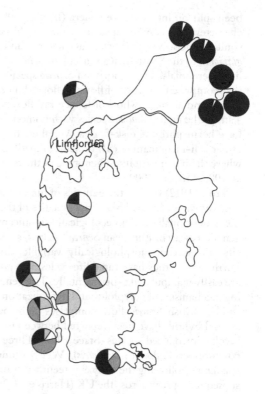

Fig. 7.10

Allozyme variation in aconitase in populations
of *Maculinea alcon* in Jutland (Denmark)
(from Gadeberg and Boomsma 1997). Populations
in the northwest of the region are less variable than
those elsewhere, with fewer alleles and less even
allele frequencies within a population.

gotes may be due to the Wahlund effect, where individuals treated as one population are actually two with hybridization between the two forms unusual.

Jutland is effectively separated by the Limfjord, the large sea inlet towards the north of the region. Populations in the northeast of the area exhibit a lower level of genetic variation than those south of the Limfjord. The results observed in Fig. 7.10 are typical of this pattern. This geographic pattern in genetic diversity may be due to different ant hosts. *Maculinea alcon* has different hosts in different geographic regions: *Myrmica rubra* in Sweden, *M. ruginodis* in Holland, and *M. scabrinodis* in France. Elmes et al. (1994) has suggested that the switch between the Swedish and Dutch host occurs in Jutland. Given that the different ants themselves require different habitats, most notably sward height via grazing regime, it is possible that this ant-host specificity or their different social forms (microgyne, with a queen smaller than the workers or macrogyne, with a larger queen than attendant workers) imposes reproductive isolation. This is then reflected in the geographic separation of the populations surveyed.

The great majority of ecological genetic work on butterflies utilizing molecular markers has been undertaken with isozymes. However, the work by Nice and Shapiro (2001) shows that the use of DNA markers is a powerful adjunct to allozymes. One consistent theme that emerges from the investigation of the ecological genetics of Lycaenid butterflies is the challenging of previously held views based solely upon field data. For example, the greater gene flow than hitherto recognized in many butterfly species and the genetic similarity between ecologically separate species.

7.3 European ragworts

Senecio (Asteraceae) is one of the largest angiosperm genera, with between 1000 and 3000 species depending on the generic concept adopted, and its members are found on all continents, except Antarctica. Jeffrey et al. (1977) divided the genus into 16 informal groups, some of which have been split off into separate genera (Bremer 1994), and Vincent (1996) proposed that six floral characters (e.g., anther apex structure) could be used to aid delimitation of *Senecio* from very similar genera. *Senecio* species have habitats and habits as diverse as the South American tree, *S. glaziovii*, to the South African succulent *S. herreianus*, although most species are annuals or short-lived perennials. The majority of *Senecio* species have heteromorphic capitula (e.g., *S. squalidus*) that comprise two types of flowers (florets): hermaphrodite disk-florets, with five short, joined petals; and an outer whorl of female ray-florets, with three of the joined petals elongated into a strap-like ligule. *Senecio* species with homomorphic capitula (e.g., *S. vulgaris* var *vulgaris*) only have hermaphrodite disk-florets. Members of the genus *Senecio*, as with many Asteraceae, are either self-incompatible (SI) or self-compatible (SC); *S. squalidus* has a sporophytic SI system, where the incompatibility phenotype of the pollen is determined by its parent plant (Brennan et al. 2002, Hiscock 2000).

Trow (1912) studied the genetics of nine floral, stem, leaf, and hair characters, although he is best known for his analysis of the genetics of the ray floret locus; one of the earliest examples of the genetic analysis of an ecologically significant plant trait. As model organisms for ecological genetic research, European *Senecio* species, such as *S. vulgaris* and *S. squalidus*, are geographically widespread, morphologically variable, and easily grown under laboratory conditions. Furthermore, they have rapid life cycles (it is possible to produce 2–3 generations per year) and are easily manipulated experimentally. The genus has been particularly important for investigating mechanisms of polyploidy and hybridization in plant speciation.

The British *Senecio* flora comprises eight native and 12 introduced *Senecio* species: seven natural hybrids have been reported between members of the British *Senecio* flora, of which five involve introduced species (Stace 1997). Three species have attracted most research attention, common groundsel (*S. vulgaris*), Welsh groundsel (*S. cambrensis*), and Oxford ragwort (*S. squalidus*). Following the early eighteenth-century introduction of *S. squalidus* from Sicily, and its subsequent spread across the UK (Harris et al. 2002; Fig. 7.11), the species has had diverse interactions with the native British *Senecio* flora, including: (i) the origin of *S. vulgaris* var. *hibernicus* via introgression of *S. squalidus* into *S. vulgaris* var. *vulgaris*; (ii) the polytopic origin of Welsh and Scottish *S. cambrensis*, between *S. squalidus* and *S. vulgaris* (Abbott, Ashton, and Forbes 1992, Ashton and Abbott 1992); and (iii) the origin of *S. eboracensis*, a fertile tetraploid derived from crosses between *S. squalidus* and *S. vulgaris* (Lowe and Abbott 2003). Moreover, Abbott et al. (2000) have argued that *S. squalidus* is itself a hybrid between two Sicilian species, *S. aethnensis* and *S. chrysanthemifolius*.

7.3.1 What is the origin of radiate *Senecio vulgaris*?

In 1875, Syme described a short-rayed variant of *Senecio vulgaris*, based on material collected in 1866 from Cork (Eire), as *S. vulgaris* var. *hibernicus*, to distinguish it from the non-radiate type, *S. vulgaris* var. *vulgaris*, and *S. vulgaris* ssp. *denticulatus*, a radiate taxon restricted to maritime dune communities and some inland montane regions of the western European seaboard (Kadereit 1984b). In addition to typical radiate *S. vulgaris*, a second type of radiate *Senecio* species was discovered in York in 1979 and is thought to have originated less than 20 years previously (Lowe and Abbott 2003). This plant is now known to be a fertile tetraploid hybrid between *S. squalidus* and *S. vulgaris*, called *S. eboracensis*.

Trow (1912) studied the radiate polymorphism in *S. vulgaris* and showed that ray florets are inherited as a single, incompletely dominant gene. When radiate (RR) *S. vulgaris* is crossed with non-radiate (rr) *S. vulgaris*, F_1 progeny (Rr) with intermediate length rays are produced. If the F_1 is selfed, three classes are found in the F_2 progeny, non-radiate (rr), intermediate (Rr), and radiate (RR), in the expected Mendelian ratios of 1:2:1. Two hypotheses have been proposed to explain the origin of ray-floret allele in radiate *S. vulgaris*: (i) hybridization with *S. squalidus* followed by backcrossing to non-radiate *S. vulgaris*; or (ii) a single gene mutation at the ray-floret locus in non-radiate *S. vulgaris*.

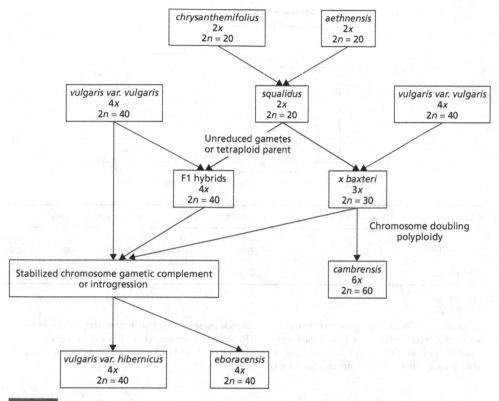

Fig. 7.11

Hybridization and introgression relationships between *Senecio* taxa involving *S. squalidus*, showing chromosome numbers and ploidy level.

Introgression is paradoxical, in that the more difficult introgression events are to detect the more evolutionarily significant effects are likely to be. Thus evidence to support an introgression hypothesis is more likely to be derived from multiple, rather than single, data sources (Arnold 1997). In the case of radiate *Senecio vulgaris* this evidence comes from: (i) the approximately parallel spread of *S. squalidus* and radiate *S. vulgaris* over the past 150 years; (ii) intermediacy of radiate *S. vulgaris* between *S. squalidus* and non-radiate *S. vulgaris* (Monaghan and Hull 1976, Richards 1975); and (iii) resynthesis of a tetraploid resembling radiate *S. vulgaris* from artificially created *S. x baxteri* followed by backcrossing to *S. vulgaris* (Ingram, Weir, and Abbott 1980; Fig. 7.11). However, Stace (1977) questioned the hybrid hypothesis and argued that a mutation at the ray-floret locus was at least as plausible. An important part of Stace's argument was based on the ease with which ray-floret mutants are found in the Asteraceae, for example, radiate to non-radiate (e.g., *Aster trifolium*, *S. cambrensis*, *S. squalidus*, and *S. jacobaea*) and non-radiate to radiate (e.g., *Bidens cernua*).

Additional evidence supporting the introgressive origin of radiate *S. vulgaris* is derived from the occurrence of *S. squalidus* genes in *S. vulgaris* populations that are polymorphic for capitulum type, but the absence of these genes from *S. vulgaris* populations monomorphic for capitulum type. In an allozyme study of 13 polymorphic *S. vulgaris* populations, 15 monomorphic *S. vulgaris* populations, and 20 *S. squalidus* populations, Abbott, Ashton, and Forbes (1992) investigated the distribution of alleles at an aspartate aminotransferase locus (*Aat*-3). Aspartate aminotransferase is an enzyme that catalyzes the conversion of α-ketoglutaric acid and aspartic

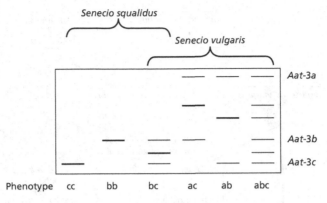

Popn. type	Capitulum phenotype	Sample	Phenotype frequency			
			ab	ac	bc	abc
Polymorphic	N	384	0.97	0.01	0.01	0.01
	R	351	0.53	0.33	0.13	0.01
Monomorphic	N	428	1.00	0.00	0.00	0.00

Fig. 7.12

Aat-3 evidence for the introgressive origin of radiate *Senecio vulgaris*. *Aat*-3 phenotype frequencies in 13 *Senecio vulgaris* populations polymorphic for radiate (R) and non-radiate (N) morphs and 15 populations monomorphic for the radiate morph (data from Abbott et al. 1992). On allozyme gels the two *Aat*-3 loci overlap each other and therefore appear as a single banding pattern.

acid to glutamate and oxaloacetate, and in *Senecio* has three isozymes (*Aat*-1, *Aat*-2, *Aat*-3). Using artificial crosses, Abbott, Ashton, and Forbes (1992) showed that in *S. vulgaris*, *Aat*-3 behaves as two independent, duplicated loci (*Aat*-3A and *Aat*-3B; Fig. 7.12), and that these loci are not linked to the ray-floret locus. Three alleles (a, b, c) occur at locus *Aat*-3. Allele *Aat*-3a is restricted to *S. vulgaris*, whilst alleles *Aat*-3b and *Aat*-3c are found in both *S. vulgaris* and *S. squalidus*.

Allele *Aat*-3c occurred in 74% of the *S. squalidus* individuals sampled. However, with duplication of the *Aat*-3 locus in *S. vulgaris*, and with each duplication sharing allele *Aat*-3a, it is difficult to genotype individuals with the *Aat*-3a allele from natural populations; hence phenotypes, rather than genotypes, are usually scored (Abbott, Ashton, and Forbes 1992). All non-radiate *S. vulgaris* plants from monomorphic populations had the *Aat*-3ab phenotype, which also occurred in 97% of non-radiate plants from polymorphic populations. In contrast, this phenotype was found in only 53% of radiate plants, the remaining plants having phenotypic combinations that included allele *Aat*-3c. Whereas allele *Aat*-3c is common in radiate *S. vulgaris*, it is rare in individuals of non-radiate *S. vulgaris* that co-occur with radiate *S. vulgaris* and is completely absent from populations monomorphic for non-radiate *S. vulgaris* (Fig. 7.12). Taken together with the historical, morphological, and cytological data, these biochemical data are convincing evidence for an introgressive origin of radiate *S. vulgaris*.

Increased genetic diversity of a recently evolved introgressant compared to the progenitor taxon is only expected for those parts of the genome involved in the introgression event; a general increase is not expected since only a small sample of the parental genetic diversity will be involved in introgression. Abbott, Irwin, and Ashton (1992), in an analysis of allozyme variation in *S. vulgaris* using 12 enzyme systems, showed that 10 of the systems were monomorphic, whilst two esterase loci (*αEst*-1 and *βEst*-3) were polymorphic. Radiate *S. vulgaris* plants contained less gene diversity at *αEst*-1 and *βEst*-3 than non-radiate *S. vulgaris* plants, irrespective of whether non-radiate *S. vulgaris* was sampled from populations monomorphic or polymorphic for capitulum

Table 7.3

Estimates of gene diversity in *S. squalidus* and radiate and non-radiate *S. vulgaris* at the loci *αEst*-1 and *βEst*-3.

Taxon	Popn. type	Popn. no.	Sample size	Locus	H_T	H_S	D_{ST}	G_{ST}
S. squalidus	–	23	822	*βEst*-3	0.143	0.132	0.011	0.077
Non-radiate *S. vulgaris*	Polymorphic	18	672	*βEst*-3	0.292	0.194	0.098	0.336
				αEst-1	0.313	0.235	0.078	0.249
				Mean	0.302	0.214	0.088	0.292
Radiate *S. vulgaris*	Polymorphic	18	594	*βEst*-3	0.112	0.070	0.042	0.375
				αEst-1	0.103	0.074	0.029	0.284
				Mean	0.109	0.072	0.037	0.337
Non-radiate *S. vulgaris*	Monomorphic	7	185	*βEst*-3	0.287	0.050	0.237	0.826
				αEst-1	0.486	0.084	0.402	0.827
				Mean	0.386	0.067	0.319	0.827

From Abbott et al. (1992).

type (Table 7.3). However, the maintenance of the difference in allele frequencies between radiate and non-radiate individuals in mixed populations is a surprise given that it is known that outcrossing between the two types can reach 35% (Marshall and Abbott 1984). Abbott, Irwin, and Ashton (1992) suggested that differential esterase diversity may reflect a breeding barrier between the two types due to reduced fitness of the products of intertype crosses compared to intratype crosses. This hypothesis was also invoked by Oxford, Crawford, and Pernyes (1996) to explain the association between capitulum type and suites of developmental, morphological, and reproductive characters in some radiate *S. vulgaris* populations. In contrast, *S. eboracensis* has a high degree of reproductive isolation from its parents, hence the decision to consider it a species rather than a variety (Lowe and Abbott 2003).

However, genetic diversity analyses alone provide little evidence as to the route by which radiate *S. vulgaris* arose through introgression. Whilst the triploid hybrid, *S. x baxteri*, has been recorded in nature and successfully raised artificially, it is unclear whether this was important in the origin of radiate *S. vulgaris*, since at least two alternative routes are feasible for the production of a fertile tetraploid F_1: (i) fusion of a normally reduced *S. vulgaris* gamete and an unreduced *S. squalidus* gamete; and (ii) fusion of a reduced *S. vulgaris* gamete and a reduced gamete of tetraploid *S. squalidus* (Fig. 7.13). Following an extensive crossing program over three generations and morphometric analysis of 16 morphological and fertility characters, Lowe and Abbott (2000) concluded that hybrid progeny similar in morphology to both radiate *S. vulgaris* and *S. eboracensis* could be produced by any of these routes, although they favored the involvement of unreduced *S. squalidus* gametes in the formation of tetraploid introgressants. Furthermore, they showed that fertile hybrid progeny could be produced within two generations and concluded that the potential exists for multiple origins of radiate *S. vulgaris* in the British Isles.

7.3.2 What is the origin of *Senecio cambrensis*?

Senecio cambrensis, one of only a few endemic plants in the British Isles, was described by Rosser in 1955 based on material collected near Wrexham (North Wales), although it had been known at the site since 1948. Based on colchicine treatment of the synthetic hybrid, *S. squalidus* x *S. vulgaris*, it has been concluded that *S. cambrensis* is a recently evolved allohexaploid, the result of hybridization between *S. squalidus* and *S. vulgaris* followed by chromosome doubling (Rosser 1955, Weir and Ingram 1980). Harris and Ingram (1992), in a restriction site study of nuclear

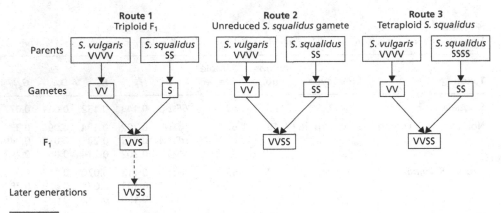

Fig. 7.13

Possible routes for the synthesis of a fertile tetraploid from the interploidy cross *Senecio vulgaris* (tetraploid; V) and *S. squalidus* (diploid; S); genomes are denoted by the letters S and V.

ribosomal DNA, showed that the restriction site profiles of *S. cambrensis* were additive between those of the two parents. Since *S. squalidus* and *S. vulgaris* did not come into contact with each other in North Wales until the early twentieth century, Ingram and Noltie (1995) concluded that *S. cambrensis* originated near Wrexham sometime between 1925 and 1948. *Senecio cambrensis* is morphologically intermediate between its two parents but has larger achenes, pollen with four, rather than three, pores, and is self-compatible. However, *S. cambrensis* is morphologically extremely plastic (Ingram and Noltie 1984), for example, capitula may be non-radiate, short-, intermediate- or long-rayed.

In 1966, *S. cambrensis* was found in Mochdre (c. 40 km NW of Wrexham), in 1982 it was discovered in Edinburgh (c. 300 km NE of Wrexham; Abbott, Ingram, and Noltie 1983) and between 1972 and 1980 it also occurred in Ludlow (c. 75 km SSE of Wrexham; Ingram and Noltie 1995). The species' disjunct distribution raised the possibility that either the Edinburgh population was established as the result of long-distance dispersal from the Welsh populations or the species had arisen more than once (Abbott, Noltie, and Ingram 1983). However, the absence of the species from suitable, intervening localities between the Scottish and Welsh sites is compelling evidence against the long-distance dispersal hypothesis.

Using a combination of biparentally inherited nuclear markers and maternally inherited chloroplast markers, hypotheses of long-distance dispersal and multiple origins can be tested. If the Edinburgh populations of *S. cambrensis* were the result of long-distance dispersal, similar nuclear and cytoplasmic genotypes would be expected at the two localities. If dual origins were responsible for the disjunct distribution, then plants from the two localities would be expected to have different genotypes due to different parental genotypes being involved in the crosses.

Ashton and Abbott (1992) studied polymorphism in allozyme phenotypes at three loci (*Acp*-1, *αEst*-1, *Aat*-3) in 16 populations of *S. cambrensis* and its putative parental taxa (Table 7.4). *Acp*-1 shows three codominant alleles, one of which is null (n; an allele that produces a non-functional allozyme), whilst *αEst*-1 has two codominant alleles. In Wales, *S. cambrensis* was monomorphic for phenotypes *Acp*-1a and *αEst*-1a, whilst in Edinburgh all the individuals had phenotypes *Acp*-1ab and *αEst*-1b. These results were interpreted as indicating differences in parentage of Welsh and Scottish *S. cambrensis*, rather than long-distance dispersal. In Scotland, *S. cambrensis* is thought to have originated from a cross between *S. vulgaris* with an *Acp*-1a allele and *S. squalidus* with an *Acp*-1b allele, whilst in Wales, *S. squalidus* donated either *Acp*-1a or *Acp*-1n. Since *S. squalidus* does not possess the *αEst*-1 locus, the phenotype for this enzyme is determined entirely

Table 7.4

Acp-1 and α*Est*-1 phenotype frequencies in six *S. cambrensis* and 10 parental populations.

Taxon	Popn. no.	Sample size	Acp-1				αEst-1		
			a	ab	b	n	a	ab	b
S. vulgaris (non-radiate)									
Edinburgh	1	48	1.00	0.00	0.00	0.00	0.63	0.04	0.33
Wales	5	213	1.00	0.00	0.00	0.00	0.71	0.01	0.28
S. squalidus									
Edinburgh	1	44	0.00	0.00	1.00	0.00	Locus absent		
Wales	3	111	0.27	0.33	0.35	0.05	Locus absent		
S. cambrensis									
Edinburgh	1	28	0.00	1.00	0.00	0.00	0.00	0.00	1.00
Wales	5	163	1.00	0.00	0.00	0.00	1.00	0.00	0.00

From Ashton and Abbott (1992).

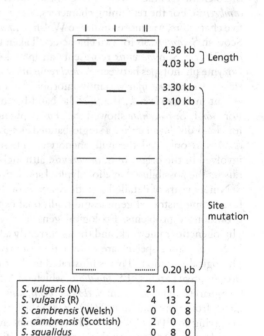

Fig. 7.14

*Cla*I digestion of *Senecio* cpDNA and polymorphism detection with the lettuce probe C6, to illustrate detection of the three most common cpDNA haplotypes in *S. vulgaris*, *S. cambrensis*, and *S. squalidus* (data from Abbott et al. 1995 and Lowe and Abbott 1996). N is non-radiate and R is radiate.

	I	II	III
S. vulgaris (N)	21	11	0
S. vulgaris (R)	4	13	2
S. cambrensis (Welsh)	0	0	8
S. cambrensis (Scottish)	2	0	0
S. squalidus	0	8	0

by the *S. vulgaris* parent. Thus, in Scotland, *S. vulgaris* must have donated the α*Est*-1b allele and in Wales the α*Est*-1a allele.

Using a heterologous cpDNA probe, derived from lettuce, three haplotypes (Types I, II, and III) can be distinguished in the maternally inherited cpDNA (Fig. 7.14; Abbott, Curnow, and Irwin 1995), based on two mutations, a 330 bp length mutation and a *Cla*I restriction site mutation (3.1 kb + 0.2 kb ↔ 3.3 kb). Harris and Ingram (1992) showed that the 330 bp length mutation occurred in the Welsh populations of *S. cambrensis* but was absent from Scottish populations.

Lowe and Abbott (1996), in a more detailed analysis of the geographical distribution of the length mutation among the parents of *S. cambrensis*, showed that it was absent from British *S. squalidus* populations (although it does occur in *S. squalidus* from southwestern Europe; Abbott, Curnow, and Irwin 1995) and from British populations of non-radiate *S. vulgaris*. However, it did occur at very low frequency in Glaswegian populations of radiate *S. vulgaris*. These data provide strong support for the dual origin of *S. cambrensis* in Wales and Scotland, although there is still no clear evidence to indicate how the length mutation was acquired in the Scottish material. Unfortunately, Scottish *S. cambrensis* is now extinct due to redevelopment of its habitat (Abbott and Forbes 2002). Ashton and Abbott (1992) provided limited evidence, based on allozyme phenotype variation at the *Aat*-3 locus, for an additional origin of *S. cambrensis* at Mochdre, and showed that natural resynthesis of the allohexaploid may be a relatively common event.

Senecio teneriffae, like *S. cambrensis*, is a hexaploid, but it is endemic to the Canary Islands. Both species are morphologically similar and may be artificially crossed to produce fertile F1 hybrids (Lowe and Abbott 1996). These data led Lowe and Abbott (1996) to make the radical suggestion that *S. cambrensis* was established following the introduction of *S. teneriffae* to the British Isles. In order to test this hypothesis, cpDNA, nuclear ribosomal DNA, morphological, and allozyme variation were surveyed in comparison to *S. cambrensis* and its parents. The cpDNAs and rDNAs of *S. teneriffae* were identical to those of *S. cambrensis*, with *S. teneriffae* possessing the 330 bp length mutation similar to Welsh *S. cambrensis*. Morphometric analysis showed that for 10 out of 26 characters there were no significant differences between the means of *S. teneriffae* and *S. cambrensis*. For the remaining characters, *S. teneriffae* differed significantly from *S. cambrensis* for five characters, was more similar to Welsh *S. cambrensis* for seven characters, and more similar to Scottish *S. cambrensis* for two characters. Taken together this data would support the hypothesis that Welsh *S. cambrensis* represents an introduction of *S. teneriffae*. However, differences in isozyme phenotypes between *S. cambrensis* and *S. teneriffae* were evident at loci βEst-3, *Aco*-1, and *Aco*-3. *Senecio teneriffae* was monomorphic for a double-banded phenotype (βEst-3a′b) that was absent in *S. cambrensis*; the βEst-3a′ band has not been found in other British *Senecio* species. For *Aco*-1, *S. teneriffae* showed the *Aco*-1b phenotype that was not present in *S. cambrensis* and for *Aco*-3 displayed either a single-banded *Aco*-3a phenotype or a null phenotype, whereas all *S. cambrensis* only had the null phenotype. These data indicate that *S. teneriffae* has not been involved in the origin of *S. cambrensis*, although it appears to have originated in a similar way, raising the possibility that allopolyploid speciation has occurred on an oceanic island.

Ninety years of detailed genetic research on European *Senecio* species has shown the value of integrating historical genetic and traditional experimental ecological studies for understanding evolutionary problems. Ecological genetic research should be undertaken within a rigorous phylogenetic framework, and this is currently a limitation of comparative evolutionary research in *Senecio*, since species are often either poorly delimited or show complex patterns of morphological variation. This is illustrated in the case of *S. aethnensis* and *S. chrysanthemifolius*, the presumptive parents of *S. squalidus*, which on Mt. Etna (Sicily) show a cline of variation through an apparent hybrid swarm: *S. chrysanthemifolius* occurs to approximately 1000 m asl and *S. aethnensis* from approximately 1600–2600 m asl, with the hybrid swarm between the two species occurring at 1300 ± 300 m asl. However, some authorities consider that a single variable species, *S. aethnensis*, encompasses this variation (e.g., Pignatti 1982). Whereas in the case of the aureoid *Senecio* complex of North America, following a cpDNA analysis, Bain and Jansen (1996) suggest that this complex of some 23 taxa may represent either a single, polymorphic biological species, or that the species limits are confounded by hybridization and introgression. These problems are amplified by the absence of a generic monograph for the whole genus and the lack of detailed, explicitly phylogenetic hypotheses of species relationships; systematic treatments of the genus are usually geographically restricted, for example, Barkley (1978) on the North American members of the genus *Senecio*. Such restrictions to the focus of research programs are necessary because of the sheer size of the genus, the apparent frequency of hybridization, and the complexity of variation patterns.

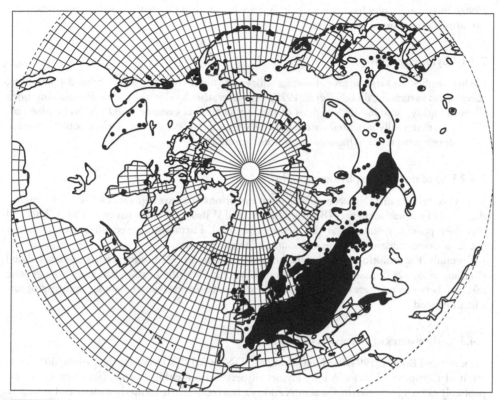

Worldwide distribution of *Ursus arctos* (brown bear).

7.4 Brown bears

Ursus arctos, the brown bear, is the most widespread of the six extant *Ursus* species. It is a Holarctic species, occurring in Europe through to Asia and North America (Fig. 7.15). Not surprisingly given this extremely broad distribution, the species is found in a wide variety of habitats, from the arid areas of China and Turkey, through temperate rainforests and boreal woodland to taiga and Arctic tundra (Nowak 1991, Paetkau et al. 1998). Over the last 150 years it has suffered a decline in number and distribution (Servheen 1989).

Decline in numbers is most extreme in western Europe where human persecution has reduced the brown bear to a few populations containing small numbers of individuals. A similar decline has been recorded in the conterminous United States. Where the species was probably widespread in the early 1800s, it is now restricted to six isolated populations in the states of Idaho, Montana, Wyoming, and Washington (Servheen 1989). The decline is mainly as a result of the western spread of European settlers from the early 1800s. By comparison, in other areas of its worldwide range, the current distribution is likely to be the same as that of recent historical time, having been little disturbed by human activity.

With the combination of its increasing rarity, its large size (it is the largest living carnivore), and general public appeal, it is not surprising that the species has been the subject of considerable research using molecular markers, including systematics (e.g., Talbot and Shields 1996), estimation of population size (e.g., Taberlet et al. 1997), and inheritance studies (e.g., Cronin et al.

1999). Phylogeography has been a significant area of study along with gene flow and it is these two areas that are summarized here.

7.4.1 Phylogeography

Earlier studies on large mammals using allozymes, including the closely related polar bear, revealed no variation (Allendorf et al. 1979). However, the development of PCR-based methods allowed highly variable DNA regions of the mitochondrial genome (mtDNA) to be assessed. While the maternally inherited nature of mtDNA needs to be considered when conclusions are being drawn (chapter 2), it has some significant advantages.

7.4.2 Use of mtDNA in phylogeography

Firstly, vertebrate mtDNA accumulates point mutations at an average rate of 5–10 times higher than that of nuclear sequences (Brown, George, and Wilson 1979, Brown et al. 1982). This will therefore provide resolution at the intraspecific level. Furthermore, recombination does not occur, so point mutations are the sole source of any variation.

Secondly, the evolutionary rate of the homologous human DNA sequence has been calculated (Vigilant et al. 1989). Assuming a similar rate of evolution in the *Ursidae*, this allows the genetic distance between two sets of data to be calibrated and an approximate date for splitting of lineages deduced.

7.4.3 Early studies of European variation

Taberlet and Bouvet (1994) used mitochondrial DNA (mtDNA) to assess the genetic differentiation of European *U. arctos*. A long history of persecution within Europe (for instance, it was probably extinct in Britain by the tenth century), has reduced the formerly widespread animal to a few populations. Remaining populations are found predominantly in mountain ranges: the Cantabrians, the Pyrenees, the Alps, and the Apennines. In some of these populations there are likely to be fewer than 10 individuals. By comparison, Scandinavia and eastern Europe (the Balkans, Carpathians, and Russia) host larger populations.

Taberlet and Bouvet (1994) sequenced the mtDNA control region of 60 brown bears, sampled from across the extant geographic range of the species in Europe. Population sizes varied between one and 13 individuals but the majority had between one to four individuals.

There was no genetic variation within individual regions. However, there is variation between regions, which could be partitioned into two distinct phylogenetic groups (or clades) corresponding to lineages distributed in western and eastern Europe. Furthermore, the western lineage could be further geographically separated into that typical of Iberia and of Italy. Calibrating the differences in the base sequence using the comparable data from humans suggested that the initial split between the western and eastern European groups occurred 0.85 Mya (million years ago), with the western lineage diverging further 0.32 Mya. It was argued that the close correspondence between genetic variation and geographic pattern was likely to be explained by events following the end of the last ice age.

Northern and central Europe was covered with an extensive ice-sheet during the last glaciation (the Pleistocene), which clearly would have precluded the persistence of any large mammals. It is likely that large animals would have survived the glaciation in refugia south of the ice-sheet, only recolonizing more northerly lands following retreat of the ice 16,000 BP, although this pattern may have been reversed in the cooler Younger Dryas period (11,000–10,000 BP). In Europe these refugia would have been in Spain (Iberia), Southern Italy, or the Balkans.

Using the molecular clock data for the homologous human DNA sequence implies that the genetic differences between the clades arose prior to the glaciation. It seems that the refugia hosted genetically different brown bear populations and the patterns now observed represent the

Fig. 7.16

Proposed European spread of *Ursus arctos* following
the end of glaciation (from Hewitt 1999).

migratory routes taken by the species as the climate ameliorated. Thus, the bears of the Iberian refugia moved into France and then radiated out into central Europe, Scandinavia, and Britain – this represents the western European (Iberian) clade (Fig. 7.16). The group forming the eastern European clade moved north from the Balkans into central Europe and then into Scandinavia (meeting the western clade). The Italian branch of the western clade is restricted in its distribution. Possibly individuals in this region were prevented by glaciation in the Alps from migrating northward until these areas had already been occupied by bears from other refugia. This is different from the pattern of post-glacial colonization observed in *Quercus* (section 7.5). However, it has been observed in other species (e.g., *Sorex araneus*, Fumagalli et al. 1996; *Fagus sylvatica*, Demesure, Comps, and Petit 1996) and is identified by Hewitt (1999) as one of the three broad European phylogeographic patterns.

7.4.4 Supporting studies of European variation

Two other studies on European *U. arctos* using mtDNA were also published shortly after the work of Taberlet and Bouvet (1994). Randi et al. (1994) surveyed genetic variation in Italian populations from the Alps and the Apennines and compared them primarily with Croatian individuals. Although the sample sizes were small (maximum 10 individuals per population, and 21 samples in all; Table 7.5), they probably represented 10–20% of the Apennine population. Kohn et al. (1995) similarly looked at a larger Italian Alpine population and compared this with Eastern European and Swedish samples, totalling 98 individuals. Like Taberlet and Bouvet (1994), both studies utilized the variation in the control region (or D-loop) of the mtDNA. Randi et al. (1994) also sequenced the cytochrome b gene. This has similar characteristics to the other mitochondrial genes previously described.

Kohn et al. (1994) were broadly in agreement with the findings of Taberlet and Bouvet and recognized a distinct western and eastern European clade, although they recorded little evidence for separation within the western clade. Moreover, they also noted that some Romanian populations contained haplotypes from both European lineages. This was the first record of such intrapopulational variation for *Ursus*, although it may simply be a product of the more intense sampling in a particular location of the study. Similarly, the monomorphy observed in the two populations studied by Randi et al. (1994) may have been due to the stochastic effects experienced by small populations, even though the phylogeographic patterns observed were concordant with that of Taberlet and Bouvet (1994).

Table 7.5

Phylogeographic studies of *Ursus arctos* using mtDNA.

Area	No. samples in survey	Region of mt genome	Authors
Europe/Scandinavia	60	Control region	Taberlet and Bouvet (1994)
Europe/Scandinavia	98	Control region	Kohn et al. (1994)
Europe (Italy and Croatia)	21	cyt-b, control region	Randi et al. (1994)
Alaska, E. Siberia, and Turkey	166	cyt-b, tRNA^{pro}, tRNA^{thr}	Talbot and Sheilds (1996)
Alaska, USA, and Canada	317	Control region	Waits et al. (1998)
Japan	2	Control region	Masuda et al. (1998)
Japan	56	Control region	Matsuhashi et al. (1999)

7.4.5 Phylogeographic variation across the rest of the species range

Further phylogeographic studies, also utilizing mtDNA, have analyzed the variation of *U. arctos* across other parts of its range including Japan, Eastern Asia, Alaska, Canada, and the conterminous United States (Table 7.5), and have revealed a broad-scale phylogeographic pattern with some finer scale complications at the local level.

7.4.6 Variation in Japanese bears

Japanese bears are very closely associated with the eastern European lineage. Migration from Beringian refugia, the ice-age landmass connecting North America and Siberia, has been cited as the source of this genetic similarity between Japan and eastern Europe (Masuda et al. 1998). Tibetan bears are also located in this clade. By comparison, Mongolian individuals have a closer affinity with the western European lineage.

Within Japan, where bears are only found on Hokkaido Island, there is variation in the samples correlated with isolation-by-distance. The three lineages found on the island correspond to different regions of forested uplands. The most recent separation between lineages is estimated to be 300,000–800,000 years ago. This is notable given that the bears had the ability to migrate freely across the island until agricultural intensification led to isolation of the uplands at the end of the nineteenth century. Matsuhashi et al. (1999) accounted for this pattern as post-glacial invasion of the three lineages across land bridges, with geographic separation maintained by low female dispersal distances.

7.4.7 Variation in North America

The clades found within North America follow a geographic pattern (Fig. 7.17). Individuals with haplotypes closely allied to the eastern European group are recorded in regions of western and central Alaska, identified as Clade II (Talbot and Shields 1996, Waits et al. 1998). In addition, three other major clades are recorded within the continent. Clade I is restricted to the Admiralty, Baranof, and Chichagof Islands of south-eastern Alaska (ABC Islands), Clade III is distributed in eastern Alaska and northern Canada, and Clade IV is located in south Canada and the three areas of the United States where *U. arctos* is still found.

Given that even the most recent separation between these two clades (245,000–310,000 years ago between Clades II and III) predates the end of the last glaciation in North America, and that *U. arctos* was probably unknown in North America until 50,000–70,000 years ago (Kurten 1968), patterns of post-glacial colonization have again been invoked to explain this distribution.

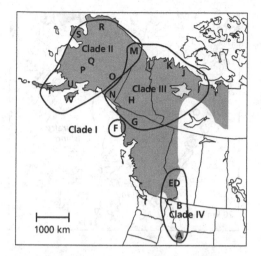

Pattern of mtDNA clade distribution of *Ursus arctos* in North America (Waits et al. 1998). Letters indicate regions where *U. arctos* was sampled. Shading represents current distribution of the species.

The first fossil evidence for the presence of *U. arctos* in North America is from an Alaskan fossil prior to the last glaciation, and it is generally considered that the cordilleran ice-sheet prevented further southward migration until the post-glacial period (Pasitschniale-Arts 1993). However, fossil bone of *U. arctos* located in the Alexander Archipelago dated from 35,000 BP, at the height of the glacial maximum supports the idea that the shorelines and coastal islands provided glacial refugia for some species. The bears of these refugia may have been ancestors of the extant ABC bears which, given the degree of divergence between this and other clades, suggests that these bears represent the oldest lineage in the New World (Talbot and Shields 1996, Waits et al. 1998). Moreover, this genetic separation is also correlated with morphological and behavioral traits.

By comparison the other Alaskan bears may be of more recent origin (suggested by the genetic similarity to the Siberian and Turkish individuals). These have migrated from the Beringian refugia. The two subclades elsewhere in Alaska also represent a broad ecological split between bears of tundra in the west and taiga in the northeast.

The view of the current pattern of mtDNA variation in North America being the product of successive waves of migration into the continent following glaciation has been questioned by the work of Leonard et al. (2000), following study of mtDNA variation in seven bears preserved in the Alaskan permafrost. In the oldest specimens (34,000–42,000 BP), predating the last glacial maximum, three clades are recorded (Clades II, III, and IV), two of which (II and IV) are not found in present-day populations in the area. By comparison samples retrieved from 15,000 BP at the end of the last ice age have only Clade III, the variant recorded in living bears. Hence, preglacially a genetically diverse population is observed that contrasts with the monomorphy of present populations. Leonard, Wayne, and Cooper (2000) have suggested that this pattern may have less to do with migration patterns but instead is due to genetic drift as populations are reduced and fragmented with the onset of glaciation followed by lineage sorting, whereby, due to stochastic processes, only one mitochondrial line survived in any particular area.

7.4.8 Gene flow

In addition to the potential role of stochastic processes influencing the global patterns of genetic diversity in *U. arctos*, there is also the possibility that the geographic patterns observed are a product of sex-biased dispersal (e.g., Paetkau, Sheilds, and Strobeck 1998), with the females moving

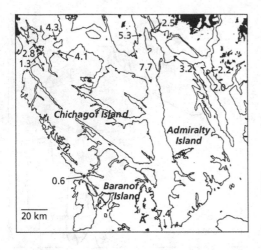

Fig. 7.18

Location of Admiralty, Baranof, and Chichagof
(ABC) islands. Figures show distances (km) between
the islands (from Paetkau et al. 1998).

little compared to the more mobile males. The maternally inherited nature of the mitochondrial
genome would ignore the contribution of males to the spread of the species. If the two sexes dispersed equal distances this would be unimportant, however, if, as suspected, this is not the case it
becomes an important factor. Thus the relative contribution of both sexes to gene flow and the
pattern of geographic distribution revealed by nuclear genes are important areas of study.

Field evidence for shorter dispersal distances and greater fidelity to a more restricted home
range in female bears was documented by Canfield and Harling (1987). Similar findings had
been revealed by Blanchard and Wright (1991). Therefore, the degree of gene flow indicated
by mtDNA will give an underestimate of the movement of the species. Consequently, a nuclear
marker inherited from both parents is required. In addition, this must have sufficient variability
in *U. arctos* to address detailed study of population structure and gene flow. Hence questions of
the relative movement of males and females can be investigated.

Microsatellites fulfill these criteria (Paetkau et al. 1997), and Paetkau et al. (1998) used the
markers to investigate further the apparent genetic separation in the Alaskan populations based
on mtDNA revealed by Waits et al. (1998). In particular, the study focused on the brown bears of
the ABC Islands and the Kodiak archipelago. As noted above the ABC bears represent a unique
mtDNA clade that diverged long before the last glaciation. The Kodiak bears were not distinct
genetically (using mtDNA) but were morphologically distinct and had been treated as a subspecies by certain authors (e.g., Kurten and Anderson 1980). In addition these two groups of
bears are found on islands with different degrees of isolation. This presents an opportunity to
investigate gene flow across water via swimming (the seas do not freeze in winter). The Kodiak
archipelago at its closest point is 35 km from the mainland. Baranof and Chichagof Islands are
less than 600 m from each other at their closest point, and 3 km from the mainland. These two
islands are divided from Admiralty Island by the Chatham straits, which is 7 km wide at its narrowest point. Admiralty Island itself is about 2 km from the mainland (Fig. 7.19). In addition to
bears sampled from the four islands, mainland populations covering the ranges of all of the
mtDNA clades were also surveyed. These came from coastal and inland populations (Fig. 7.18).
In total 55 individuals were assayed.

All populations exhibited significant differences in allele frequency except for Baranof and
Chichagof Islands. While Baranof and Chichagof bears can effectively be treated as one group,
they are distinct from Admiralty Island bears. Thus the microsatellite data provides a greater
degree of resolution of the bears of these islands than the mtDNA. In addition, nuclear alleles
exhibited by bears on all three islands are also recorded in mainland coastal populations. This
suggests gene flow across the water between these islands and the mainland.

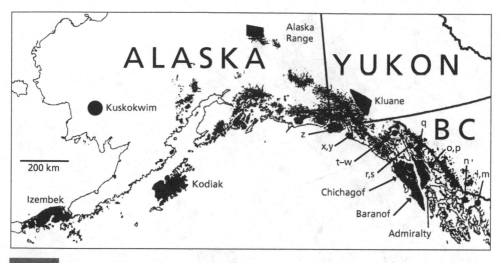

Sites of sampling of *Ursus arctos* in Alaska.

Further evidence for gene flow between the mainland and the ABC Islands is found in the high levels of genetic diversity relative to the number of bears on the islands. For instance, there is much greater variation recorded on Baranof and Chichagof together than on Kodiak Island, areas of comparable size (Table 7.6). It seems most likely that the difference in levels of variation is due to gene flow between the mainland and ABC Islands, which does not occur between the mainland and the Kodiak Island. Given the mtDNA results, it therefore seems that males are responsible for gene flow. However, the clear separation of Admiralty Island from Baranof and Chichagof Islands implies that they will not swim as far as 7 km, by comparison a distance of 600 m presents no obstacle. This is reflected in the genetic similarity between Baranof and Chichagof bears, which is comparable to the genetic similarity between mainland populations separated by similar distances, which are walked rather than swum.

The differentiation of the bears of the Kodiak archipelago from all other island and mainland populations is unsurprising given the 35 km distance between the group of islands and the mainland. The most likely explanation is that individuals colonized the area towards the end of the last (Wisconsin) glaciation. The subsequent isolation and possibly small gene pool allowing drift resulting in their morphological and molecular distinctiveness.

7.4.9 Gene flow in Scandinavian bears

Further discrepancy between male and female movements has been revealed in studies of Scandinavian bears (Waits et al. 2000) using the same microsatellites as the Alaskan study. However, a key difference between the Alaskan and Scandinavian bears is the likely available habitat for migrating individuals. Alaska represents an area where bears are found at a density that is relatively uninfluenced by humans. By comparison the Scandinavian bears are recovering from near extinction due to vigorous hunting programs. Numbers fell from close to 5000 bears in the 1800s to a minimum of 130 bears in 1930 before recovering to its current level of 1000 and which is continuing to expand (Swenson et al. 1994). These bears survived in four Swedish subpopulations (Fig 7.20), the Norwegian representatives being completely eliminated. Consequently movement by bears away from home territories is likely to be into unoccupied areas.

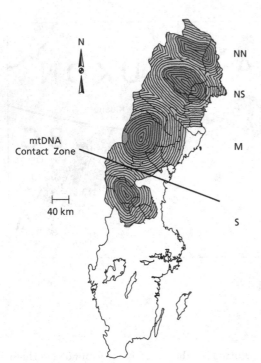

N

NN

NS

mtDNA
Contact Zone

M

⊢—⊣
40 km

S

Fig. 7.20

Location of Scandinavian bear populations
(NN, NS, M, and S) (Waits et al. 2000). The line
between the M and S populations represents the
contact zone between the western haplotype (S)
and the eastern haplotype (M, NS, and NN, see
section 7.4.3).

The expansion of the species into suitable habitats in Scandinavia may be close to the pattern
of spread of the species following the last glaciation. Scandinavia also represents the area where
the European western clade meets the eastern clade (Taberlet and Bouvet 1994), the southernmost Scandinavian population representing the western clade, the other three representing the
eastern clade. A detailed study of the mtDNA contact zone (Taberlet et al. 1995) had shown individuals crossing the zone in both directions. However, the 19 microsatellite loci showed an equal
amount of genetic divergence between the four areas where the bears had persisted. There was
no greater distinction between two groups of different mtDNA clades than comparison between
populations from the same mtDNA clade. Hence this lack of accord between mtDNA and
nuclear DNA markers is further evidence for the greater movement of males than females.

Despite the severe bottleneck which the Scandinavian population had experienced, there is
still a high level of genetic variation within the four populations (measured against the expected
heterozygosity) comparable with levels of variation present in the Alaskan and Canadian populations that have been relatively free of human influence (Table 7.6).

7.4.10 Ecological genetic results and conservation

With the increasing scarcity of the brown bear, conservation is an implicit part of the genetic
work undertaken on the species to date. The geographic patterns revealed by mtDNA within
Europe have led Taberlet and Bouvet (1994) to suggest that individuals from separate clades
should be prevented from breeding by zoos to maintain genetic purity of lineages. This will
prevent any genetic problems for potential future reintroduction programs. Paradoxically this
contradicts the frequently stated modern aim of conservation of conserving genetic variation
(Paabo 2000).

The finer scale variation revealed by microsatellite variation has led to a reconsideration of the
genetic treatment for *in situ* conservation. This in turn is driven by the aims of conservation

Table 7.6

Sample size (N) and expected heterozygosity (H_E) in North American and Scandinavian populations of *U. arctos*. In particular note the levels of variation in small Scandinavian populations, Baranof and Chichagof Islands and Kodiak Island.

Population	N	H_E
Alaska range, Alaska	228	0.78
Kluane, Yukon	50	0.76
Richardson Mountains, NWT	119	0.75
Brooks Range, Alaska	148	0.75
Scandinavia – NS	**108**	**0.70**
Scandinavia – NN	**29**	**0.69**
Flathead River, BC/MT	40	0.69
Scandinavia – S	**156**	**0.68**
Kuskoskim Range, Alaska	55	0.68
Scandinavia – M	**88**	**0.67**
East Slope, Alberta	45	0.67
Admiralty Island, Alaska	30	0.63
Coppermine, NWT	36	0.61
Yellowstone, MT/WY	57	0.55
Baranof and Chichagof Is, Alaska	**35**	**0.49**
Kodiak Island, Alaska	**34**	**0.27**

BC, British Columbia; Is, Island; MT, Montana; NWT, Northwest Territories; WY, Wyoming (from Waits et al. 2000).

genetics raised by Moritz (1994a, b). Two distinct conservation units have been proposed: evolutionarily significant units (ESUs) and management units (MUs). ESUs have been described as populations or groups of populations that have been subject to long-term evolutionary separation (Ryder 1986). In practice, defining evolutionary separation has been difficult. MUs identify a population or populations with distinct allele frequency distributions. As such they identify the area to be considered for monitoring and management of a species. Essentially, ESUs identify the past history of a species and MUs reflect the current genetic pattern.

From the previously described work on brown bear European phylogeography (based on mtDNA), the western and eastern European clades could be viewed as ESUs. However, this approach was not adopted by Waits et al. (2000), who considered the separation of the two mtDNA lineages was as likely to be a product of genetic drift and lineage sorting rather than as a result of separation prior to the last glaciation. Similarly they considered that the microsatellite data should also lead the Scandinavian brown bear to be considered as a single ESU despite fulfilling the criterion of clear allelic differentiation between populations. This differentiation is due to a combination of the high resolution of microsatellites and the recent genetic bottleneck.

Similarly the allelic differences would support a number of separate MUs. Allocating distinct MU status to all or some of these subpopulations implies that they are to be prevented from mixing. However, the genetic data needs to be considered along with a fuller consideration of the broader conservation picture. This is reflected in the current Swedish policy of allowing the expansion of the bears to recover continuous coverage over a major part of its former range. Protection from hunting has allowed the four remaining Scandinavian subpopulations to increase in numbers with an attendant increase in size and range. As this process progresses, increasing mixing between subpopulations will occur initially through male movements, but eventually, it is hoped, by female expansion.

Ursus arctos shows distinct large-scale phylogeographic patterns in mtDNA variation across its current global range. This may be due to expansion of the species range following the last ice age (16,000 BP). In Europe spread has occurred northwards from preglacial refugia in the south of the continent. Similarly a close genetic relationship between some Asian and North America populations may be due to movement across the Beringian land bridge. The possibility, at least in some areas of Alaska, of some regions maintaining populations in ice-free areas adjacent to other areas experiencing glaciation needs consideration. Similarly populations in some areas may have contained more variation preglacially than at present, the loss being due to stochastic processes. Further analysis of individuals preserved in permafrost would also provide more information on patterns of variation.

Finally, the potential bias of the molecular marker, the maternal inheritance of mtDNA in this case, needs to be considered when interpreting phylogeographic patterns. This has led to the utilization of highly variable nuclear markers (microsatellites) for fine-scale gene flow studies. Along with ecological studies these have revealed the low dispersal of female brown bears. This suggests that the bias of mtDNA may be significant in understanding the phylogeographic patterns of the species. The challenge is to incorporate the results of large-scale maternally inherited studies with finer scale results from nuclear markers.

Microsatellite studies have also shown that previously seriously threatened populations may continue to maintain levels of variation commensurate with populations that have not suffered similar reductions in size due to human persecution. The conservation implications (both *ex situ* and *in situ*) of genetic results have been the subject of some discussion (Paabo 2000, Waits et al. 2000). This may be partly due to the aims of conservation genetics itself being developed. While elucidation and explanation of patterns of genetic variation remain important in their own right, the utilization of such information in conservation remains difficult while the aims of the discipline are at a nascent state.

7.5 European oaks

Oaks belong to the genus *Quercus*, which contains more than 500 species and is the most speciose group in the Fagaceae family. About half the species are evergreen and their center of diversity and presumed origin is Asia. All species bear recognizable acorns and these are the oaks' predominant means of propagation. The majority of seed dispersal tends to be within the vicinity of the mother tree and is mediated by gravity, but for many species a sizable proportion of acorns are dispersed over medium to long distances via small mammals and birds engaged in caching behavior. Oaks are monoecious, obligate outcrossing trees with unspectacular wind-pollinated flowers.

Oaks are known as one of the social hardwood species (together with beech), primarily for their tendency to occur in clumped distributions in forest or woodland. Their large dominating tree form and integral position in many wooded ecosystems has meant that many other species (e.g., insects and fungi) have become associated with oaks and rely on them for ecological niches and survival. This central ecosystem role is matched by their economic and social importance to human populations. Oaks have extensive histories of usage both for their timber (house and shipbuilding) and acorns (animal and human food sources) in many modern and ancient civilizations. As an indication of their central anthropological status, oaks are the subject of numerous myths and folklore and were sacred to many ancient peoples including the Greeks and Celts.

In terms of studying genetic variation, phylogeography, and gene flow, oaks are probably the most extensively investigated tree species in the world. This formidable volume of literature, which spans more than 50 years, has shed light on many aspects of historical dispersion, colonization, and contemporary gene flow for oak species. This chapter will predominantly review the literature of European oaks but considers case studies of oak species from other continents where relevant. The subject areas covered in this chapter relate directly to the theoretical and analytical topics described in previous chapters of this book, and are split into three review sections as follows: (i) The historical development of the literature on range-wide genetic variation and phylogeography studies together with implications for postglacial colonization and dispersal dynamics of European oaks. (ii) Studies of population genetic structure at the population level, together with direct estimates of gene flow via pollen and seed. These studies are set in the context of theoretical predictions of genetic structure under different dispersal and gene flow models. (iii) Finally the level of hybridization between oak species, which appears to be a common phenomenon for many species pairs, is reviewed. The level of genetic differentiation and genomic location of markers is discussed in the context of hybridization, ecotypic differentiation, and species concepts.

7.5.1 Range-wide genetic variation and phylogeography

Early analyses of genetic structure over large geographic areas indicated that there was a significant difference in the level and partitioning of variation between the nuclear and organelle genomes of European white oaks. Kremer et al. (1991) found that for nuclear-encoded allozyme markers, only 4% of the total variation was partitioned between French populations of *Quercus robur* L. (pedunculate oak) and *Q. petraea* (Matt.) Liebl. (sessile oak), whereas 91% of the variation was allocated to the intrapopulation component, and indicated that extensive gene flow occurs between populations over a wide geographic scale. For chloroplast DNA variation, which is exclusively maternally inherited in oaks (Dumolin, Demesure, and Petit 1995), 88% of the total variation was partitioned between populations, and indicated little exchange of seeds between populations. Together these estimates indicate that there is a significant imbalance between the levels of pollen- and seed-mediated gene flow between oak populations and were calculated as a proportion of 250:1 respectively by Ennos (1994).

Certainly this was an interesting ecological genetic observation, but the discovery of cpDNA variation within oak populations (largely due to the development of universal PCR markers, e.g., Demesure, Sodzi, and Petit 1995, Taberlet, Gielly, and Bouvet 1991, see chapter 2) and refinements in statistical analysis (e.g., Pons and Petit 1995, 1996), have made possible detailed description of the population structure of organelle genomes. These methodological developments together with the phylogenetically interpretable pattern of variation within cpDNA (Wolfe, Li, and Sharp 1987) appeared to offer a marker with the phylogeographic potential for unravelling species history. Thus, at least for oaks, there was a parallel marker to the mitochondrial genome (mtDNA) of animals, which has proved so useful for phylogeographic studies (see chapter 6, Avise 1994).

Fossil pollen evidence of species history

For plant species, fossil pollen records also offer the possibility to determine the historical geographic distribution of a species, and they can be of tremendous use together with phylogeographic data to provide independent lines of evidence for post-glacial migration patterns. The fossil pollen record for the European flora (including oaks) is particularly good, and has allowed detailed description of vegetation dynamics since the last glacial maximum (LGM, 18,000 BP). Much of the northern part of the continent (British Isles and Scandinavia) was covered in ice sheets, and fossil pollen data suggests that areas south of the ice fields were treeless with abundant Arctic-alpine and steppe vegetation (Godwin and Deacon 1974). Whilst it is not possible to differentiate the pollen of oak species beyond major groupings (e.g., deciduous white oaks, which includes *Q. robur*, *Q. petraea*, and *Q. pubescens*), fossil pollen evidence indicates that oak species were confined to three main Pleistocene refugia in southern Europe, in Iberia (Spain and Portugal), Italy, and the Balkans (Huntley and Birks 1983). Some or all of these areas were also major refugia for much of Europe's other flora and fauna (Taberlet et al. 1998). It is believed that oak started to emerge from these refuge areas as the ice caps began to retreat about 12,000 BP, and were able to colonize as far north as southern Britain and Scandinavia by approximately

10,000 BP (Birks 1989, Godwin 1975). However, the direction of colonization and the influence that the different refugial areas have had on the contemporary oak range is not discernible from pollen data, and requires combination with phylogeographic studies.

Early investigations of cpDNA phylogeographic pattern

Two phylogeographic studies of oaks published simultaneously in 1993 indicated that the pattern of chloroplast variation within Europe has probably been strongly influenced by the postglacial history of the species (Ferris et al. 1993, Petit, Kremer, and Wagner 1993). Both studies identified a major phylogeographic split in white oaks through central France, which the authors ascribed to a contact zone between colonization waves from two separate refugia. In addition, both studies indicated that chloroplast variation appeared to be independent of species boundaries and confirmed that a large amount of "chloroplast capture" had occurred between these oak species due to extensive and recent hybridization (Whittemore and Schaal 1991).

Further refinements of the European Pleistocene history of oaks were made at local (e.g., Britain, Ferris et al. 1995, Denmark, Johnk and Siegismund 1997) and regional scales (Ferris 1996). In addition, two studies of particular importance outlined the pattern of Pleistocene colonization of oak across the whole of western and central Europe (Dumolin-Lapègue et al. 1997, Ferris et al. 1998). In combination, these studies described cpDNA variation from almost 400 populations of eight white oak species, and appeared to corroborate the picture of species history painted by earlier pollen core analysis, described above. Populations sampled from areas identified as refugia by pollen core analysis (Iberian and Italian Peninsulas, and the Balkans) harbored higher levels of cpDNA diversity than more northern populations. Such a pattern of cpDNA variation suggests that refugial areas have remained populated by oaks for a longer time than more northerly regions. In addition, the three refugial areas were genetically differentiated and it was possible to trace these separate refugial lineages across Europe to indicate the routes of colonization that oaks had taken following the retreat of the ice sheets. Oaks from the Iberian Peninsula had colonized western France, the British Isles, and western Scandinavia; oaks from the Italian Peninsula had colonized eastern France, Germany, and central Scandinavia; and material from the Balkans had colonized southern France, eastern European countries (e.g., Hungary, Czech Republic, and Poland), the Baltic countries, eastern Scandinavia (mainly Finland), and probably western Russia.

Recent integration of pollen core literature

At the time, these studies were an innovation in the field of plant phylogeography. However, these results have since been incorporated into a more comprehensive study involving a truly European-wide team of researchers and producing an almost saturated survey of cpDNA variation together with a reanalysis of the available pollen core data from across central and western Europe. The pollen core analysis was based on 600 profiles (Brewer et al. 2002), and supports the earlier findings of Huntley and Birks (1983) that during the glacial maximum oaks were restricted to the southern peninsulas of Iberia, Italy, and the Balkans. The analysis by Brewer et al. (2002) also highlighted some particularly interesting insights into temporal and geographic variation in oak colonization (see Fig. 7.21). The initial stages of colonization took place in the late-glacial interstadial (13–11,000 BP) when oaks spread into central Europe from primary refugia of the peninsulas. It appears that this range expansion was rapid and extensive, and oaks could even have reached areas of northern France and Germany at the end of this period. However, this expansion was halted by a recent cold period, between 8000 and 10,000 BP (known as the Younger Dryas), which appears to have caused a halt in range expansion and even extinction of oak populations in northern areas. During this time oaks appear to have survived in mountainous regions of central Europe (e.g., Alps and Carpathian mountains) outside of the primary refugia and these locations have been identified as secondary refugia. Oak spread into

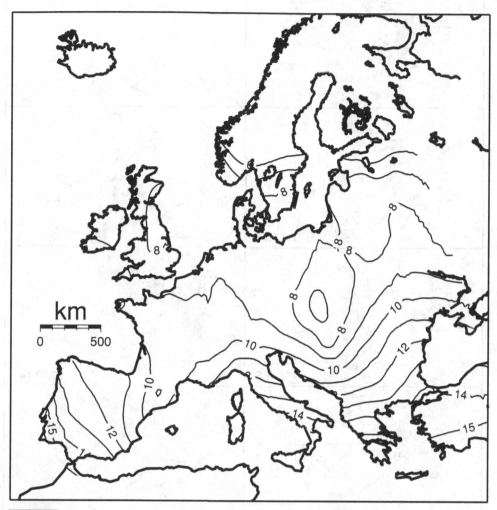

Fig. 7.21

Pollen core record indicating the date that oak pollen was first recorded at sustained levels between 13,000 and 6000 BP (from Brewer et al. 2002). During the last glacial maximum (>13,000 BP) the Iberian, Italian, and Balkan peninsulas are clearly identified as refugia, and following post-glacial climatic warming oaks recolonized central and northern European areas from these refugia. While the pollen core record indicates the timing of events it is not a good indicator of the route the particular oaks took during recolonization (apart from distinguishing overall south/north range change and indicating a leading western edge of colonization).

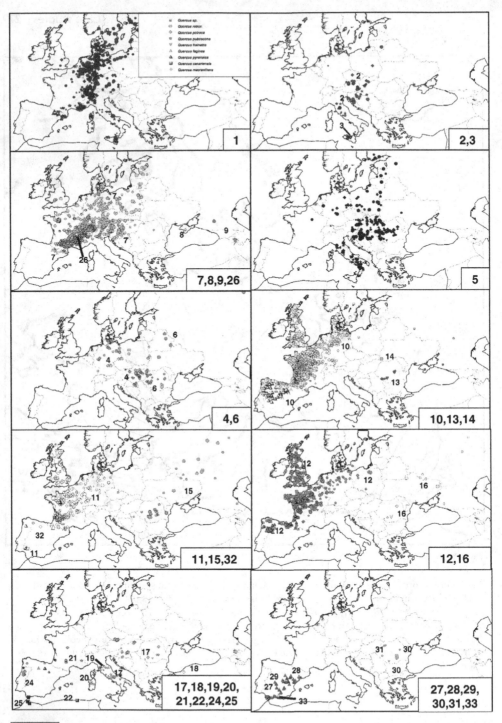

Fig. 7.22

Distribution of chloroplast DNA variation within a total of 2593 oak populations (11,828 individual trees from eight white oak species) were sampled and genotyped from 38 countries in western and central Europe (from Petit et al. 2002b). A total of 45 chloroplast variants were detected and the distribution of the most common types is presented.

central and northern European regions, from its primary refugia and secondary brief interstadial refugia, following climatic stabilization in the Holocene (since 8000 BP). Range expansion was noted to be faster in western compared to central regions, probably due to more rapid warming in coastal areas and physical barriers (e.g., mountain ranges) in central areas. Shifts in temperature and competition with other species are both thought to play a role in the colonization dynamics of oak during this time, but by 6000 BP, oak had reached its maximum European extension (i.e., northern Britain and Scandinavia).

Recent phylogeographic developments

Onto this detailed temporal picture of oak colonization, the results of a combined analysis of cpDNA from 16 European laboratories were superimposed (results were produced mostly as part of a European-funded project called fairoak). The phylogeographic analysis is based on a sample of 2613 populations (12,214 individuals trees and eight oak species) from 37 countries, analyzed for four cpDNA loci (highlighting 45 variants) (Petit et al. 2002a, 2002b). The distribution of cpDNA variants across Europe is shown in Fig. 7.23. Besides adding a tremendous

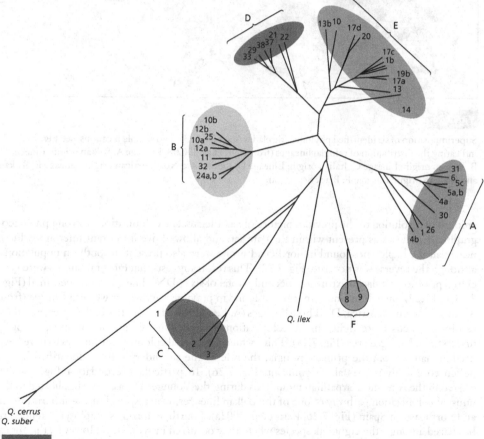

Fig. 7.23

Phylogeny (based on Fitch–Margoliash method) of 42 haplotypes and six main lineages identified for cpDNA variants found during screening of 11,828 individuals from 2593 oak populations (from Petit et al. 2002b). The lineages are labeled A to F (Fig. 7.24) and the distribution of the individually numbered cpDNA variants can be seen in Fig. 7.22.

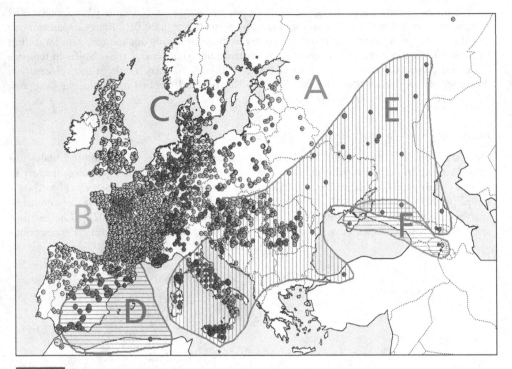

Fig. 7.24

Superimposition of six identified phylogenetic clades (see Fig. 7.23) onto sample locations (see Fig. 7.22), indicating the distribution of refugial lineages (from Petit et al. 2002b). Lineage A: Balkan origin; Lineage B: Iberian origin; Lineage C: Italian origin; Lineage D: Iberian and North African origin; Lineage E: Balkan and Russian origin; Lineage F: Black Sea origin.

amount of resolution to the previous picture of oak Pleistocene colonization, a strong phylogeographic structure was present within the data set, and allowed five important inferences. First, nearly all the haplotypes found in northern Europe were also present in southern populations, although the reverse was not true (Fig. 7.22). This result suggests that most mutations were generated prior to postglacial expansion. Second, a total of six cpDNA lineages were identified (Figs. 7.23 and 7.24), including a new lineage not found in previous surveys for western Europe (from Spain and northern Africa, D). Three lineages (A, E, and F) were from the Balkan area, and their resolution sheds more light on the colonization dynamics in this important and previously understudied refugial area (Fig. 7.25). Third whilst postglacial colonization appears to have been predominantly from the primary refugia, the role of the secondary refugia, identified by the pollen core analysis, was also significant (Fig. 7.26). In particular, secondary refugia, which existed in the Alps and Carpathian mountains during the Younger Dryas, were implicated in the unusual colonization pathway of one of the Balkan lineages, which spread into southern France and northeastern Spain (Fig. 7.26, Petit et al. 2002a). Fourth, chloroplast haplotypes tended to be shared amongst the eight oak species where they occurred in sympatry. However, the ability of certain cpDNA lineages to spread out of refugia may be linked to the colonization and establishment ability of the species possessing those haplotypes. A fifth and final important observation concerned the age of chloroplast variation and arose from analysis of the cpDNA phylogeny (Fig. 7.24). Although a clock for the rate of mutation accumulation in the chloroplast DNA of plants is notoriously difficult to calibrate (Wolfe, Li, and Sharp 1987), it appears that it was not

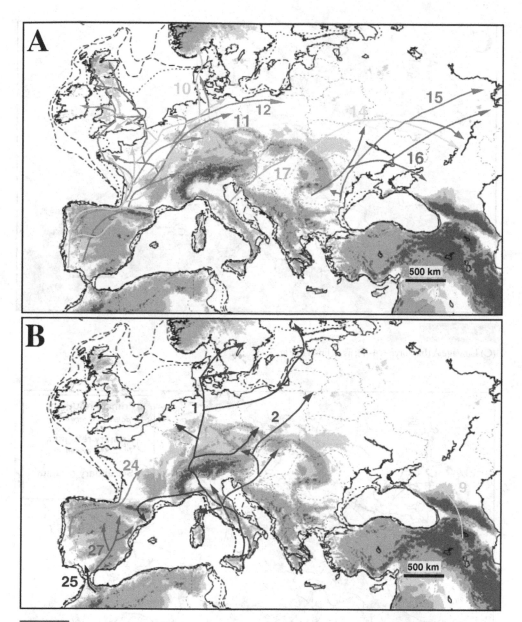

Fig. 7.25

Inferred colonization routes for all frequent haplotypes (from Petit et al. 2002a). Routes for each haplotype (numbered arrows) are influenced by combining timing of events from pollen cone data (Fig. 7.21) and the geographic distribution of haplotypes (Fig. 7.22). (A) Lineages B (haplotypes 10–12) and E (haplotypes 14–17). (B) Lineages B (haplotypes 24–25), C (haplotypes 1–2), D (haplotype 27), and F (haplotypes 9).

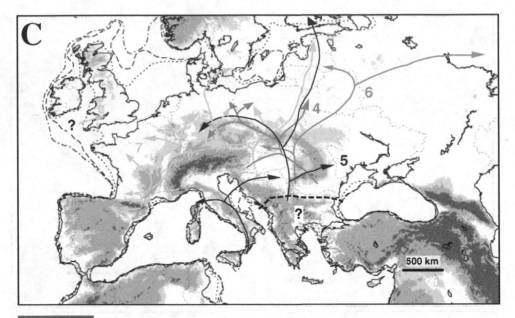

Fig. 7.25 cont.

(C) Lineage A (haplotypes 4–7), as defined in Fig. 7.23.

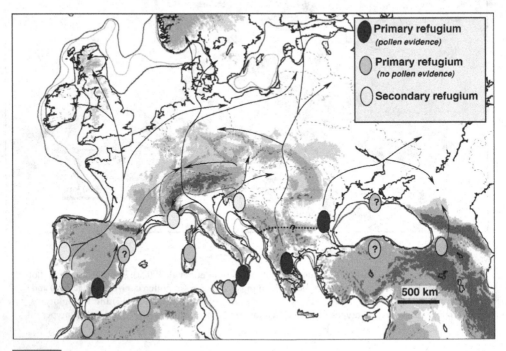

Fig. 7.26

Distribution of primary and secondary oaf refugia, inferred from pollen and/or chloroplast haplotype distribution data, and major post-glacial movements (Petit et al. 2002a).

possible for all the observed variation to have accumulated whilst regions were separated during the last glacial maximum (approximately 100,000 BP). Some of the observed lineage splits are so deep that they must have been isolated for millions rather than tens of thousands of years, and are expected to be the product of isolation during multiple glacial cycles. Vegetation changes before the onset of glacial periods are predicted to be very rapid and lead mainly to the extinction of northern populations rather than invasion of saturated, climax communities already present in southern refugial locations (Hewitt 1996). Under such dynamics, there is expected to be little mixing of different cpDNA lineages during interglacial periods. It should be noted that the maintenance of ancient, ancestral haplotypes alongside more newly arisen ones, has enabled the fine-scale analysis of the postglacial colonization dynamics of oaks, and is a situation that may not be shared by all plant species (Schaal et al. 1998). In addition, from studies of other tree species that have combined molecular, palynological, and computer simulation studies (e.g., alder, King and Ferris 1998; and beech, Demesure, Comps, and Petit 1996), it appears that the routes of colonization followed and the pattern of genetic variation established during migration are particular to the life history of individual species (Comps et al. 2001, Taberlet et al. 1998).

Local-scale pattern of chloroplast DNA variation

The resolution of the European-wide pattern of cpDNA variation was accompanied by a suite of papers detailing patterns of cpDNA at the country or local scales, for example, descriptions of the British Isles (Cottrell et al. 2002), France (Petit et al. 2002c), Italy (Fineschi et al. 2002), Scandinavia (Jensen et al. 2002), and the Iberian Peninsula (Olalde et al. 2002). The results of these papers are too detailed to be summarized here, but many discuss the influence of human activity on the contemporary pattern of cpDNA variation. Despite this caveat, most of the country-wide studies identify large areas of forest that are fixed for a single chloroplast haplotype (some over several hundred square kilometers, Fig. 7.27), and this pattern of cpDNA distribution is predominantly responsible for the high values of differentiation observed for oaks. This pattern of cpDNA structure was also noted in a study by Petit et al. (1997) who speculated that the phenomenon was due to the dynamics of seed dispersal, and referred to these dominant haplotype patches as "footprints of colonization". A series of simulation studies modeling dispersal strategies by Hewitt (1996), Ibrahim, Nichols, and Hewitt (1996), Le Corre et al. (1997), Petit et al. (2001, Fig. 7.28), describe the results of a mode of dispersal called leptokurtic, where most seed dispersal is local but with occasional very long-distance events. After several hundred generations, large patches of the recolonized landscape are dominated by a single genotype and mimic the empirical results of Petit et al. (1997). In addition, haplotype diversity was found to reduce only slowly across the colonized landscape and helps explain why relatively high cpDNA diversity is maintained across northern Europe. Such a model of seed dispersal is realistic for oaks that have seeds that are predominantly gravity dispersed but for which additional dispersal can occur over very long distances by birds (e.g., jays). Thus once long-distance dispersal events have established a small patch of trees possessing a single haplotype ahead of an advancing wave of colonization, local dispersal would increase the size and density of the patch to such an extent that by the time the advancing wave had reached the single haplotype patch it would be almost impenetrable to the influence of external seed influx (although not pollen influx). The establishment and maintenance of this patchy distribution of cpDNA variation thus appear to be a consequence of the peculiar combination of rare long-distance dispersal together with heavy local dispersal and the long generation time of oaks (Le Corre et al. 1997).

Other advances for phylogeographic study

In addition to chloroplast DNA variation, mitochondrial DNA analysis has been undertaken for oaks (Dumolin-Lapègue, Kremer, and Petit 1999, Dumolin-Lapègue, Pemonge, and Petit 1999). Whilst the analysis is not nearly so extensive as for chloroplast DNA, results indicate that

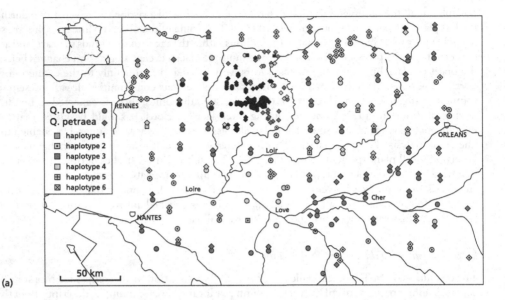

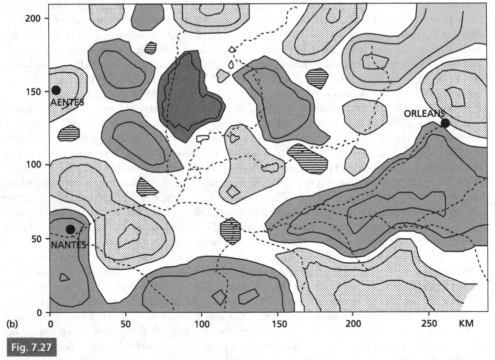

Fig. 7.27

Evidence for "footprints of colonization" in contemporary oak populations. (a) Distribution of six cpDNA haplotypes within forests sampled in an area of north-western France (Petit et al. 1997). (b) Averaging statistics (a method known as kriging) applied to haplotype frequency data in relation to geographic distribution. Areas dominated by a single haplotype (>60%) are indicated at the >60%, >80%, and 100% level (from Petit et al. 1997).

Pure diffusion

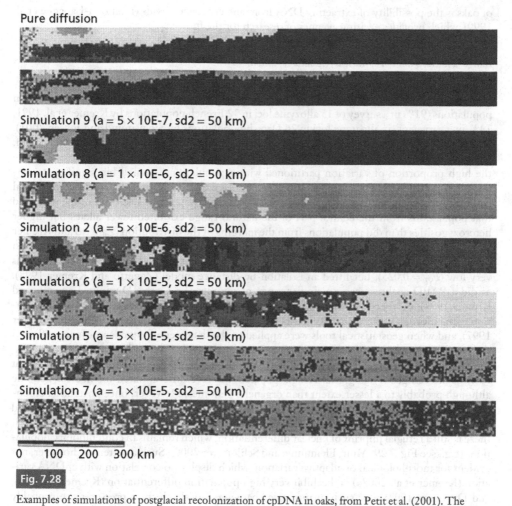

Simulation 9 (a = 5 × 10E-7, sd2 = 50 km)

Simulation 8 (a = 1 × 10E-6, sd2 = 50 km)

Simulation 2 (a = 5 × 10E-6, sd2 = 50 km)

Simulation 6 (a = 1 × 10E-5, sd2 = 50 km)

Simulation 5 (a = 5 × 10E-5, sd2 = 50 km)

Simulation 7 (a = 1 × 10E-5, sd2 = 50 km)

0   100   200   300 km

Fig. 7.28

Examples of simulations of postglacial recolonization of cpDNA in oaks, from Petit et al. (2001). The program is based on cellular automata, with 5 km cells, and 1000 trees at carrying capacity, and was tested against the individual-based model of Le Corre et al. (1997). The simulation plot measures 100 × 1320 km, and the initial condition was an equal mixture of four cpDNA variants in the first 20 km (diversity of 0.75), whereas the remaining cells were empty. The simulations were carried out using the stratified dispersal approach of Nichols and Hewitt (1994), where the probability of the seed density in a given distance is given by: $F = (1 - a)N[0, sd1] + aN[0, sd2]$. The first normal distribution represents the short distance dispersal ($sd2 = 250$ m) and the second represents the long distance dispersal ($sd2 = 50$ km). The parameter a accounts for the proportion between these two distributions. In the pure diffusion model, $a = 0$. In the examples shown, fixation for a single haplotype occurs at 1200 km in the two pure diffusion cases as well as in the simulations characterized by low values of a ($<5 \times 10^{-6}$), but diversity is maintained and a strong patch-like structure is present greater than distances of 1200 km for dispersal profiles with a higher proportion of long-distance (>50 km) events ($>5 \times 10^{-6}$).

mitochondrial variation arises independently of chloroplast variation and is also only maternally inherited. Thus analyses of both genomes may be useful in mapping postglacial colonization pathways, although interpretation of variation in the mtDNA genome may be prone to homoplasy due to the structural characteristics and potential for recombination within this genome (see chapters 2 and 6). Another recent development that may shed light on the historical spread of oaks is the possibility of extracting DNA from ancient macrofossils (Dumolin-Lapègue et al. 1999), which provides exciting avenues of research for the future.

Range-wide distribution of nuclear-encoded variation

In contrast to surveys of organelle DNA, range-wide analysis of nuclear-encoded variation indicates that oaks maintain high levels of diversity but that most variation is partitioned within populations (91% in a survey of 15 allozyme loci in 32 French populations by Kremer et al. 1991; 74% in a survey of six allozyme loci in 26 Danish populations by Siegismund and Jensen 2001). Indirect estimates of gene flow (i.e., Nm, see chapter 4) are consequently extremely high, presumably due to the highly efficient wind-pollinated reproductive mode of the species. Despite the high proportion of variation partitioned within populations, a small but significant component is partitioned between populations and appears to have been influenced by postglacial colonization dynamics. Allozyme analysis of 81 *Q. petraea* populations across Europe indicated that populations from the central part of the natural range exhibited fewer alleles but higher heterozygosities than did populations from the margins of the distribution (Zanetto and Kremer 1995), and the authors speculated that these patterns were due to Pleistocene colonization dynamics. Population differentiation across the range of the sample (Iberia to the Balkans) was very low ($F_{ST} = 0.025$), but fitted an isolation-by-distance model (Zanetto and Kremer 1995), as did a RAPD analysis across a similar range of samples (Le Corre, Dumolin-Lapègue, and Kremer 1997). Further analysis of allozyme and RAPD variation found that there was a correlation with chloroplast variation (Kremer et al. 2002a, Le Corre, Dumolin-Lapègue, and Kremer 1997), and when geostatistical tools were applied to the data, gene frequency variance reflected the postglacial pathways of colonization (Le Corre et al. 1998). These observations suggest that the nuclear genome must have been differentiated between populations inhabiting different Pleistocene refugia. Thus glacial refugia populations were differentiated for nuclear genome although probably to a lesser extent than organelle genomes. However, once postglacial colonization began, extensive gene flow would have led to contact between refugia areas and caused extensive mixing of nuclear genomes (Kremer et al. 2002a). Despite this mixing, it appears that there is still a refugial imprint of nuclear differentiation, which remains in contemporary populations (e.g., see Fig. 7.29, Muir, Flemming, and Schlötterer 2000). Such a pattern is, however, not evident for morphological or adaptive variation, which display no correlation with cpDNA variation (Kremer et al. 2002a) and exhibit very high population differentiation (Kremer, Zanetto, and Ducousso 1997). For such quantitative characters, local selection pressures are probably responsible for the contemporary distribution of genetic variation.

7.5.2 Population genetic structure and contemporary gene dispersion

Alongside almost exhaustive phylogeographic description, oak has been the subject of extensive population genetic study. At the scale of single populations, local gene flow dynamics (seed and pollen dispersal mechanisms), population density (Gram and Sork 1999), regeneration ecology, and selection, all influence genetic structure. Whilst several studies have documented the level of diversity within and between tree populations, only a few have examined fine-scale genetic structure. Part of the reason for the lack of such studies is that sufficiently polymorphic markers and easily applicable statistical methods of analysis have only recently been available (e.g., Berg and Hamrick 1995, Degen 2000, Epperson and Allard 1989, Perry and Knowles 1991).

Patterns of fine-scale genetic structure within oak populations

For oak species, more than 50 polymorphic microsatellite loci have now been isolated (e.g., *Q. macrocarpa*, Dow, Ashley, and Howe 1995; *Q. petraea*, Steinkellner et al. 1997; *Q. robur*, Kampfer et al. 1998). In a study of an intensively sampled plot in France (Petite Charnie), 355 mature *Q.*

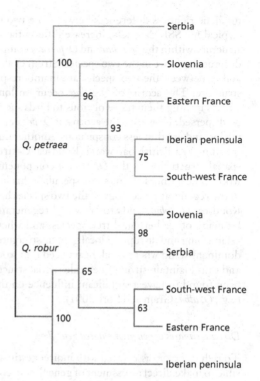

Fig. 7.29

UPGMA dendrogram of 10 European *Q. robur* and *Q. petraea* populations (162 individuals) based on the proportion of shared alleles at 20 microsatellite loci (Muir, Flemming, and Schlötterer 2000). One population from each species was sampled at five locations. Numbers are bootstrap support values. In accordance with the assumption that *Q. robur* and *Q. petraea* are separate taxonomic units, all populations of the same species group together. The separation between the two species is low but supported by high bootstrap values (100%). Similarly, *F*-statistics also indicated a significant differentiation between the two species ($P < 0.01$), but was mainly influenced by allele frequency differences at only three of the 20 loci surveyed. There was also a significant component of variation due to the geographic origin of the populations ($P < 0.01$), and distribution partially corresponds to glacial refugial regions.

(a)

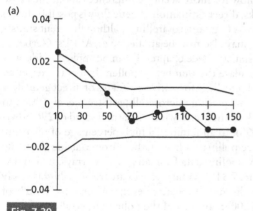

(b)

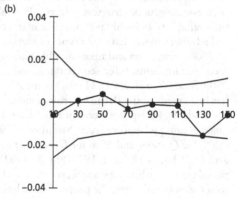

Fig. 7.30

Fine-scale spatial genetic structure and pollen-mediated gene flow within the French forest stand, Petite Charnie of 166 *Q. petraea* trees, 183 *Q. robur* trees and six intermediate mapped individuals (Strieff et al. 1998). Spatial autocorrelation analysis within the wood, where black circles indicate observed values for ordered allele analysis and thin lines are 95% confidence limits for (a) *Q. petraea* (significant spatial structure detected up to 40 m). (b) *Q. robur* (no spatial structure detected).

robur and *Q. petraea* trees were screened for six microsatellite and four allozyme loci (Bacilieri, Labbé, and Kremer 1994, Streiff et al. 1998). For both species, microsatellites highlighted stronger spatial structure than did allozymes, but for both marker sets, the *Q. petraea* component of the wood exhibited significantly higher spatial structure than did the *Q. robur* component (Fig. 7.30). The discrepancy in results obtained using the two marker sets is believed to be mainly due to allelic richness differences between the two marker types, as higher numbers of rare alleles (typical for SSR data sets) increase differentiation estimates. Comparison of fine-scale genetic structure within the *Q. robur* and *Q. petraea* components of the wood hint at interesting biological differences between the two species. Streiff et al. (1998) suggest that differences in seed dispersal ability between the two species are mainly responsible for the observed differences in genetic structure. The acorns of *Q. robur* occur on long peduncles, which project them clear of leaf whorls making them more obvious to birds such as jays (Bossema 1979) and thus more likely to be dispersed. However, the acorns of *Q. petraea* are nestled within the leaf whorls on short peduncles, and are thus less conspicuous. Another microsatellite study of fine-scale genetic structure, this time in a British oak wood (Roudsea, Cottrell et al. 2003), also found significantly stronger spatial structuring in the *Q. petraea* component compared to the *Q. robur* component of the wood. The authors of this study speculate that in addition to seed dispersal differences, variation in the regeneration ecology of the two species has a significant influence on genetic structure. At Roudsea, *Q. robur* tends to grow and regenerate at low density in open woodland surrounded by many other herb and tree species, and where non-local colonizers have a good chance of establishing and disrupting local genetic structure. However, *Q. petraea* tends to occur as a mono-dominant stand where local, heavy seed rain would be likely to swamp out non-local colonizers, and thus maintain strong fine-scale spatial structure. Habitat type and forest structure have also been found to have a significant influence on the spatial genetic structure in other oak species (e.g., *Q. alba*, Gram and Sork 2001).

Direct estimates of pollen-mediated gene flow

To reduce errors associated with indirect estimates of gene flow (see chapters 3 and 4), it is possible to make direct assessment of gene flow events using highly polymorphic molecular markers. Such improvements in gene flow estimation allow for more accurate inferences about the cause of fine-scale genetic structure patterns. For oaks, direct estimation of gene flow is possible due to the availability of highly polymorphic marker sets (e.g., microsatellites), although recent statistical advances mean that other types of markers may also now be applied (e.g., AFLPs, Gerber et al. 2000). Exclusion and maximum likelihood statistics (see chapter 4) can be applied to identify seed contaminants, infer seed parents, and calculate the number of pollen donors (Lexer et al. 2000, Smouse et al. 2001). One of the most biologically informative estimates of direct gene flow has been made at the Petite Charnie stand in France using microsatellites (Streiff et al. 1999). In the study, 984 offspring were collected from 13 mother trees within a natural stand of 296 previously genotyped mature trees (Streiff et al. 1998). In all families, a high percentage of offspring (65% for *Q. robur* and 69% for *Q. petraea*) were pollinated from fathers from outside the study area (5.76 ha; Streiff et al. 1999, Fig. 7.31a). By pooling data from all progeny arrays, pollen dispersal curves within the stand were inferred (Fig. 7.31b). Whilst proximate trees are highly likely to act as pollen donors, the proportion of long-distance pollination events is very high (>60% of pollination events involved fathers more than 200 m away; and the pollen dispersal curves suggest that pollination events could be over several kilometers, Streiff et al. 1999, Fig. 7.31). Overall direct estimates of pollen-mediated gene flow for oaks correlate well with the indirect estimates made from population genetic structure (Zanetto and Kremer 1995). Both types of study indicate a substantial level of pollination involving very long-distance transfer (km not m), which is probably the main mechanism maintaining connectivity between even widely separated populations.

Microsatellite analysis has also been used to examine parentage of seedlings within populations of *Q. macrocarpa* (Dow and Ashley 1996). Of the 100 saplings analyzed, 94 matched at least one of 62 adult bur oaks within the sample stand. In addition, matched saplings tended to grow in dense half-sib clusters around the presumed maternal parent, although some long-distance seed dispersal events were recorded. The mast-seeding (Sork 1993, Sork and Bramble 1993) and secondary transfer of acorns by birds (Bossema 1979, Kollman and Schill 1996) and small mammals

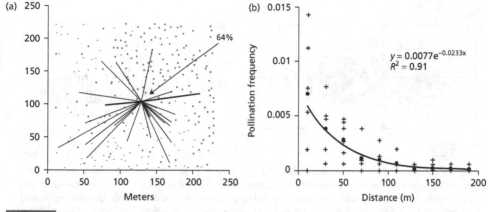

(a) Direction, distance, and frequency (indicated by thickness of line) of pollinations inferred by paternity analysis of open-pollinated progeny derived from a single *Q. petraea* tree. Despite neighboring trees being the most common fathers of the progeny, 64% of fathers could not be identified with the 5-hectare stand and therefore pollinations must be occurring over distances much greater than 100 m. (b) Distribution of mating frequencies as a function of the distance between parental trees. Mating frequencies, that is, number of matings observed over the total number of matings analyzed, are plotted as a function of the distance between the maternal and paternal parents. Black circles represent average values of mating frequencies for each maternal tree (crosses).

(Kikuzawa 1988, Miyaki and Kikuzawa 1988, Sork 1984, Sork, Stacey, and Averett 1983), all have a role in producing this effective reproductive and dispersal strategy for oaks. Seed dispersal estimates, inferred from the genetic structure of maternally inherited markers (cpDNA), and actual measures of seed-mediated gene flow in natural populations appear to be concordant for oaks. Such empirical validation provides further support for results of simulation models on the influence of dispersal on genetic structure.

7.5.3 Species differentiation and hybridization

Despite background knowledge on historical and population processes of European deciduous oaks, the differentiation and classification of species remain problematic and controversial (see chapter 5 for discussion of species definition). Extensive historical and contemporary hybridization between white oaks is indicated from the genetic data, yet species remain morphologically distinct and taxonomically identifiable. The biological species concept defines a species by genetic isolation, that is, those entities that do not exchange genes with one another (Mayr 1963), and appears difficult to apply in the case of oaks. Even Darwin noted the almost continuous nature of variation between European deciduous oaks and for him, as for us today, oaks are a group that represent considerable taxonomic difficulty.

Morphological and ecological differentiation of species

Early work using morphological and genetic differences to distinguish white oak species found such a range of overlap that some workers classified sessile and pedunculate oak as belonging to the same biological species, but representing different ecotypes (Cousens 1963, 1965, Kleinschmit et al. 1995, Rushton 1983). However, recent work has shown that it is possible to distinguish between white oaks using a combination of morphological characters. For example,

(a)

(b)

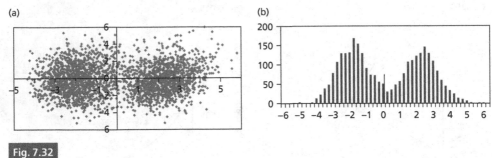

Fig. 7.32

Range of variation for 13 morphometric characters analyzed in 1346 *Q. petraea* trees and 1519 *Q. robur* trees taken from nine exhaustively sampled sympatric stands across Europe. (a) Plot of first two principal components showing morphometric clustering and range of variation of individuals: *Q. robur* (right) and *Q. petraea* (left). (b) Frequency distribution of individuals along first principal component. Axis indicating comparative lack of intermediate morphological types/hybrids between bimodal distribution of *Q. robur* (right) and *Q. petraea* (left).

Bruschi et al. (2000) found that it was possible to distinguish between *Q. petraea* and *Q. pubescens* based solely on micromorphological features (e.g., stomatal and stellate trichomes characteristics). In addition, Kremer et al. (2002b), found that morphological differences for a range of leaf characteristics for *Q. petraea* and *Q. robur* were stable throughout nine sympatric populations sampled from across Europe (Fig. 7.32). The deciduous oak species also differ in their ecological tolerance, and environmental conditions largely determine the distribution of species on a local scale. For example, *Q. robur* tends to occur on poorly drained nutrient-rich soils on the lower portion of slopes, whereas *Q. petraea* occurs on well-drained, nutrient-poor soils, generally found on the upper portion of slopes. Differences in wood anatomy between the two species is also discernible (Feuillat et al. 1997) and may have an ecophysiological and economic relevance due to its relationship with stem shake (Mather, Kanowski, and Savill 1991). Similar environmental adaptations have been seen for other oak species, for example, water availability and herbivory in *Q. rubra* (Sork, Stowe, and Hochwender 1993, Stowe, Sork, and Farrell 1994).

Evidence for reproductive isolation of species

In addition to morphological differences, there are a number of breeding system differences that distinguish white oaks. For example, *Q. robur* and *Q. petraea* are predominantly outcrossing, and whilst interspecific crosses can be forced, natural gene flow between the two species was found to be relatively low and asymmetric. Dengler (1941 in Aas 1991) was the first to note the asymmetric nature of crosses between these two species when he found a greater success rate when pollinating *Q. robur* flowers with *Q. petraea* pollen (15%) than the reciprocal cross (6%). These early observations of low and assymmetric gene flow between these two species were confirmed in later crossing studies (Aas 1991, Bacilieri et al. 1996, Steinhof 1993) and estimates from wild populations (Streiff et al. 1999). Such a pattern of gene flow may reinforce succession between the two species where pedunculate oak is considered a pioneer species and is progressively replaced by sessile oak. The unidirectional nature of gene flow appears to be a consequence of differences in flowering time and probably contributes to the prezygotic isolation of the two species (Aas 1991). Postzygotic breeding barriers for the two species have also been suggested. Rushton (1978, 1979) found that the fertility of morphologically intermediate individuals was significantly decreased relative to pure species types and Kremer et al. (2002b) suggested that ecological selection might remove interspecific hybrids, reinforcing and maintaining pure species types (see chapter 5 for discussion).

Level and partitioning of diversity within oak species

Several studies using molecular markers have noted differences in the level of diversity partitioned between populations of *Q. robur* and *Q. petraea*. Genetic diversity within and differentiation between populations of *Q. petraea* are generally higher than for *Q. robur*, independent of marker type used (allozymes, Kremer et al. 1991; allozymes, Zanetto, Roussel, and Kremer 1994; allozymes and RAPDs, Kleinschmitt et al. 1995; SSRs, Siegismund and Jensen 2001; SSRs and AFLPs, Mariette et al. 2002; cpDNA, Petit et al. 2002a). These differences are explained by dispersal differences and the asymmetric nature of gene flow between the species, where the *Q. petraea* genome can be enriched by effective capture of genetic diversity from *Q. robur* due to pollen swamping.

Differentiation of species using molecular markers

Differences in morphology, ecology, or diversity between these oak species have not, however, translated into consistent differences in molecular marker frequency. Analyses of cpDNA indicate that oak species restricted to common refugia during the last glacial maximum probably hybridized extensively and thus contemporary populations cannot be distinguished on the basis of cpDNA type alone (Petit et al. 2002a, 2002b). Indeed there are several examples of "chloroplast capture" between sympatric oak species that have come into recent contact (Ferris et al. 1993, 1998, Whittemore and Schaal 1991). Perhaps more important is the difficulty in finding nuclear-based molecular markers that differentiate species. An allozyme study conducted on French populations, indicated that only 5% of the variation within the data set was partitioned between the two species, and that species-specific markers were not identified (Kremer et al. 1991), a situation reflected in a single population analysis (Bacilieri, Ducousso, and Kremer 1994). An allozyme study of Danish material (Siegismund and Jensen 2001) indicated that differentiation between the two species was relatively high (F_{ST} was 0.235), although the error rate in assigning species based solely on allozyme variation was also high (10–14%). One of the most extensive searches for genomic regions that differentiate the two species involved screening 2800 RAPD fragments (Bodénès et al. 1997). Only 2% exhibited significant frequency differences and none was specific to either species. Other allozyme, RAPD, AFLP, and SSR studies (e.g., Mariette et al. 2002, Moreau, Kleinschmit, and Kremer 1994, Muir, Fleming, and Schlötterer 2000, Samuel 1999, Zanetto, Roussel, and Kremer 1994) have also found a similar situation, that is, that a small proportion of surveyed markers exhibit significant frequency differences between morphologically classified groups but none were exclusive when applied to a wide range of material (see Fig. 7.29).

The application of genomics to differentiate oak species

The taxonomic debate surrounding *Q. robur* and *Q. petraea* continues. The advent of genomics research has led to genetic maps being constructed using all available molecular and morphological markers, and has resulted in some interesting perspectives on the debate. For *Q. robur*, Barreneche and coworkers (1998) have mapped rDNA, 271 RAPDs, 10 SCARs, 18 SSRs, minisatellites, and six allozymes, and Bakker (2001) has mapped six microsatellites, 343 AFLPs, and 35 morphological and ecological traits. From the results of Bakker (2001), morphological traits were predominantly grouped on only three linkage groups, and large parts of the genome contributed little to the morphological and ecological expression of *Q. robur* (at least for the surveyed characters). It is expected that those "neutral" molecular markers that do distinguish species are probably associated with morphological or ecophysiological characters. Thus, whilst a proportion of the genome (both non-coding and coding) appears to be shared between species due to interspecific hybridization, particular combinations of adaptive genes remain associated either due to fertility consequences (e.g., Rieseberg et al. 1996, Rushton 1978, 1979) or ecological selection pressures (Kremer et al. 2002b). Doubtless the taxonomic debate over white oaks will continue but the reinforcement of taxonomically identifiable types within natural populations favors the maintenance of white oak taxa as true species.

REFERENCES

Aas, G. 1991. Crossing experiments on pedunculate and sessile oak (*Quercus robur* L. and *Q. petraea* (Matt.) Leibl.). *Allgemeine Forst und Jagdzetung*, **162**: 141–5.

Abbott, R.J., Ashton, P.A., and Forbes, D.G. 1992. Introgressive origin of the radiate groundsel, *Senecio vulgaris* L. var. *hibernicus* Syme: *Aat*-3 evidence. *Heredity*, **68**: 425–35.

Abbott, R.J., Curnow, D.J., and Irwin, J.A. 1995. Molecular systematics of *Senecio squalidus* L. and its close relatives. In D.J.N. Hind, C. Jeffery, and G.V. Pope, eds. *Advances in Compositae Systematics*. Kew, UK: Royal Botanic Gardens, pp. 223–37.

Abbott, R.J. and Forbes, D.G. 2002. Extinction of the Edinburgh lineages of the allopolyploid neospecies, *Senecio cambrensis* Rosser (Asteraceae). *Heredity*, **88**: 267–9.

Abbott, R.J., Ingram, R., and Noltie, H.J. 1983. Discovery of *Senecio cambrensis* Rosser in Edinburgh. *Watsonia*, **14**: 407–8.

Abbott, R.J., Irwin, J.A., and Ashton, P.A. 1992. Genetic diversity of esterases in the recently evolved stabilized introgressant, *Senecio vulgaris* L. var. *hibernicus* Syme, and its parental taxa *S. vulgaris* L. var. *vulgaris* L. and *S. squalidus* L. *Heredity*, **68**: 547–56.

Abbott, R.J., James, J.K., Irwin, J.A., and Comes, H.P. 2000. Hybrid origin of the Oxford Ragwort, *Senecio squalidus* L. *Watsonia*, **23**: 123–38.

Abbott, R.J., Noltie, H.J., and Ingram, R. 1983. The origin and distribution of *Senecio cambrensis* Rosser in Edinburgh. *Transactions of the Botanical Society, Edinburgh*, **44**: 103–6.

Allendorf, F.W., Christiansen, F.B., Dobson, T., Eaves, W.F., and Frydenberg, O. 1979. Electrophoretic variation in large mammals, 1 The Polar Bear, *Thalarctos maritimus*. *Heredita*, **91**: 19–22.

Arnold, M.L. 1997. *Natural Hybridisation and Evolution*. Oxford: Oxford University Press.

Ashton, P.A. and Abbott, R.J. 1992. Multiple origins and genetic diversity in the newly arisen allopolyploid species, *Senecio cambrensis* Rosser (Compositae). *Heredity*, **68**: 25–32.

Avise, J.C. 1994. *Molecular Markers, Natural History and Evolution*. New York: Chapman and Hall.

Avise, J.C. 2000. *Phylogeography. The history and formation of species*. Cambridge, Massachusetts: Harvard University Press.

Bacilieri, R., Ducousso, A., and Kremer, A. 1994. Genetic, morphological, ecological and phenological differentiation between *Quercus petraea* (Matt.) Liebl. and *Quercus robur* L. in a mixed stand of northwest France. *Silvae Genetica*, **44**: 1–10.

Bacilieri, R., Ducousso, A., Petit, R.J., and Kremer, A. 1996. Mating system and asymmetric hybridisation in a mixed stand of European oaks. *Evolution*, **50**: 900–8.

Bacilieri, R., Labbé, T., and Kremer, A. 1994. Intraspecific genetic structure in a mixed population of *Quercus petraea* (Matt.) Liebl. and *Q. robur* L. *Heredity*, **73**: 130–41.

Bain, J.F. and Jansen, R.K. 1996. Numerous chloroplast DNA polymorphisms are shared among different populations and species in the aureoid *Senecio* (*Packera*) complex. *Canadian Journal of Botany*, **74**: 1719–28.

Bakker, E.G. 2001. *Towards molecular tools for management of oak forests: genetic studies on indigenous* Quercus robur L. *and* Q. petraea *(Matt.) Liebl. populations*. PhD thesis, University of Wageningen, Alterra Scientific Contributions, Wageningen.

Barkley, T.M. 1978. *Senecio. North American Flora, Ser. II*, **10**: 50–139.

Barreneche, T., Bodénès, C., Lexer, C., Trontin, J.F., Fluch, S., Streiff, R., Plomion, C., Roussel, G., Steinkellner, H., Burg, K., Favre, J.M., Glössl, J., and Kremer, A. 1998. A genetic linkage map of *Quercus robur* L. (pedunculate oak) based on RAPD, SCAR, microsatellite, minisatellite, isozyme and 5S rDNA markers. *Theoretical and Applied Genetics*, **97**: 1090–103.

Berg, E.E. and Hamrick, J.L. 1995. Fine-scale genetic structure of a Turkish oak forest. *Evolution*, **49**: 110–20.

Birks, H.J.B. 1989. Holocene isochrone maps and patterns of tree-spreading in the British Isles. *Journal of Biogeography*, **16**: 503–40.

Blanchard, B.M. and Wright, R.K. 1991. Movements of Yellowstone Grizzly Bears. *Biological Conservation*, **58**: 41–67.

Bodénès, C., Joandet, S., Laigret, F., and Kremer, A. 1997. Detection of genomic regions differentiating two closely related oak species *Quercus petraea* (Matt.) Liebl. and *Quercus robur* L. *Heredity*, **78**: 433–44.

Bossema, I. 1979. Jays and oaks: an eco-ethological study of a symbiosis. *Behaviour*, **70**: 1–117.

Bremer, K. 1994. *Asteraceae: Cladistics and Classification*. Portland, Oregon: Timber Press.

Brennan, A.C., Harris, S.A., Tabah, D.A., and Hiscock, S.J. 2002. The population genetics of sporophytic self-incompatibility in *Senecio squalidus* L. (Asteraceae) I: *S* allele diversity in a natural population. *Heredity*, **89**: 430–8.

Brewer, S., Cheddadi, R., Beaulieu, J.L., Reille, M., and data contributors. 2002. The spread of deciduous *Quercus* throughout Europe since the last glacial period. *Forest Ecology and Management*, **156**: 27–48.

Brown, W.M., George, M. Jr., and Wilson, A.C. 1979. Rapid evolution of animal mitochondrial DNA. *Proceedings of the National Academy of Sciences USA*, **76**: 1967–71.

Brown, W.M., Prager, E.M., Wang, A., and Wilson, A.C. 1982. Mitochondrial DNA sequences of primates: tempo and mode of evolution. *Journal of Molecular Evolution*, **18**: 225–39.

Brummitt, J.M. 1971. Plant records. *Watsonia* **8**: 408.

Bruschi, P., Vendramin, G.G., Bussotti, F., and Grossoni, P. 2000. Morphological and molecular differentiation between *Quercus petraea* (Matt.) Liebl. and *Quercus pubescens* Willd. (Fagaceae) in Northern and Central Italy. *Annals of Botany*, **85**: 325–33.

Bush, G.L. 1994. Sympatric speciation in animals: new wine in old bottles. *TREE*, **9**: 285–8.

Canfield, J. and Harting, A.L. 1987. Home range and movements. In M.N. Le Franc, M.B. Moss, U.A. Patnode, and W.C. Sugg, eds. *Grizzly Bear Compendium*. Washington DC: National Wildlife Federation, pp. 27–35.

Comps, B., Gömöry, D., Letouzey, J., Thiébaut, B., and Petit, R.J. 2001. Diverging trends between heterozygosity and allelic richness during postglacial colonization in the European beech. *Genetics*, **157**: 389–97.

Costa, J.T., McDonald, J.H., and Pierce, N.E. 1996. The effect of ant association on the population genetics of the Australian butterfly *Jalmenus evagoras* (Lepidoptera: Lycaenidae). *Biological Journal of the Linnean Society*, **58**: 287–306.

Cottrell, J.E., Munro, R.C., Tabbener, H.E., Gillies, A.C.M., Forrest, G.I., Deans, J.D. and Lowe, A.J. 2002. Distribution of chloroplast DNA variation in British oaks (*Quercus robur* and *Q. petraea*): the influence of postglacial colonisation and human management. *Forest Ecology and Management*, **156**: 181-96.

Cottrell, J.E., Munro, R.C., Tabbener, H.E., Milner, A.D., Forrest, G.I., and Lowe, A.J. 2003. Comparison of fine-scale genetic structure within two British oakwoods using microsatellites: consequences of colonisation dynamics and past management. *Forest Ecology and Management*, **176**: 287–303.

Cousens, J.E. 1963. Variation of some diagnostic characters of the sessile and pedunculate oaks and their hybrids in Scotland. *Watsonia*, **5**: 273–86.

Cousens, J.E. 1965. The status of pedunculate and sessile oaks in Britain. *Watsonia*, **6**: 161–76.

Cronin, M., Sheilder, R., Hechtel, J., Strobeck, C., and Paetkau, D. 1999. Genetic relationships of Grizzly Bears (*Ursus arctos*) in the Prudhoe Bay Region of Alaska: inference from microsatellite DNA, mitochondrial DNA, and field observations. *Journal of Heredity*, **90**: 622–8.

Degen, B. 2000. *SGS: Spatial Genetic Software. Computer program and user's manual.* http://Kourou.cirad.fr/genetique/software.html.

Demesure, B., Comps, B., and Petit, R.J. 1996. Chloroplast phylogeography of the common beech (*Fagus sylvatica* L.) in Europe. *Evolution*, **50**: 2515–20.

Demesure, B., Sodzi, N., and Petit, R.J. 1995. A set of universal primers for amplification of polymorphic non-coding regions of mitochondrial and chloroplast DNA in plants. *Molecular Ecology*, **4**: 129–31.

Dow, B.D. and Ashley, M.V. 1996. Microsatellite analysis of seed dispersal and parentage of saplings in bur oak, *Quercus macrocarpa*. *Molecular Ecology*, **5**: 615–27.

Dow, B.D., Ashley, M.V., and Howe, H.F. 1995. Characterization of highly variable (GA/CT)n microsatellites in the bur oak, *Quercus macrocarpa*. *Theoretical and Applied Genetics*, **91**: 137–41.

Dumolin-Lapègue S., Demesure, B., and Petit, R.J. 1995. Inheritance of chloroplast and mitochondrial genomes in pedunculate oak investigated with an efficient PCR method. *Theoretical and Applied Genetics*, **91**: 1253–6.

Dumolin-Lapègue, S., Demesure, B., Fineschi, S., Le Corre, V., and Petit, R.J. 1997. Phylogeographic structure of white oaks throughout the European continent. *Genetics*, **146**: 1475–87.

Dumolin-Lapègue, S., Kremer, A., and Petit, R.J. 1999. Are chloroplast and mitochondrial variation species independent in oaks? *Evolution*, **53**: 1406–14.

Dumolin-Lapègue, S., Pemonge, M.-H., Gielly, L., Taberlet, P., and Petit, R.J. 1999. Amplification of oak DNA from ancient and modern wood. *Molecular Ecology*, **8**: 1–8.

Dumolin-Lapègue, S., Pemonge, M.-H., and Petit, R.J. 1999. Association between chloroplast and mitochondrial lineages in oaks. *Molecular Biology and Evolution*, **15**: 1321–31.

Elmes, G.W., Thomas, J.A., Hammerstedt, O., Muguria, M.L., Martin, J., and van Der Made, J.G. 1994. Differences in host-ant specificity between Spanish, Dutch and Swedish populations of the endangered butterfly, *Maculinea alcon* (Denis et Schiff.) (*Lepidoptera*). *Memorabillia Zoologica*, **48**: 55–68.

Ennos, R. 1994. Estimating the relative rates of pollen and seed migration among plant populations. *Heredity*, **72**: 250–9.

Epperson, B.K. and Allard, R.W. 1989. Spatial autocorrelation analysis of the distribution of genotypes within populations of lodgepole pine. *Genetics*, **121**: 369–77.

Ferris, C. 1996. Ancient history of the common oak. *Tree News (Autumn)*: 12–13.

Ferris, C., King, R.A., Väinölä, R., and Hewitt, G.M. 1998. Chloroplast DNA recognizes three refugial sources of European oaks and suggests independent eastern and western immigrations to Finland. *Heredity*, 80: 584–93.

Ferris, C., Oliver, R.P., Davy, A.J., and Hewitt, G.M. 1993. Native oak chloroplasts reveal an ancient divide across Europe. *Molecular Ecology*, 2: 337–44.

Ferris, C., Oliver, R.P., Davy, A.J., and Hewitt, G.M. 1995. Using chloroplast DNA to trace postglacial migration routes of oaks into Britain. *Molecular Ecology*, 4: 731–8.

Feuillat, F., Dupouey, J.L., Sciama, D., and Keller, R. 1997. A new attempt at discrimination between *Quercus petraea* and *Quercus robur* based on wood anatomy. *Canadian Journal of Forest Research*, 27: 343–51.

Figurny-Puchalska, E., Gadeberg, R.M.E., and Boomsma, J.J. 2000. Comparison of genetic population structure of the large blue butterflies *Maculinea nausithous* and *M. teleius*. *Biodiversity and Conservation*, 9: 419–32.

Fineschi, S., Taurchini, D., Grossoni, P., and Vendramin, G.G. 2002. Chloroplast DNA variation of white oaks in Italy. *Forest Ecology and Management*, 156: 103–4.

Fumagalli, L., Hauser, J., Taberlet, P., Gielly, L., and Steward, D.T. 1986. Phylogenetic structures of the Holarchic *Sorex araneus* group and its relationship with *S. samniticus*, as inferred from mt DNA sequences. *Hereditas*, 125: 191–9.

Gadeberg, R.M.E. and Boomsma, J.J. 1997. Genetic population structure of the large blue butterfly *Maculinea alcon* in Denmark. *Journal of Insect Conservation*, 1: 99–111.

Gerber, S., Mariette, S., Streiff, R., Bodénès, C., and Kremer, A. 2000. Comparison of microsatellites and amplified fragment length polymorphism markers for parentage analysis. *Molecular Ecology*, 9: 1037–48.

Godwin, H. 1975. *The History of the British Flora*. 2nd edn. Cambridge: Cambridge University Press.

Godwin, H. and Deacon, J. 1974. Flandrian history of oak in the British Isles. In M.G. Morris and F.H. Perring, eds. *The British Oak*. Cambridge: The Botanical Society of the British Isles, pp. 51–61.

Gram, W.K. and Sork, V.L. 1999. Population density as a predictor of genetic variation for woody plant species. *Conservation Biology*, 13: 1079–87.

Gram, W.K. and Sork, V.L. 2001. Association between environmental and genetic heterogeneity in forest tree populations. *Ecology*, 82: 2012–21.

Harris, S.A. 2002. Introduction of Oxford ragwort, *Senecio squalidus* L. (Asteraceae), to the United Kingdom. *Watsonia*, 24: 31–43.

Harris, S.A. and Ingram, R. 1992. Molecular sytematics of the genus *Senecio* L. I: Hybridisation in a British polyploid complex. *Heredity*, 69: 1–10.

Hewitt, G.M. 1996. Some genetic consequences of ice ages and their role in divergence and speciation. *Biological Journal of the Linnean Society*, 58: 247–76.

Hewitt, G.M. 1999. Post-glacial re-colonization of European biota. *Biological Journal of the Linnean Society*, 68: 87–112.

Hiscock, S.J. 2000. Genetic control of self-incompatibility in *Senecio squalidus* L. (Asteraceae): a successful colonizing species. *Heredity*, 85: 10–19.

Huntley, B. and Birks, H.J.B. 1983. *An Atlas of Past and Present Pollen Maps of Europe, 0–13,000 Years Ago*. Cambridge: Cambridge University Press.

Ibrahim, K.M., Nichols, R.A., and Hewitt, G.M. 1996. Spatial patterns of genetic variation generated by different forms of dispersal during range expansion. *Heredity*, 77: 282–91.

Ingram, R. and Noltie, H.J. 1984. Ray floret morphology and the origin of variability in *Senecio cambrensis* Rosser, a recently established allopolyploid species. *New Phytologist*, 96: 601–7.

Ingram, R. and Noltie, H.J. 1995. *Senecio cambrensis* Rosser. *Journal of Ecology*, 83: 537–46.

Ingram, R., Weir, J., and Abbott, R.J. 1980. New evidence concerning the origin of inland radiate groundsel, *S. vulgaris* L. var. *hibernicus* Syme. *New Phytologist*, 84: 543–6.

Irwin, J.A. and Abbott, R.J. 1992. Morphometric and isozyme evidence for the hybrid origin of a new tetraploid radiate groundsel in York, England. *Heredity*, 69: 431–9.

Jeffrey, C., Halliday, P., Wilmot-Dear, M., and Jones, S.W. 1977. Generic and sectional limits in *Senecio* (Compositae) I. Progress report. *Kew Bulletin*, 32: 47–67.

Jensen, J.S., Gillies, A.C.M., Csaikl, U., Munro, R.C., Madsen, S.F., Roulund, H., and Lowe, A.J. 2002. Chloroplast DNA variation within Scandinavia. *Forest Ecology and Management*, 156: 167–80.

Johnk, N. and Siegismund, H.R. 1997. Population structure and post-glacial migration routes of *Quercus robur* and *Quercus petraea* in Denmark, based on chloroplast DNA analysis. *Scandinavian Journal of Forest Science*, 12: 130–7.

Kaaber, S. 1964. Studies on *Maculinea alcon* (Shiff.) – *rebeli* (Hir) (Lep. Lycaenidae) with reference to the taxonomy, distribution and phylogeny of the group. *Entomologiske Meddelelser*, 38: 277–319.

Kadereit, J.W. 1984. Studies on the biology of *Senecio vulgaris* L. ssp. *denticulatus* (O.F. Muell.) P.D. Sell. *New Phytologist*, 97: 681–9.

Kampfer, S., Lexer, C., Glössl, J., and Steinkelner, H. 1998. Characterization of $(GA)_n$ microsatellite loci from *Quercus robur*. *Hereditas*, 129: 183–6.

Kikuzawa, K. 1988. Dispersal of *Quercus mongolica* acorns in a broadleaved deciduous forest 1. Disappearance. *Forest Ecology and Management*, 25: 1–8.

King, R.A. and Ferris, C. 1998. Chloroplast DNA phylogeography of *Alnus glutinosa* (L.) Gaertn. *Molecular Ecology*, **7**: 1151–61.

Kleinschmitt, J.R.G., Bacilieri, R., Kremer, A., and Roloff, A. 1995. Comparison of morphological and genetic traits of pedunculate oak (*Q. robur* L.) and sessile oak (*Q. petraea* (Matt.) Liebl.). *Silvae Genetica*, **44**: 256–69.

Kohn, M., Knauer, F., Stoffella, A., Schroder, W., and Paabo, S. 1995. Conservation genetics of the European brown bear – a study using excremental PCR of nuclear and mitochondrial sequences. *Molecular Ecology*, **4**: 95–103.

Kollman, J. and Schill, H.P. 1996. Spatial patterns of dispersal, seed predation and germination during colonization of abandoned grassland by *Q. petraea* and *C. avellana. Vegetatio*, **125**: 193–205.

Kremer, A., Dupouey, J.L., Deans, J.D., Cottrell, J., Csaikl, U., Finkeldey, R., Espinel, S., Jensen, J., Kleinschmit, J., Van Dam, B., Ducousso, A., Forrest, I., Lopez de Heredia, U., Lowe, A.J., Tutkova, M., Munro, R.C., Steinhoff, S., and Badeau, V. 2002b. Leaf morphological differentiation between Quercus robur and Quercus petraea is stable across Western European oak stands. *Annals of Forest Science*, **59**: 777–87.

Kremer, A., Kleinschmidt, J., Cottrell, J., Cundall, E.P., Deans, J.D., Ducousso, A., König, A.O., Lowe, A.J., Munro, R.C., Petit, R.J., and Stephan, B.R. 2000a. Is there a correlation between chloroplastic and nuclear divergence, or what are the roles of history and selection on genetic diversity in European oaks? *Forest Ecology and Management*, **156**: 75–87.

Kremer, A., Petit, R.J., Zanetto, A., Fougère, V., Ducousso, A., Wagner, D., and Chauvin, C. 1991. Nuclear and organelle gene diversity in *Quercus robur* and *Q. petraea*. In G. Müller-Starck and M. Ziehe, eds. *Genetic Variation in European Populations of Forest Trees*. Frankfurt-am-Main: Sauerländer-Verlag, pp. 141–66.

Kremer, A., Zanetto, A., and Ducousso, A. 1997. Multilocus and multitrait measures of differentiation for gene markers and phenotypic traits. *Genetics*, **145**: 1229–41.

Kurten, B. 1968. *Pleistocene mammals of Europe*. Chicago: Aldine.

Kurten, B. and Anderson, E. 1980. *Pleiostocene Mammals of North America*. New York: Columbia University Press.

Le Corre, V., Dumolin-Lapègue, S., and Kremer, A. 1997. Genetic variation at allozyme and RAPD loci in sessile oak *Quercus petraea* (Matt.) Liebl. The role of history and geography. *Molecular Ecology*, **6**: 519–29.

Le Corre, V., Machon, N., Petit, R.J., and Kremer, A. 1997. Colonization with long-distance seed dispersal and genetic structure of maternally inherited genes in forest trees: a simulation study. *Genetic Research, Cambridge*, **69**: 117–25.

Le Corre, V., Roussel, G., Zanetto, A., and Kremer, A. 1998. Geographical structure of gene diversity in *Quercus petraea* (Matt.) Liebl. III Patterns of variation identified by geostatistical analyses. *Heredity*, **80**: 464–73.

Leonard, J.A., Wayne, R.K., and Cooper, A. 2000. Population genetics of ice-age brown bears. *Proceedings of the National Academy of Sciences USA*, **97**: 1651–4.

Lexer, C., Heinze, B., Gerber, S., Macalka-Kampfer, S., Steinkellner, H., Kremer, A., and Glössl, J. 1999. Microsatellite analysis of maternal half-sib families of *Quercus robur*, pedunculate oak: II inferring the number of pollen donors from the offspring. *Theoretical and Applied Genetics*, **99**: 185–91.

Lowe, A.H. and Abbott, R.J. 1996. Origins of the new allopolyploid species *Senecio cambrensis* and its relationship to the Canary Islands endemic *Senecio teneriffae* (Asteraceae). *American Journal of Botany*, **83**: 1365–72.

Lowe, A.J. and Abbott, R.J. 2000. Routes of origin of two recently evolved hybrid taxa: *Senecio vulgaris* var. *hibernicus* and York radiate groundsel (Asteraceae). *American Journal of Botany*, **87**: 1159–67.

Lowe, A.J. and Abbott, R.J. 2003. A new British species, *Senecio eboracensis* (Asteraceae), another hybrid derivative of *S. vulgaris* L. and *S. squalidus* L. *Watsonia*, **24**: 375–88.

Mariette, S., Cottrell, J., Csaikl, U., Goikoetxea, P., Konig, A., Lowe. A., Van Dam, B., Barreneche, T., Bodénès, C., Streiff, S., and Kremer, A. 2002. Comparison of levels of genetic diversity detected with AFLP and microsatellite markers within and among mixed *Q. petraea* (Matt.) Leibl. and *Q. robur* L. stands. *Silvae Genetica*, **51**: 72–9.

Marshall, D.F. and Abbott, R.J. 1984. Polymorphism for outcrossing at the ray floret locus in *Senecio vulgaris* L. II. Confirmation. *Heredity*, **52**: 331–6.

Masuda, R., Murata, K., Aiurzaniin, A., and Yoshida, M.C. 1998. Phylogenetic status of brown bears *Ursus arctos* of Asia: a preliminary result inferred from mitochondrial DNA control region sequences. *Hereditas*, **128**: 277–80.

Mather, R.A., Kanowski, P.J., and Savill, P.S. 1991. Genetic determination of vessel area in oak (*Quercus robur* L. and *Q. petraea* Liebl.): a characteristic related to the occurrence of stem shakes. *Annales des Sciences Forestieres, Supplement*, **50**: 395s–8s.

Matsuhashi, T., Masuda, R., Mano, T., and Yoshida, M.C. 1999. Microevolution of the mitochondrial DNA control region in the Japanese brown bear (*Ursus arctos*) population. *Molecular Biology and Evolution*, **16**: 676–84.

Mayr, E. 1963. *Animal Species and Evolution*. Cambridge, MA: Harvard University Press.

Miyaki, M. and Kikuzawa, K. 1988. Dispersal of *Quercus mongolica* acorns in a broadleaved deciduous forest 1. Scatterhoarding by mice. *Forest Ecology and Management*, **25**: 9–16.

Monaghan, J. and Hull, P. 1976. Differences in vegetative characteristics among four populations of *Senecio vulgaris* L. possibly due to interspecific hybridization. *Annals of Botany*, **40**: 125–8.

Moreau, F., Kleinschmit, J., and Kremer, A. 1994. Molecular differentiation between *Q. petraea* and *Q. robur* assessed by random amplified DNA fragments. *Forest Genetics*, **1**: 51–64.

Moritz, C. 1994a. Application of mitochondrial DNA analysis in conservation: a critical review. *Molecular Ecology*, 3: 401–11.

Moritz, C. 1994b. Defining "evolutionarily significant units" for conservation. *Trends in Ecology and Evolution*, 9: 373–5.

Muir, G., Fleming, C.C., and Schlötterer, C. 2000. Species status of hybridizing oaks. *Nature*, 405: 1016.

Nei, M. 1972. Genetic distance between populations, *Am. Nat.*, 106: 283–92.

Nice, C.C. and Shapiro, A.M. 1999. Molecular and morphological divergence in the butterfly genus *Lycaeides* (Lepidoptera: Lycaenidae) in North America: evidence of recent separation, *Journal of Evolutionary Biology*, 12, 936–50.

Nice, C.C. and Shapiro, A.M. 2001. Population genetic evidence of restricted gene flow between host races in the butterfly genus *Mitoura* (Lepidoptera: Lycaenidae), *Annals of the Entomological Society Of America*, 94, 257–267.

Nichols, R.A. and Hewitt, G.M. 1994. The genetic consequences of long distance dispersal during colonisation. *Heredity*, 72: 312–17.

Niegel, J.E. and Avise, J.C. 1986. Phylogenetic relationships of mitochondrial DNA under various demographic models of speciation. In E. Nevo and S. Karlin, Eds. *Evolutionary Processes and Theory*. New York: Academic, pp. 515–34.

Nowak, R.M. 1991. *Walker's Mammals of the World, 5th ediction, Volume II*. Baltimore, Maryland: Johns Hopkins University Press.

Olalde, M., Herran, O.M., Espinel, H., and Goicoechea, P.G. 2002. White oaks phylogeography in the Iberian peninsula. *Forest Ecology and Management*, 156: 89–102.

Oxford, G.S., Crawford, T.J., and Pernyes, K. 1996. Why are capitulum morphs associated with other characters in natural populations of *Senecio vulgaris* (groundsel)? *Heredity*, 76: 192–7.

Paabo, S. 2000. Of bears, conservation genetics, and the value of time travel. *Proceedings of the National Academy of Sciences USA*, 97: 1320–1.

Paetkau, D., Sheilds, G.F., and Strobeck, C. 1998. Gene flow between insular, coastal and interior populations of brown bears in Alaska. *Molecular Ecology*, 7: 1283–92.

Paektau, D., Waits, L.P., Clarkson, P.L., Craighead, L., and Strobeck, C. 1997. An empirical evaluation of genetic distance statistics using microsatellite data from bear (*Ursidea*) populations. *Genetics*, 147: 1943–57.

Paetkau, D., Waits, L.P., Clarkson, P.L., Craighead, L., Vyse, E., Ward, R., and Strobeck, C. 1998. Variation in genetic diversity across the range of North American brown bears. *Conservation Biology*, 12: 418–29.

Pasitschniale-Arts, M. 1993. *Ursus arctos*. *Mammalian Species*, 439: 1–10.

Perry, J. and Knowles, P. 1991. Spatial genetic structure within three sugar maple (*Acer saccharum* Marsh.) stands. *Heredity*, 66: 137–42.

Peterson, M.A. 1995. Phenological isolation, gene flow and developmental differences among low and high elevation populations of *Euphilotes enoptes* (Lepidoptera: Lycaenidae). *Evolution*, 49: 446–55.

Peterson, M.A. 1996. Long distance gene flow in the sedentary butterfly, *Euphilotes enoptes* (Lepidoptera: Lycaenidae). *Evolution*, 50: 1990–9.

Petit, R.J., Bialozyt, R., Brewer, S., Cheddadi, R., and Comps, B. 2001. From spatial patterns of genetic diversity to postglacial migration in forest trees. In J. Silvertown and J. Antonovics, eds. *Integrating Ecology and Evolution in a Spatial Context*. Oxford: Blackwell Science.

Petit, R.J., Brewer, S., Bordacs, S., Burg, K., Cheddadi, R., Coart, E., Cottrell, J., Csaikl, U., Fineschi, S., Finkelday, R., Goicoechea, P.G., Jensen, J.S., Konig, A., Lowe, A.J., Madsen, S.F., Matyas, G., Oledska, I., Popescu, F., Slade, D., Van Dam, B., de Beaulieu, J.-L., and Kremer, A. 2002a. Identification of postglacial colonisation routes of European white oaks based on chloroplast DNA and fossil pollen evidence. *Forest Ecology and Management*, 156: 49–74.

Petit, R.J., Csaikl, U., Bordacs, S., Burg, K., Coart, E., Cottrell, J., Deans, D., Dumolin-Lapègue, S., Fineschi, S., Finkelday, R., Gillies, A., Goicoechea, P.G., Jensen, J.S., Konig, A., Lowe, A.J., Madsen, S.F., Matyas, G., Oledska, I., Pemonge, M.-H., Popescu, F., Slade, D., Taurchini, D., Van Dam, B., Ziegenhagen, B., and Kremer, A. 2002b. Chloroplast DNA variation in European white oaks: phylogeography and patterns of diversity based on data from over 2600 populations. *Forest Ecology and Management*, 156: 5–26.

Petit, R.J., Kremer, A., and Wagner, D.B. 1993. Geographic structure of chloroplast DNA polymorphisms in European oaks. *Theoretical and Applied Genetics*, 87: 122–8.

Petit, R.J., Latouche-Hallé, C., Pemonge, M.-H., and Kremer, A. 2002c. Chloroplast DNA variation of oaks in France and the influence of forest fragmentation on genetic diversity. *Forest Ecology and Management*, 156: 115–29.

Petit, R.J., Pineau, E., Demesure, B., Bacilier, R., Ducousso, A., and Kremer, A. 1997. Chloroplast DNA footprints of postglacial recolonization by oaks. *Proceedings of the National Academy of Sciences USA*, 94: 9996–10,001.

Pignatti, S. 1982. *Flora d'Italia. Vol. 3*. Bologna: Edagricole.

Pons, O. and Petit, R.J. 1995. Estimation, variance and optimal sampling of gene diversity. I. Haploid locus. *Theoretical and Applied Genetics*, 90: 462–70.

Pons, O. and Petit, R.J. 1996. Measuring and testing differentiation with ordered versus unordered alleles. *Genetics,* **144**: 1237–45.

Randi, E., Gentile, L., Boscagli, G., Huber, D., and Roth, H.U. 1994. Mitochondrial DNA sequence divergence among some West-European brown bear (*Ursus arctos* L.) populations – lessons for conservation. *Heredity,* **73**: 480–9.

Richards, A.J. 1975. The inheritance and behaviour of the rayed gene complex in *Senecio vulgaris. Heredity,* **34**: 95–104.

Rieseberg, L.H., Sinervo, B., Linder, C.R., Ungerer, M.C., and Arias, D.M. 1996. Role of gene interactions in hybrid speciation: evidence from ancient and experimental hybrids. *Science,* **272**: 741–5.

Rosser, E. 1955. A new British species of *Senecio. Watsonia,* **3**: 228–32.

Rushton, B.S. 1978. *Quercus robur* and *Q. petraea:* a multivariant approach to hybrid problems. I. Data acquisition, analysis and interpretation. *Watsonia,* **12**: 81–101.

Rushton, B.S. 1979. *Quercus robur* and *Q. petraea:* a multivariant approach to hybrid problems. II. The geographical distribution of population types. *Watsonia,* **12**: 209–224.

Rushton, B.S. 1983. The analysis of variation of leaf characters in *Quercus robur* L. and *Quercus petraea* (Matt.) Liebl. Population samples from Northern Ireland. *Irish Forestry,* **40**: 52–77.

Ryder, O.A. 1986. Species conservation and systematics: the dilemma of subspecies. *Trends in Ecology and Evolution,* **1**: 9–10.

Samuel, R. 1999. Identification of hybrids between *Quercus petraea* and *Quercus robur* (*Fagaceae*): results obtained with RAPD markers confirm allozyme studies based on the Got-2 locus. *Plant Systematics and Evolution,* **217**: 137–46.

Schaal, B.A., Hayworth, D.A., Olsen, K.M., Rauscher, J.T., and Smith, W.A. 1998. Phylogeographic studies in plants: problems and prospects. *Molecular Ecology,* **7**: 465–74.

Scott, J.A. 1986. *The Butterflies of North America. A Natural History Field Guide.* Stanford, CA: Stanford University Press.

Servheen, C. 1989. Status of the world's bears, 2nd International Conference of Bear Research and Management, *Monograph* **2**.

Siegismund, H.R. and Jensen, J.S. 2001. Intrapopulation and interpopulation genetic variation of *Quercus* in Denmark. *Scandinavian Journal of Forest Science,* **16**: 103–16.

Slatkin, M. 1977. Gene flow in natural populations. *Annual Review of Ecology and Systematics,* **16**: 393–430.

Slatkin, M. 1993. Isolation by distance in equilibrium and non-equilibrium populations. *Evolution,* **47**: 264–79.

Smouse, P.E., Dyer, R.J., Westfall, R.D., and Sork, V.L. 2001. Two-generation analysis of pollen flow across a landscape I. Male gamete heterogeneity among females. *Evolution,* **55**: 260–71.

Sork, V.L. 1984. Examination of seed dispersal and survival in red oak, *Quercus rubra* (*Fagaceae*), using metal-tagged acorns. *Ecology,* **65**: 1020–2.

Sork, V.L. 1993. Evolutionary ecology of mast-seeding in temperate and tropical oaks (*Quercus* spp.). *Vegetatio,* **107/108**: 133–47.

Sork, V.L. and Bramble, J. 1993. Ecology of mast-fruiting in three species of North American deciduous oaks. *Ecology,* **74**: 528–41.

Sork, V.L., Stacey, P., and Averett, J.E. 1983. Utilization of red acorns in non-bumper crop year. *Oecologia (Berlin),* **59**: 49–53.

Sork, V.L., Stowe, K.A., and Hochwender, C. 1993. Evidence for local adaptation in closely adjacent subpopulations of northern red oak (*Quercus rubra* L.) expressed as resistance to leaf herbivores. *The American Naturalist,* **142**: 928–36.

Stace, C.A. 1977. The origin of radiate *Senecio vulgaris* L. *Heredity,* **39**: 383–8.

Stace, C.A. 1997. *The New Flora of the British Isles.* Cambridge: Cambridge University Press.

Steinhoff, S. 1993. Results of species hybridisation with *Quercus robur* L. and *Quercus petraea* (Matt.) Liebl. *Annales des Sciences Forestieres, Supplement,* **50**: 137s–43s.

Steinkellner, H., Fluch, S., Turetschek, E., Lexer, C., Streiff, R., Kremer, A., Burg, K., and Gloessl, J. 1997. Identification and characterization of (GA/CT)n microsatellite loci from *Quercus petraea. Plant Molecular Biology,* **33**: 1093–6.

Stowe, K.A., Sork, V.L., and Farrell, A.W. 1994. Effect of water availability on phenotypic expression of herbivore resistance in northern red oak seedlings (*Quercus rubra* L.). *Oecologia,* **100**: 309–15.

Streiff, R., Ducousso, A., Lexer, C., Steinkellner, H., Glössl, J., and Kremer, A. 1999. Pollen dispersal inferred from paternity analysis in a mixed oak stand of *Quercus robur* L. and *Quercus petraea* (Matt.) Liebl. *Molecular Ecology,* **8**: 831–41.

Streiff, R., Labbe, T., Bacilieri, R., Steinkellner, H., Glössl, J., and Kremer, A. 1998. Within-population genetic structure in *Quercus robur* L. and *Q. petraea* (Matt.) Liebl. Assessed with allozymes and microsatellites. *Molecular Ecology,* **7**: 317–28.

Swenson, J.E., Sandegren, F., Bjarvall, A., Soderberg, A., Wabakken, P., and Franzen, R. 1994. Size, trend, distribution and conservations of the Brown Bear, *Ursus arctos,* populations in Sweden. *Biological Conservation,* **70**: 9–17.

Syme, J.T.B. 1875. *Senecio vulgaris* L. var. *hibernica mihi. Botl. Exch. Club, Rep. Curators,* **1972–74**: 27–8.

Taberlet, P. and Bouvet, J. 1994. Mitochondrial DNA polymorphism, phylogeography and conservation genetics of the brown bear *Ursus arctos* in Europe. *Proceedings of the Royal Society of London Series B – Biological Sciences,* **225**: 195–200.

Taberlet, P., Camarra, J.J., Griffin, S., Uhres, E., Hanotte, O., Waits, L.P., Dubois Paganon, C., Burke, T., and Bouvet, J. 1997. Noninvasive genetic tracking of the endangered Pyrenean Brown Bears population. *Molecular Ecology,* **6**: 869–76.

Taberlet, P., Fumagalli, L., Wust-Saucy, A.-G., and Cossons, J.-F. 1998. Comparative phylogeography and post-glacial colonisation routes in Europe. *Molecular Ecology,* **7**: 453–64.

Taberlet, P., Gielly, L., and Bouvet, J. 1991. Universal primers for amplification of three non-coding regions of chloroplast DNA. *Plant Molecular Biology,* **17**: 1105–9.

Taberlet, P., Swenson, J.E., Sandegren, F., and Bjarvall, A. 1995. Localization of a contact zone between two highly divergent mitochondrial DNA lineages of the Brown Bear (*Ursus arctos*) in Scandinavia. *Conservation Biology,* **9**: 1253–61.

Talbot, S.L. and Sheilds, G.F. 1996. Phylogeography of brown bears (*Ursus arctos*) of Alaska and paraphyly within the *Ursidae. Molecular Phylogenetics and Evolution,* **5**: 477–94.

Thomas, J.A., Elmes, G.W., Wardlaw, J.C., and Woyciechowski, M. 1989. Host specificity among *Maculinea* butterflies in *Myrmica* nests. *Oecologia,* **79**: 452–7.

Tilden, J.W. and Smith, A.C. 1986. *A Field Guide to Western Butterflies.* Boston: Houghton-Mifflin.

Trow, A.H. 1912. On the inheritance of certain characters in the Common Groundsel – *Senecio vulgaris,* Linn. – and its segregates. *Journal of Genetics,* **2**: 239–76.

Vigilant, L., Pennington, R., Harpending, H., Kocher, T.D., and Wilson, A.C. 1989. Mitochondrial DNA sequences in single hairs from a southern African population. *Proceedings of the National Academy of Sciences, USA,* **86**: 9350–4.

Vincent, P.L.D. 1996. Progress on clarifying the generic concept of *Senecio* based on an extensive world-wide sample of taxa. In D.J.N. Hind and H.J. Beentje, eds. *Compositae: systematics.* Kew, UK: Royal Botanic Gardens, pp. 597–611.

Waits, L.P., Talbot, S.L., Ward, R.H., and Sheilds, G.F. 1998. Mitochondrial DNA phylogeography of the North American brown bear and implications for conservation. *Conservation Biology,* **12**: 408–17.

Waits, L., Taberlet, P., Swenson, J.E., Sandegren, F., and Franzen, R. 2000. Nuclear DNA microsatellite analysis of genetic diversity and gene flow in the Scandinavian brown bear (*Ursus arctos*). *Molecular Ecology,* **9**: 421–31.

Weir, J. and Ingram, R. 1980. Ray morphology and cytological investigations of *Senecio cambrensis* Rosser. *New Phytologist,* **86**: 237–41.

Whittemore, A.T. and Schaal, B.A. 1991. Interspecific gene flow in sympatric oaks. *Proceedings of the National Academy of Sciences USA,* **88**: 2540–4.

Wolfe, K.H., Li, W.-H., and Sharp, P.M. 1987. Rates of nucleotide substitution vary greatly among plant mitochondrial, chloroplast and nuclear DNA. *Proceedings of the National Academy of Sciences USA,* **84**: 9054–8.

Zanetto, A. and Kremer, A. 1995. Geographical structure of gene diversity in *Quercus petraea* (Matt.) Liebl. I. Monolocus patterns of variation. *Heredity,* **75**: 506–17.

Zanetto, A., Roussel, G., and Kremer, A. 1994. Geographic variation of inter-specific differentiation between *Quercus robur* L. and *Quercus petraea* (Matt.) Liebl. *Forest Genetics,* **1**: 111–23.

Data analysis software

Programme name	Operating system	Location	Basic descriptive stats, e.g. HW, disequilibrium	Genetic diversity	Genetic differentiation /inbreeding coefficient
AFLP-SURV	Windows & Mac	http://www.ulb.ac.be/ sciences/lagev/aflp-surv.html	Dominant	Dominant	Dominant
ARELQUIN (supersedes WINAMOVA + other packages)	Windows, Mac & Linux	http://lgb.unige.ch/arlequin/	Codominant, dominant, haploid	Codominant, dominant, haploid	Codominant, dominant, haploid
BOTTLENECK	Windows	http://www.montpellier.inra.fr/ URLB/bottleneck/bottleneck.html		Codominant data, detects population bottlenecks	
CERVUS	Windows	http://helios.bto.ed.ac.uk/ evolgen/cervus/cervus.html			
CONTRIB & RAREFAC	Windows & Linux	http://www.pierroton.inra.fr/ genetics/labo/Software/	Codominant	Codominant	
FaMoZ	Windows & Linux	http://www.pierroton.inra.fr/ genetics/labo/Software/Famoz/			
FSTAT	Dos & Windows	http://www.unil.ch/izea/ softwares/fstat.html	Codominant	Codominant	Codominant
GDA	Windows	http://lewis.eeb.uconn.edu/ lewishome/software.html	Codominant	Codominant	Codominant
GenAlEx	Windows & Mac	http://www.anu.edu.au/ BoZo/GenAlEx/	Codominant, dominant, haploid	Codominant, dominant, haploid	Codominant, dominant, haploid
GENECLASS	Windows	http://www.montpellier.inra.fr/ URLB/geneclass/geneclass.html			
GENEPOP	DOS & web	http://wbiomed.curtin.edu.au/genepop/	Codominant		Codominant
GENETIX	Windows	http://www.univ-montp2.fr/ ~genetix/genetix/genetix.htm.	Codominant	Codominant	Codominant
GENSTAT	Windows	http://www.nag.co.uk/ Commercially available			
GEODIS	Windows & Mac	http://inbio.byu.edu/Faculty/ kac/cr&all_lab/geodis.htm			
HAPLODIV & HAPLONST	Windows & Linux	http://www.pierroton.inra.fr/ genetics/labo/Software/		Haploid	Haploid
IBD	Windows & Mac	http://www.bio.sdsu.edu/ pub/&y/IBD.html			
KINSHIP	Mac	http://www.gsoftnet.us/GSoft.html			

Genetic distance	Spatial autocorrelation/ phylogeographic structure	Indirect gene flow	Outcrossing rate, correlated matings, neighborhood size	Direct gene flow, parentage analysis	Tree building/ cluster analysis	Confidence/ permutation testing	Additional features
Dominant							
Codominant, dominant, haploid					Codominant, dominant, haploid	Codominant, dominant, haploid	
			Codominant	Codominant		Codominant & dominant	
						Codominant	
			Codominant & dominant	Codominant & dominant		Codominant & dominant	
Codominant					Codominant	Codominant	
Codominant, dominant, haploid	Codominant, dominant, haploid	Codominant, dominant, haploid					
						Mulilocus data for assignment	
	Codominant	Codominant					
Codominant	Codominant	Codominant				Codominant	
				Continuous or discrete data	Continuous or discrete data	Continuous or discrete data	Many other statistical functions
Ordered data	Ordered data				Ordered data		
Haploid	Haploid						
	Codominant & distance matrix						
Kinship for codominant data							

Programme name	Operating system	Location	Basic descriptive stats, e.g. HW, disequilibrium	Genetic diversity	Genetic differentiation /inbreeding coefficient
LAMARC	Windows	http://evolution.genetics.washington.edu/lamarc.html	Sequence, SNPS & SSRs		
LEA	Windows & Linux	http://www.cnrs-gif.fr/pge/bioinfo/lea/index.php?lang=en			
LINKDOS	Web	http://wbiomed.curtin.edu.au/genepop/linkdos.html	Disequilibrium loci test		
MEGA	Windows	http://www.megasoftware.net			
MICROSAT	Windows & Mac	http://hpgl.stanford.edu/projects/microsat/	Microsatellite	Microsatellite	Microsatellite
MSA	Windows & Linux	http://i122server.vu-wien.ac.at/MSA/MSA_download.html	Codominant	Codominant	Codominant
MINITAB	Windows & Mac	http://www.minitab.com/ price > US$1000			
MLTR (& others)	Windows & DOS	http://genetics.forestry.ubc.ca/ritl&/programs.html			
NTSYS	Windows	Exeter Software (US$250) http://www.exetersoftware.com/cat/ntsyspc/ntsyspc.html 100 North Country Road, Setauket, New York 11733, USA			
PAPA	Windows	http://www.bio.ulaval.ca/contenu-fra/professeurs/prof-l-bernatchez.html			
PARENTE	Windows	http://www2.ujf-grenoble.fr/leca/membres/manel.html			
PATRI		http://www.biom.cornell.edu/Homepages/Rasmus_Nielsen/index.html			
PAUP (plus MACCLADE)	Windows & Mac	US$85–150 from http://www.sinauer.com/formpurch.htm, & e-mail at orders@sinauer.com			
PHYLIP	Windows & Mac	http://evolution.genetics.washington.edu			
PHYLOGENY	Windows & Mac	http://evolution.genetics.washington.edu/phylip/software.html			
POPGENE	Windows	http://www.ualberta.ca/~fyeh/index.htm	Codominant & dominant	Codominant & dominant	Codominant & dominant
POPULATIONS	Windows & Linux	http://www.cnrsgif.fr/pge/bioinfo/populations/index.php?lang=en			
RAPDistance	DOS	http://life.anu.edu.au/molecular/software/rapid.htm			
RSTCALC	Windows	http://helios.bto.ed.ac.uk/evolgen/rst/rst.html	Microsatellite data		Microsatellite data
RELATEDNESS	Mac	http://www.gsoftnet.us/GSoft.html			
SGS	Windows	http://kourou.cirad.fr/genetique/software.html			

Genetic distance	Spatial autocorrelation/ phylogeographic structure	Indirect gene flow	Outcrossing rate, correlated matings, neighborhood size	Direct gene flow, parentage analysis	Tree building/ cluster analysis	Confidence/ permutation testing	Additional features
					Sequence, SNPS & SSRs	Sequence, SNPS & SSRs	
						Assignment test of hybridization independent nuclear loci	
Sequence & discrete data					Sequence & discrete data	Sequence & discrete data	Many
Microsatellite						Microsatellite	
Codominant						Codominant	
					Continuous or discrete data	Continuous or discrete data	Many other statistical functions
			Codominant & dominant				
					Continuous or discrete data	Continuous or discrete data	Many other statistical functions
				Codominant			
				Codominant			
Sequence & discrete data					Sequence & discrete data	Sequence & discrete data	Many
Sequence & discrete data					Sequence & discrete data	Sequence & discrete data	Many
					Comprehensive list of phylogeny prgraames		
Codominant & dominant		Codominant & dominant			Codominant & dominant		
Codominant							
Dominant					Dominant		
Microsatellite data							
Relatedness using codominant data							
	Haploid, codominant & dominant						

Programme name	Operating system	Location	Basic descriptive stats, e.g. HW, disequilibrium	Genetic diversity	Genetic differentiation /inbreeding coefficient
SPLITSTREE	Windows & Mac	ftp://ftp.uni-bielefeld.de/pub/math/splits			
SPSS	Windows	http://www.spss.com/ available as separate models US$300–700			
STATISTICA	Windows	http://www.statsoftinc.com/ addition.html price >US$1000			
STRUCTURE	Windows & Linux	http://pritch.bsd.uchicago.edu/ software.html			
TFPGA & MANTEL-STRUCT	Windows	http://bioweb.usu.edu/mpmbio/index.htm	Codominant & dominant		Codominant & dominant
TREE-PUZZLE	Windows	http://www.tree-puzzle.de			
TREEVIEW	Windows, Mac & Linux	http://taxonomy.zoology.gla.ac.uk/rod/ treeview.html			

Note: This list is not comprehensive & it does not necessarily imply recommendation by the authors.

Genetic distance	Spatial autocorrelation/ phylogeographic structure	Indirect gene flow	Outcrossing rate, correlated matings, neighborhood size	Direct gene flow, parentage analysis	Tree building/ cluster analysis	Confidence/ permutation testing	Additional features
					Sequence & discrete data		
					Continuous or discrete data	Continuous or discrete data	Many other statistical functions
					Continuous or discrete data	Continuous or discrete data	Many other statistical functions
						Codominant, dominant, haploid assignment test	
Codominant & dominant	Codominant & dominant	Codominant			Codominant & dominant		Hierarchical analysis
Sequence & discrete data					Sequence & discrete data	Sequence & discrete data	
					Standard distance data & NEXUS format		

B

Which distance algorithm should be used and when?

There are a large number of distance algorithms that can be used for representing similarity or distance between individuals or groups of individuals (Legendre and Legendre 1998). The suitability of each will depend on the type of data available for analysis and thus the distance measures described below are listed according to the data type with which they would most typically be used.

Morphometric or quantitative data

Individual-based estimates

Euclidean distance (d_{ab}) is the simplest distance algorithm, and is widely used for many purposes including morphometric data. The best way to visualize the relationship between two entities (a and b) is within a two-dimensional plot, where the locations of a and b are described according to their coordinates along two axes (i.e., a_1, a_2 and b_1, b_2, respectively) that represent the values for two differentiating characters (character 1 and character 2).

Using Pythagoras' theorem, the distance between the two points in a two-dimensional space is:

$$d_{ab} = \sqrt{(a_1 - b_1)^2 + (a_2 - b_2)^2} \qquad (1)$$

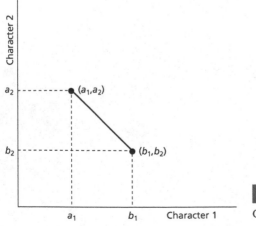

Fig. B1

Genetic distance.

Thus, d_{ab} for many characters (i.e., z dimensions) is:

$$d_{ab} = \sqrt{\sum_{i=1}^{i=z} (a_i - b_i)^2} \tag{2}$$

Population-based estimates

Several measures have been proposed to examine distances between populations, where data are available on character means, variances, and covariances. One widely used measure is by Mahalanobis (1948), D^2, which takes into account correlations between characters and is therefore not violated if non-independent variables are used.

$$D_{ij}^2 = \sum_{r=1}^{r=n} \sum_{s=1}^{s=n} (\mu_{ri} - \mu_{rj})(\mu_{si} - \mu_{sj}) v^{rs} \tag{3}$$

where μ_{ri} and μ_{si} are the means of variables, x_r and x_s, in the ith population, n is the number of individuals, and v^{rs} is the inverse of the covariance between the two variables being compared. The covariance between the variables x_r and x_s is defined as:

$$c_{jk} = \sum_{i=1}^{i=n} \frac{(x_{ir} - \bar{x}_r)(x_{is} - \bar{x}_s)}{(n-1)} \tag{4}$$

Codominant data from single loci, unordered alleles

Individual-based estimates

For data derived from isozymes, microsatellites, or other single locus nuclear loci for diploid organisms, a simple relatedness measure can be applied, where the genetic relatedness (R) between two individuals (a and b) is calculated based on the number of alleles that two individuals share:

$$R = \frac{n_{ab}}{2} \tag{5}$$

where n_{ab} is the number of alleles at a locus shared between individuals a and b. Clearly, for individuals that share both alleles, $R = 1$ and for those that only share one allele, $R = 0.5$ (e.g., parent–sibling relationship). This relatedness measure can be summed across loci.

For microsatellite data, a stepwise mutation model can be applied based on the size of the repeat motif (chapter 3). Under this model, two individuals that have alleles similar in repeat number are presumed to have diverged more recently than two individuals that have alleles with very different repeat numbers. For individuals, the average repeat number can be calculated for two alleles at a locus and compared:

$$\Delta\mu = (\mu_a - \mu_b)^2 \tag{6}$$

where, μ_a and μ_b are the mean number of repeats for alleles at a specified locus in individuals a and b, respectively. For multiple loci, the arithmetic mean is calculated across loci.

For data where the number of mutations separating alleles is available (e.g., RFLP), then a simple distance metric between two individuals, e.g., is the average number of mutations separating alleles at a locus can be applied in a similar manner.

Population-based estimates

Euclidean distance (Eqs. (1) and (2)) can be used to describe the genetic distance between populations or groups, as well as individuals. In which case, alleles are the characters and allele frequencies are the coordinate values to be used in Eqs. (1) and (2).

A refinement of the linear Euclidean distance is to use the square root of the allele frequencies as coordinates, and therefore populations lie on the surface of a hypersphere of unit radius (Weir 1990). The angle (θ) between the lines connecting the origin to the points representing the two populations (P and Q) is then calculated as:

$$\cos(\theta) = \sum_i \sqrt{(p_i q_i)} \tag{7}$$

where p_i and q_i are the frequencies of the ith alleles in populations P and Q, respectively. If $1 - \cos(\theta)$ is used as the distance estimator, then the distance between populations with identical allele frequencies will be 0, and those with no alleles in common will have a value of 1. In addition, the chord length between coordinates along the arc may be used and equals $\sqrt{[2 - 2\cos(\theta)]}$ (Cavalli-Sforza and Bodmer 1971).

None of the distances measured used so far incorporate genetic processes. However, genetic distance is presumed to be positively related to the time, since populations diverge from a common ancestral population. Thus, a genetic model specifying processes, such as mutation and drift, causing the populations to diverge is required.

Roger's (1972) distance is defined as:

$$D = \sqrt{\left(0.5 \sum_{i=1}^{a} (p_i - q_i)^2\right)} \tag{8}$$

where p_i and q_i are the frequencies of the ith allele in populations P and Q, respectively, and l is the number of alleles. For more than one locus, the mean value across loci is used. Values range from 0, where the allele frequencies are the same for two populations, to 1, where no alleles are shared. A similarity index is defined as $S = 1 - D$.

Nei's (1972) standard genetic identity is commonly used and defined as:

$$I = \frac{\sum_{i=1}^{a} p_i q_i}{\sqrt{\left(\sum_{i=1}^{a} p_i^2 \sum_{i=1}^{a} q_i^2\right)}} \tag{9}$$

For multiple loci, the overall identity is:

$$I = \frac{\mathcal{J}_{pq}}{\sqrt{(\mathcal{J}_p \mathcal{J}_q)}} \tag{10}$$

where, $\mathcal{J}_{pq}$, $\mathcal{J}_p$, and $\mathcal{J}_q$ are the arithmetic means across loci for $\sum p_i q_i$, $\sum p_i^2$, and $\sum q_i^2$, respectively. Values range between 0 and 1, although the standard genetic distance calculated from the identity ($D = -\ln I$) ranges between 0 and infinity, and is interpreted as the mean number of codon substitutions per locus, corrected for multiple hits.

Nei's (1978) unbiased genetic identity should be used for unequal population sizes and is defined as:

$$ I = \frac{2n - 1}{2n} \left[\frac{\displaystyle\sum_{i=1}^{a} p_i q_i}{\sqrt{\left(\displaystyle\sum_{i=1}^{a} p_i^2 \sum_{i=1}^{a} q_i^2 \right)}} \right] \tag{11} $$

Several other refinements to genetic distance measures have been developed (e.g., Cockerham 1984) and may include (or assume) a specified mutation rate for calculation of time since divergence. Where no mutation rate is specified then an *infinite alleles model* is assumed.

The measure of coancestry, F_{ST}, can also be used as a pairwise population comparison to examine the amount of drift between populations, without assuming a mutation model (chapter 3). The measure of coancestry, R_{ST} (chapter 3), which assumes a stepwise mutation model for microsatellite data, can also be used as a pairwise population comparisons.

The average repeat motif number for a locus can also be calculated for a population and applied to interpopulation comparisons using Eq. (6).

Multilocus dominant genetic data

Individual-based estimates

Several algorithms are available to calculate the similarity of pairs of samples which are characterized by presence–absence data. Most of these estimates make no assumption of a genetic model and are solely phenetic estimates of similarity based on band sharing. For data types, such as RAPD and AFLP, where the basis of band absence is unknown and therefore cannot be assumed to have an identical cause for all individuals, genetic distance estimates which remove the shared absence component from calculations are preferred (Weising et al. 1995).

Two measures are most commonly used. Jaccard's (1908) similarity coefficient ($\mathcal{J}_{ij}$), which only takes positive matches into account, is defined for two individuals, i and j, as:

$$ \mathcal{J}_{ij} = \frac{C_{ij}}{(n_i + n_j - C_{ij})} \tag{12} $$

where C_{ij} is the number of positive matches between individuals i and j, and n_i and n_j are the total number of bands in individuals i and j, respectively.

Dice's similarity coefficient (S; Nei and Li 1979) is similar and defined as:

$$ S = \frac{2n_{ij}}{(n_i + n_j)} \tag{13} $$

The shared absence of bands may indicate genetic similarity between closely related individuals, as there is a high probability that the absence of a fragment is due to ancestry rather than convergent evolution (Adams 1995). In such a case, a distance measure such as Gower (1971) can be used:

$$ S = \frac{1}{p} \sum_p \frac{C_{ij}}{(n_i + n_j - 2C_{ij})} \tag{14} $$

where p is the number of bands scored.

Gower's distance estimate should be restricted to very similar individuals, for cases where similarities are unknown Jaccard's or Dice's algorithms (Eqs. (12) and (13)) are recommended.

Gower (1985) has discussed the relationships between similarity measures. Most of the metrics give similarity values between 0 to +1 and can be converted to distances as [1 − (similarity)] or √[1 − (similarity)]. In addition, most are very similar in their ability to distinguish relative differences between individuals, although the magnitude of scores is quite different.

Population-based estimates

The problem with dominant data is that there is no direct estimate of the number of heterozygotes in a population. One way to perform population analysis is to assume each fragment represents a locus and that the two alleles (i.e., fragment presence or absence) are in Hardy–Weinberg equilibrium. In this way the proportion of heterozygotes at each locus can be estimated from the frequency of fragment absence. However, this argument may be flawed, for example, fragments may be length variants of a single locus (non-independent data) and the population may not be in Hardy–Weinberg equilibrium. Even if a direct measure of gene diversity is available from another marker system, Lynch and Milligan (1994) make several recommendations over the use of dominant marker data in population-based estimates: (i) two to 10 times more individuals need to be analyzed for dominant data than codominant data; and (ii) to avoid bias in parameter estimation, marker alleles should be of relatively low frequency (0.2 to 0.5). However, despite these steps, bias cannot be completely eliminated (Isabel et al. 1999).

Once these assumptions have been made, and appropriate action taken, it is possible to calculate population genetic distance estimates using different measures, including Nei's (1972, 1978) and Roger's distances.

To avoid making these assumptions for dominant data, it is possible to calculate an ANOVA-based estimate of distance taking into account the proportion of variation within and between populations. A procedure called AMOVA allows a Φ_{ST} (Phi$_{ST}$) estimate of population distance to be calculated (Excoffier et al. 1992, Huff et al. 1993; chapter 3). Individual distance estimates for dominant data (e.g., Jaccard or Dice coefficients) should be used with this analysis.

In addition, new Bayesian methods have been proposed for use with dominant data (chapter 3).

DNA restriction site and nucleotide sequence data, haploid, and diploid genomes

Individual-based estimates

For restriction fragment data, Upholt's (1977) distance (F) is based on the concept of the overall proportion of shared fragments:

$$F = \frac{2N_{XY}}{(N_X + N_Y)} \tag{15}$$

where N_X, N_Y, and N_{XY} are the number of fragments observed in sequences X and Y and shared by X and Y, respectively. The number of base substitutions per nucleotide is then estimated as:

$$p = 1 - \left\{0.5\left[-F + \sqrt{(F^2 + 8F)}\right]\right\}^{\frac{1}{r}} \tag{16}$$

where r is the number of bases in a restriction enzyme's recognition site.

For restriction sites, the proportion of shared sites (S) is derived in the same way as F in Eq. (16), and the number of base substitutions is calculated as the mean of values calculated separately for enzymes having different length recognition sites as:

$$p = \frac{-\ln(S)}{r} \tag{17}$$

For nucleotide sequence data that are the same length and unambiguously aligned, the percent sequence difference can be calculated as:

$$p = \frac{z_d}{z_t} \tag{18}$$

where z_d is the number of nucleotides which differ between two sequences and z_t is the total number of nucleotides compared. This estimate is fine for closely related sequences but for more divergent sequences, a multiple substitution correction should be applied (e.g., Jukes and Cantor 1969, Kimura 1980).

Population-based estimates

Using the measures above, mean genetic distance can be estimated for a population of haploid genomes by:

$$\text{mean}(p) = \sum_i \sum_j f_i f_j p_{ij} \tag{19}$$

where f_i and f_j are the frequencies of the ith and jth sequences in the sample and p_{ij} is the estimated sequence divergence between the ith and jth sequences (Nei 1978).

For data derived from two or more populations, estimates of net sequence divergence (p_{corr}) between populations can be calculated as:

$$p_{corr} = p_{xy} - 0.5(p_x + p_y) \tag{20}$$

where p_{xy} is the mean genetic distance in pairwise comparisons of individuals between populations X and Y, and p_x and p_y are the mean genetic distances among individuals of each respective populations. The estimate is thus based on a correction for within-population polymorphism.

REFERENCES

Adams, P.P. 1995. *PCO3D, Principal Coordinate Analysis in 3D. A Computer Program.* Gruver: Baylor University.

Cavalli-Sforza, L.L. and Bodmer, W.F. 1971. *The Genetics of Human Populations.* San Francisco: Freeman.

Cockerham, C.C. 1984. Drfit and mutation with a finite number of allelic states. *Proceedings of the National Academy of Sciences USA*, **81**: 530–4.

Excoffier, L., Smouse, P.E., and Quattro, J.M. 1992. Analysis of molecular variance inferred from metric distances among DNA haplotypes: application to human mitochondrial DNA restriction data. *Genetics*, **131**: 479–91.

Gower, J.C. 1971. Statistical methods of comparing multivariate analyses of the same data. In R. Hodson, D.G. Kendall, and P. Tatu, eds. *Mathematics in the Archaeological and Historical Sciences.* Edinburgh: Edinburgh University Press, pp. 138–49.

Gower, J.C. 1985. Measures of similarity, dissimilarity, and distance. In S. Klotz and N.J. Johnson, eds. *Encylopedia of Statistical Sciences.* New York: Wiley, pp. 397–405.

Huff, D.R., Peakall, R., and Smouse, P.E. 1993. RAPD variation within and among natural populations of outcrossing buffalo grass [*Buchloe dactyloides* (Nutt.) Engelm.]. *Theoretical and Appplied Genetics*, **86**: 927–34.

Isabel, N., Beaulieu, J., Theriault, P., and Bousquet, J. 1999. Direct evidence for biased gene diversity estimates from dominant random amplified polymorphic DNA (RAPD) fingerprints. *Molecular Ecology*, **8**: 477–83.

Jaccard, P. 1901. Etude comparative de la distribution florale dans une portion des Alpes et des Jura. *Bull. Soc. Vaudoise Sci. Nat.*, **37**: 547–79.

Jukes, T.H. and Cantor, C.R. 1969. Evolution of protein molecules. In H. Munro, ed., *Mammalian Protein Metabolism.* New York: Academic Press, pp. 21–132.

Kimura, M. 1980. A simple method for estimating evolutionary rate of base substitutions through comparative studies of nucleotide sequences. *Journal of Molecular Evolution*, **16**: 111–20.

Legendre, P. and Legendre, L. 1998. *Numerical Ecology*. Amsterdam: Elsevier.

Lynch, M. and Milligan, B.G. 1984. Analysis of population genetic structure with RAPD markers. *Molecular Ecology*, **3**: 91–9.

Mahalanobis, P.C. 1948. Historic note on D^2 statistics. *Sankhya*, **9**: 237.

Nei, M. 1972. Genetic distance between populations. *American Naturalist*, **106**: 283–92.

Nei, M. 1978. Estimation of average heterozygosity and genetic distance from a small number of individuals. *Genetics*, **89**: 583–90.

Nei, M. and Li, W.-H. 1979. Mathematical model for studying genetic variation in terms of restriction endonucleases. *Proceedings of the National Academy of Sciences USA*, **76**: 5269–73.

Upholt, W.B. 1977. Estimation of DNA sequence divergence from comparison of restriction endonuclease digests. *Nucleic Acids Research*, **4**: 1257–65.

Weir, B.S. 1996. *Genetic Data Analysis*, 2nd edn. Sunderland, MA: Sinauer.

Weising, K., Nybom, H. Wolff, K., and Meyer, W. 1995. *DNA Fingerprinting in Plants and Fungi*. London: CRC Press.

Glossary

Accuracy A measure of how close a hypothesized tree is to the true tree (in a phylogenetic context).

Adaptation The evolution of a particular character or characters in response to a specific selective pressure.

Adaptive gene complex A suite of tightly linked genes that code for an array of interlinked characters of high adaptive value. For instance, the genes covering style length, anther type, and pollen shape in the genus *Primula*.

Allele One of two or more alternative forms of a gene, locus, or DNA sequence.

Allele frequency The abundance of a particular allele in a population or species (expressed as a proportion of one).

Allopatry Geographically separate populations or taxa.

Allopolyploidy Speciation where hybridization between two taxa is followed by chromosome doubling to confer fertility. Alternatively allopolyploid species can arise by fusion of two sets of unreduced gametes (*see also* autopolyploidy).

Allozyme The alternative forms of a particular protein visualized on a gel as bands of different mobility. Each separate allozyme is coded for by a different allele of the same gene.

Amino acid The basic components of proteins identified by possession of an amino group (-NH$_2$) and a carboxylic acid (-COOH) both joined to a central carbon atom. This carbon atom is also joined to a hydrogen molecule and a variable side chain.

Amplified fragment length polymorphism (AFLP) The selective amplification, using PCR, of a set of DNA fragments produced by a frequently cutting (usually 4 bp recognition site) and rarely cutting restriction enzyme (usually 6 bp recognition site).

Anonymous marker A genetic marker where the inheritance pattern of the marker is unknown.

Assay The processing of plant or animal material to visualize patterns of genetic variation.

Assortative mating Deviation from mating patterns expected by random assortment of genes due to the increased incidence of particular phenotype/genotype mating with an individual of the same phenotype/genotype (positive assortative mating), or a different phenotype/genotype (negative assortative mating).

Autapomorphy Unique derived character state.

Autopolyploidy The origin of a new species or race through the spontaneous doubling of chromosomes (*see also* allopolyploidy).

Autapomorphy Unique derived character state.

Average gene diversity A measure of the expected level of heterozygosity calculated from allele frequencies revealed by assay, averaged across all loci.

Backcrossing Crossing between an interspecific hybrid and one or both parental taxa (*see also* hybrid zone and introgression).

Base pair A pair of DNA nucleotides joined between the two bases by hydrogen bonds.

Bias A method of selecting units from a population such that some units have a greater chance of being selected than others.

Biparental inheritance Pattern of inheritance where each parent contributes one set of chromosomes to the zygote (*see also* maternal and paternal inheritance).

Bootstrapping Method of estimating variance parameters (particularly confidence intervals) by repeated random sampling, with or without replacement, of subsets of the data set.

Bottlenecks   Sudden reduction in population size.

Bulk collection   Sampling without maintaining a distinction between different genotypes/individuals in a population.

Coalescence   The principle of using ordered data to trace allele changes back through time.

cpDNA   *see* chloroplast DNA.

Chloroplast DNA   Double-stranded, haploid circular DNA molecule found in the chloroplast organelle of plants. Usually maternally inherited in angiosperms and paternally inherited in gymnosperms.

Chromosome number   The number of chromosomes found in a cell or organism, expressed as either mitotic ($2n$) or gametic (n) numbers.

Chromosome pairing   The positioning of homologous chromosomes alongside each other at prophase I of meiosis.

Chromosome structure   The size and shape (e.g. position of the centromere) of a chromosome.

Cluster   A group of genetically similar samples positioned together in a pictorial representation such as a dendrogram or a multivariate plot.

Coding region   A length of DNA whose sequence codes for either an amino acid or RNA sequence.

Codominant inheritance   A pattern of inheritance where heterozygous individuals express a different phenotype to either of the two homozygous forms.

Codominant marker   A genetic marker in which both alleles are expressed, thus heterozygous individuals can be distinguished from either homozygous state.

Co-efficient of relatedness   A method of determining relationships between families/populations based upon the proportion of shared genetic material.

Co-factor   A substance required for the catalytic function of an enzyme.

Confidence interval   A description of the range within which a variable is typically found.

C-value   The total amount of DNA in a haploid genome.

Deletion   A mutation involving the loss of a section of DNA.

Dendrogram   A method of representing genetic similarities by a branching tree. The greater the degree of similarity between individuals, the shorter the branch lengths separating them.

Diallelic locus   A polymorphic loci with two alleles present.

Dioecy   A breeding system in which male and female gametes are borne in separate individuals.

Diploid   A nucleus or individual having two copies of each chromosome.

DNA marker   *see* Genetic marker

Dominant inheritance   An inheritance system where an allele masks the expression of another (recessive) allele at the same locus, thus the phenotype of heterozygotes is indistinguishable from those of homozygotes.

Dominant marker   A marker that shows dominant inheritance with homozygous dominant individuals indistinguishable from heterozygous individuals.

Drift-migration equilibrium   The maintenance of genetic diversity in a population whereby the loss of alleles by genetic drift is offset by the gain of alleles into the population by migration.

Effective population size   The number of individuals in a population reproducing and contributing to the alleles present in the next generation.

Enzyme   A protein molecule responsible for catalyzing biological reactions.

Evenness   The relative abundance of alleles in a polymorphic population.

Exon   The coding region of a gene.

Ex situ   Conservation of an organism away from its natural habitat, e.g. in zoos or botanic gardens.

Family collection   Collection of samples where individuals with the same maternal parent are kept together.

Fitness   The ability of an individual, relative to other members of the same species, to produce viable offspring which themselves survive to reproduce and leave viable offspring.

Fitness components   The two aspects of fitness fecundity, the number of offspring produced; and survivorship, the proportion of offspring surviving to reproductive age.

Fixation   The loss of alleles from a polymorphic population until only one remains, i.e., becomes monomorphic.

Gamete   A typically haploid cell produced by meiosis which may unite with another such cell to form a zygote.

Gene   Specific nucleotide sequence of DNA that codes for a particular protein, tRNA, or rRNA.

Gene bank   Stored samples of a plant, typically seeds, collected to represent a proportion of the genetic variation found in the species.

Gene diversity   A measure of the genetic variation found in a population or species based on the mean expected heterozygosity.

Gene duplication   A mutation in which a sequence of DNA is copied and incorporated into the genome.

Gene pool   The genes of all the breeding individuals in a discrete population.

Gene tree   The evolutionary history of a gene represented as a branching diagram.

Genetic bottleneck   Loss of genetic variation in a population due to a fall in population numbers. While population size may recover relatively quickly, genetic variation remains low until restored by mutation and/or gene flow.

Genetic cross   Interbreeding of individuals of known genotype with the aim of determining the inheritance pattern of the locus under investigation.

Genetic differentiation   A measure of the allocation of genetic variation within a species and among populations. Species with a high level of genetic differentiation show high variation between populations.

Genetic distance   The degree of genetic similarity between a pair of individuals, populations, or species. Values typically vary between 0 (identical) and 1 (completely different).

Genetic drift   Change in allele frequencies within a population over time due to the sampling effects of small population size.

Genetic load   A measurement of the amount of deleterious mutation in a population leading to a reduction in fitness (*see also* Inbreeding depression).

Genetic marker   A sequence of DNA or protein that can be screened to reveal key attributes of its state or composition and thus used to reveal genetic variation.

Genome   The full complement of genes present in a haploid set of chromosomes in an organism.

Genome size   A measure of the total amount of DNA found in an organism's genome (recorded in kilobases, Kb, of DNA).

Genomics   The process of accessing gene function or relative gene position or interactions.

Genotype   State (e.g., allelic composition) for a particular genetic locus of an organism.

Germplasm   Collection of a species' genetic material, usually composed of material with reproductive potential.

Haploid   One copy of each chromosome, typically found in a gamete.

Haplotype   A set of tightly linked genes inherited together as a block.

Heteroplasmic cells   Cells from an individual containing copies of the same organelle with different haplotypes.

Heterozygote   An individual with two different alleles at a locus.

Highly repetitive DNA   *see* Repetitive DNA

Histochemical stain   A stain utilized to reveal the distribution of specific substances in tissues.

Homologous   Identical characters that are similar due to descent from a common ancestor.

Homoplasmic cell   A cell where organellar genomes are identical.

Homoplasy   Identical characters that have evolved separately in independent evolutionary lineages.

Homozygote   An individual with two copies of the same allele at a locus.

Hybrid zone   Area where two closely related species are in reproductive contact producing a range of hybrid and backcrossed entities.

Inbreeding   Reproduction between closely related individuals; includes self-fertilization.

Inbreeding depression   Reduction of fitness due to expression of deleterious alleles in homozygous state following breeding with close relatives.

Indel   A form of mutation where a sequence of DNA is inserted and/or deleted.

Infinite allele model (IAM) Mutation pattern where an allele can spontaneously mutate into an allele of any size or character (*see also* Stepwise mutation model).

Information content The amount of information that a particular marker yields about polymorphism in a population.

Insertion Mutation where a sequence of DNA is incorporated into a chromosome.

Interspersed repeats A form of repetitive DNA where units of a particular sequence are scattered across the genome (*see also* Tandem repeats).

Introgression The transfer of genes or alleles from one taxon to another due to hybridization between the taxa followed by repeated back-crossing of hybrids with one of the parents.

Intron Section of a gene that is transcribed but removed before translation.

Inversion A mutation where a DNA sequence is removed from a chromosome and then reinserted in the opposite orientation.

Invert repeat Either of a pair of DNA sequences that are identical (or nearly identical) and are orientated in opposite directions.

Isoelectric focusing The concentration of proteins in a gel matrix with a pH gradient causing them to migrate to the point where their net charge is neutral (isoelectric point).

Isolation by distance Reduction of gene flow in species with low dispersal powers due to spatial separation of one population from another.

Isozyme Alternative forms of a protein coded by different loci.

Jacknifing Method of estimating variance parameters (particularly standard errors) by calculating derived values using the full data set, followed by removing each datum in turn and recalculating desired characters.

Kilobase (Kb) 1000 base pairs of DNA or 1000 bases of RNA.

Lineage A particular monophyletic group of genotypes (taxa, populations, or individuals) that are related by descent more closely than other such groups.

Lineage sorting The reduction of the number of lineages within a taxon or population due to the action of natural selection and chance.

Linkage Two loci found on the same chromosome, identified by having a lower rate of recombination between them than if they were independently assorted. The closer the position of the two loci, the lower the rate of recombination between them.

Linkage disequilibrium Two alleles from different loci on the same chromosome co-occurring at a significantly greater frequency than that expected by random association.

Locus (*plural:* Loci) A specific region or position on the genome or chromosome.

Marker An observable, typically discrete, gentically controlled trait (*see also* Genetic marker).

Marker index A measure for comparing techniques of generating genetic diversity incorporating both information content (q.v.) and the multiplex ratio (q.v.).

Maternal inheritance Inheritance of a trait or genome solely through the female gamete, usually of organelle origin.

Metapopulation A population consisting of a network of partially isolated smaller populations.

Microsatellites Short tandem repeats of a short sequence of (typically two to four) nucleotides randomly distributed throughout the genome.

Minimum viable population (MVP) The minimum effective population required to persist despite genetic drift, demographic, and environmental stochasticity.

Minisatellites Highly repetitive sequences, typically occurring near telomeres, consisting of nucleotides 10–100 bp long repeated in tandem arrays.

Mitochondrial DNA (mtDNA) Haploid, double-stranded circular DNA molecule located in the mitochondria, typically maternally inherited.

Moderately repetitive DNA *see* Repetitive DNA.

Molecular marker *see* Genetic marker.

Monoecy Breeding system in which male and female gamete-producing structures are in different locations on the same organism.

Monomorphic Absence of more than one allele at a particular locus or gene, leading to a uniform phenotype in the population.

Monophyletic Group of genotypes/individuals or species which is an inclusive combination of an ancestor and all its descendents.

Multiallelic locus Presence of two or more alleles at a particular locus.

Multiplex ratio The number of loci that can be combined in a single analysis.

Mutation Change in nucleotide sequence of an organism.

Mutation rate The frequency at which a particular mutation occurs in a genome.

Non-coding region Section of the genome that is non-transcribed.

Nuclear DNA Sequence of DNA nucleotides located in the nucleus.

Nucleotide Compound containing a pentose sugar, phosphate, and either a purine or pyrimidine base that constitutes the monomer of DNA and RNA molecules.

Nucleotide substitution Form of mutation whereby one nucleotide is changed for another within a DNA sequence. This may be synonymous (substitutions typically at the third base of a codon, where the mutation does not influence amino acid sequence in the protein) or non-synonymous (where each mutation does lead to an alteration of the amino acid sequence).

Null allele An allele that is not expressed and therefore does not produce a band on a gel.

Ordered data Data in which the evolutionary direction can be identified, with original characters (ancestral) mutating into novel characters (derived, *see also* Unordered data).

Organelle Membrane-bound structure within an eukaryotic cell responsible for a particular function or functions.

Orthology Divergence between two homologous characters following speciation.

Outbreeding depression Reduction in fitness due to break up of adaptive gene complexes following reproduction between widely disparate individuals.

Outcrossing Breeding between two individuals that do not share a close genetic relationship.

Panmixia Mating between individuals in a population which is random with respect to phenotype/genotype (*see also* Assortative mating).

Parapatric Occupying adjacent contiguous areas.

Paraphyletic A grouping of genotypes/individuals/species that does not include all descendents of an ancestor.

Parology Divergence between two homologous characters or loci following genome or locus duplication.

Paternal inheritance Transfer of an inherited characteristic through the male line/gamete.

Peptide A compound of two or more amino acids.

Periodic variation Variation that is regular and predictable.

Phenotype The observable characteristics of an organism, often an interaction between the genotype and environment.

Phylogeny A hypothesized evolutionary relationship of descent and ancestry between genotypes, individuals, or species, often presented as a dichotomous tree.

Phylogeography The study of the geographical distribution or pattern of evolutionary lineages.

Private alleles The possession of unique alleles within a sampling unit (e.g., population).

Plastid Double membrane-based organelle within the cytoplasm.

Pollen core analysis Stratigraphic sampling and identification of pollen to reveal environmental information of past habitats.

Polymerase chain reaction A technique for increasing the number of target DNA sequences by several orders of magnitude by repeated cycles of denaturation, primer annealing, and fragment extension using natural properties of a thermostable DNA polymerase.

Polymorphic Loci with more than one allele leading to different phenotypes in the population (*see also* monomorphic).

Polypeptide Single-chain polymer of many amino acids.

Polyploid Cell or individual containing more than two copies of the haploid genome complement (*see also* Allopolyploidy and Autopolyploidy).

Polytopic Having more than one origin (of a species).

Population Ecological population; a group of individuals of the same species within the same habitat at the same time. Statistical population:

all the items under study. Genetical population: all the individuals connected by gene flow, i.e., the gene pool.

Precision A measure of how many possible options are excluded from a particular analysis (in a phylogenetic context).

Progeny array Group of progeny derived from a single maternal parent, often used to infer outcrossing or inbreeding rates in plant populations.

Protein Large polymer of one or more polypeptide chains.

Protein marker *see* Genetic marker.

Provenance The origin of a sample individual.

Qualitative variation Discrete variation with phenotypes being classified into clear distinctions without intermediates.

Quantitative variation Continuous variation with phenotypes falling on a continuum.

Random amplified polymorphic DNA (RAPD) The selective amplification, using PCR, of a set of DNA fragments produced by amplification of genomic regions delineated by primers of random sequence (usually decamers).

Random sample A method of selecting units from a population such that each unit has an equal chance of being selected.

Rarefaction Method of addressing unequal sample sizes by recording the frequency of alleles in a large sample and estimating the number of each allele that would occur at these frequencies in smaller samples.

Replication The copying of both strands within a DNA molecule to produce two identical pairs of DNA sequence.

Recombination Exchange of genetic material between homologous chromosomes to break up linkage groups and yield allelic combinations not recorded in parental generations.

Relatedness A value (r) that records the degree of shared genetic material between individuals.

Repetitive DNA Repeated nucleotide sequences found throughout the genome. Distinguished into two types corresponding to the frequency with which the sequences appear in the genome; moderately repetitive DNA (present in a few to about 10^5 copies in

the genome), and highly repetitive DNA (present in about 10^5 to 10^7 copies in the genome).

Reproductive clique A group of individuals within which breeding is largely confined.

Reproductive isolating mechanism A barrier to reproduction between groups of organisms that prevents gene flow between them. These may be pre-zygotic (operating before fertilization can occur) or post-zygotic (operating after fertilization, e.g. infertile offspring).

Restriction fragment length polymorphism (RFLP) Fragments of DNA of varying size produced by cutting DNA with restriction enzymes.

Restriction site variation The basis of the generation of polymorphisms revealed by restriction enzymes. Such enzymes cleave DNA where specific DNA sequences are recognized. Thus mutations within restriction sites will no longer be cut.

Richness A measure of the absolute number of variants in a sample.

Secondary product A product of a secondary metabolic pathway (often termed a secondary metabolite).

Segregation pattern The observed banding pattern and its correspondence to a Mendelian ratio.

Selection The influence of the environment (in its broadest sense) in determining which individuals will breed and pass their genes on to the next generation and those who will not breed.

Selective sweep A mutation which confers a fitness advantage, such that all other forms of the allele are lost and the locus becomes monomorphic.

Self-incompatibility The inability of a hermaphrodite or monoecious individual to fertilize itself and produce viable offspring.

Selfing Successful fusion of male and female gametes produced by the same individual.

Sequence analysis Determination of the sequence of DNA bases in a specific region.

Silent nucleotide A mutation in DNA sequence that does not alter the amino acid sequence of a protein.

Speciation The process of generating new species due to reproductive isolation, ecological differentiation, and/or character change.

Standard deviation   A measure of spread of data from the mean.

Stepwise mutation model (SMM) Mutation causing loss or gain of a single tandem repeat within microsatellites (*see also* Infinite alleles model).

Stochastic process Random fluctuating factor.

Stratification   The dividing of a population into discrete non-overlapping sections.

Structural mutations   Mutations, such as deletions, that alter the structure of a chromosome.

Subpopulation   Spatially distinct unit of a statistical population.

Sympatry   Occupying the same geographic area.

Symplesiomorpy   Shared ancestral character states.

Synapomorphy   Shared derived character states.

Tandem repeats   A form of repetitive DNA with units of usually short sequences repeated hundreds or thousands of times (*see also* Interspersed repeats).

Taxon (*plural:* Taxa)   The organisms comprising a particular taxonomic entity (e.g., species or sub-species).

Transcription   Production of an mRNA molecule from the coding strand of a DNA molecule.

Translation   Production of a peptide with a particular amino acid sequence from the nucleotide sequence in an mRNA molecule.

Unequal sex ratio   A population with a disproportionate number of males or females.

Unordered data   Data in which the evolutionary direction of genetic change cannot be identified (*see also* Ordered data).

Uniparental inheritance   A genetic character that is passed on solely along the male or female lines as often found for organellar genomes (*see also* Biparental inheitance).

Variance   A measure of spread of data from the mean.

Wahlund effect   The sampling of a single population that is actually two or more genetic populations with limited gene flow between them. Such sampling leads to a deficit of heterozygotes for alleles that differentiate the two genetic populations.

Zymogram   A stained starch or agarose gel revealing the location of isozyme bands.

Index

Page numbers in *italics* refer to figures and in **bold** refer to tables/boxes.